# Myocardial Perfusion, Reperfusion, Coronary Venous Retroperfusion

To my wife, Nora

S. Meerbaum (Ed.)

# Myocardial Perfusion, Reperfusion, Coronary Venous Retroperfusion

Springer-Verlag
Berlin Heidelberg GmbH

The Editor:
Samuel Meerbaum, Ph. D., F.A.C.C.
5741 El Canon Avenue
Woodland Hills
California 91367
USA

CIP-Titelaufnahme der Deutschen Bibliothek

**Myocardial perfusion, reperfusion, coronary venous retroperfusion** / S. Meerbaum (ed.).

ISBN 978-3-662-12558-8    ISBN 978-3-662-12556-4 (eBook)
DOI 10.1007/978-3-662-12556-4
NE: Meerbaum, Samuel [Hrsg.]

Copyright © 1990 by Springer-Verlag Berlin Heidelberg
Originally published by Dr. Dietrich Steinkopff Verlag GmbH & Co. KG, Darmstadt in 1990
Softcover reprint of the hardcover 1st edition 1990


Editor: James C. Willis − Medical Editorial: Juliane K. Weller-Ratka −
Production: Heinz J. Schäfer

# Preface

The primary goal of this book is to survey issues pertaining to coronary venous retroperfusion, a potentially beneficial treatment of myocardial ischemia. Since attention will be focused on the myocardium, its normal or deranged antegrade perfusion will be discussed first, along with factors affecting blood supply to regionally ischemic tissue. After pointing to coronary reperfusion and its rapidly expanding applications, the principle of coronary venous interventions will be discussed.

Recent anatomic observations are presented to clarify features of the coronary venous sytems, some of which remain inadequately defined, yet play a crucial role in determining effectiveness of all retrograde methods. The remainder of the text concentrates on the development of retroperfusion sytems designed for retrograde treatment of myocardium jeopardized by deficient antegrade blood delivery, secondary to coronary artery obstruction. Retroinfusion of contrast is also considered as a potential diagnostic tool. The final chapter of the book reports on recent efforts aimed at a mathematical modeling of mechanisms and effects of coronary venous interventions.

Coronary artery bypass revascularization is a commonly practiced surgical procedure, while thrombolysis and angioplasty are now established clinical interventions for emergency treatment of evolving acute infarction and for dilatation of severe coronary stenoses. To enhance the success rate of the primary interventions, effective means are sought to 1) extend the viability of ischemic myocardium, 2) minimize reperfusion injury and accelerate return of function, and 3) reduce the incidence of coronary artery restenosis or reocclusion. Retrograde techniques are a potentially useful complementary support and may in some instances constitute an alternative treatment. Such techniques are now being refined and initial clinical evaluation has begun. Since most of these systems deliberately obstruct and significantly modify the coronary venous circulation, questions can be raised about safety as well as effectiveness. It is therefore essential that application rationales be based upon a satisfactory understanding of physiologic principles, and that they take into account the available retroperfusion research experience. This text provides a description of the current state of the art, and also offers an outline of criteria to be kept in mind when applying the coronary venous techniques.

Woodland Hills, California, 1990                                    Samuel Meerbaum

# Acknowledgment to co-authors:

As editor and author of many chapters of this book, I wish to specially acknowledge the contributions of my co-authors. Thus, Drs. Pakalska and Von Luedinghausen present information on coronary venous anatomy, which was heretofore lacking, and must be considered as the basis for all efforts aimed at retroperfusion development and application. I believe that those engaged in formulating rationales for new coronary venous systems and protocols for retrograde treatments should study these anatomic data and guidelines. The pressure-controlled intermittent coronary occlusion or PICSO method is described and results are discussed by Drs. Mohl and Neumann. Along with the clinically oriented synchronized retroperfusion (SRP) system, PICSO has also been applied, e.g., intraoperatively after reperfusion and in patients undergoing thrombolytic treatment. Dr. Drury describes the recent SRP studies during percutaneous transluminal coronary angioplasty in one of the largest series of clinical retroperfusion investigations. Feasibility and safety of the system was demonstrated, and data on effectiveness during PTCA balloon inflations are being gathered. Drs. Schreiner and Neumann contribute the final chapter of the book, describing new approaches to a mathematical modeling and computer computational analysis of coronary venous interventions in a variety of parametric settings.

To all the above co-authors, I wish to express my gratitude. I trust that the reader of this book will find a logical continuity of subjects, starting with the chapters on myocardial perfusion and reperfusion, continuing through anatomic descriptions toward consideration of the retrograde interventions. The latter are first dealt with in terms of general principals, proceeding to effects of coronary vein occlusion alone, to the PICSO chapter, a chapter on coronary vein arterialization and surgical retroperfusion, to clinically oriented SRP and drug retroinfusion, and a chapter on clinical SRP. There is also a description of a retrograde coronary venous myocardial contrast echo method, and, finally, a chapter on modeling approaches. I believe that the text, illustrations, and bibliographies are sufficiently clear and comprehensive to be of interest to all those engaged in or wishing to acquaint themselves with coronary venous intervention techniques.

# Contents

# Authors' addresses:

Dr. Samuel Meerbaum, Ph.D., F.A.C.C.
5741 El Canon Avenue
Woodland Hills, California 91367, USA

Dr. I. K. Drury, M.D., F.A.C.C.
17641 Tarzana St.
Encino, California 1316, USA

Prof. Dr. med. M. v. Lüdinghausen
Anatomisches Institut
Koellikerstraße 6
8700 Würzburg, FRG

Dr. Werner Mohl, M.D., Ph.D.
2nd Surgical Clinic
University of Vienna
Spitalgasse 23
A-1090 Vienna, Austria

Dr. Friderike Neumann
2nd Dept. of Surgery
University of Vienna
Spitalgasse 23
A-1090 Vienna, Austria

Dr. Wolfgang Schreiner, Ph.D.
2nd Dept. of Surgery
University of Vienna
Spitalgasse 23
A-1090 Vienna, Austria

Dr. Eva Ratajczyk-Pakalska, M.D.
Institute of Cardiology
Medical Academy in Lodz
Sterlinga 1/3
PL-91-425 Lodz, Poland

# Myocardial perfusion

S. Meerbaum

Comprehensive reviews of the myocardial circulation's structure and dynamic function have been provided by Berne [1], Feigl [2], Schaper [3], Marcus [4], Klocke [5], Hoffman [6], and others. In this chapter, we wish to highlight the important myocardial perfusion factors which must be considered in any antegrade or retrograde treatment of myocardial ischemia.

## Introduction

Understandably, major emphasis in the past has been placed on examination of the coronary arterial system, primarily its epicardial vessels, which are the usual site of atherosclerotic lesions underlying ischemic heart disease. There has also been substantial progress in understanding the intramyocardial coronary circulation, although more remains to be learned about such features as anastomosing myocardial vessels, sinusoids, arteriovenous communications, and Thebesian connections with the cardiac chambers.

The relationships of normal epicardial and myocardial coronary pressures, flows and blood volumes have been modeled and studied in recent investigations. Unresolved microcirculatory mechanisms, along with difficulties of performing quantitative measurements compound the current descriptions of myocardial circulation dynamics, in both normal and diseased states. Thus, investigators are still seeking a better understanding of arterio-arterial, veno-venous and arterio-venous anastomoses and the diversity of coronary control mechanisms. Other issues relate to various vascular "waterfall" models of effective coronary backpressure, the newly appreciated role of intramural vascular capacitances, and details of the phasic "squeezing" of the microvasculature, producing antegrade as well as backward blood displacements.

It should not be surprising to find additional circulatory complexities when one superimposes a steady or intermittent coronary vein obstruction, along with retroperfusion. There is as yet little systematic modeling and only limited understanding of myocardial mechanisms during coronary venous interventions. These retrograde manipulations modify the coronary flows and pressures in normal or ischemic states. Among significant factors to be considered are the coronary venous capacitances and resistances, the function of Thebesian vessel drainage into cardiac chambers, the frequently very significant epicardial coronary veno-venous shunts, and alterations in the distribution between endocardial and epicardial per-

1

fusion. These and other factors must be considered to assure oneself of both safety and effectiveness of retrograde coronary venous interventions.

We begin this chapter by a brief recapitulation of the coronary vasculature and its function, and follow-up by a survey of reported observations concerning myocardial perfusion. We also describe several methods for coronary blood flow measurement.

## 1. Vascular Anatomy

*Epicardial coronary arteries* [7, 8]: The right and left coronary arteries course and branch on the outer surface of the heart (Fig. 1), escaping most of the systolic compression effects normally experienced by the intramyocardial blood vessels. Variations in detailed vasculature and perfusion patterns are encountered within and between species. In the human, the right coronary artery descends to the diaphragmatic surface of the heart, where it issues the acute marginal branch, which supplies the right ventricular wall. Apical collaterals connect with terminal branches of the left anterior descending artery. On the posterior aspect of the heart, the right coronary artery turns toward the apex in the posterior descending and smaller atrioventricular branches. Occlusion of the right coronary artery results in posterior inferior infarction.

The left main coronary artery divides basically into the left circumflex and left anterior descending branches, The latter supplies major branches of the interventricular septum, and features as many as five diagonal branches. The left anterior descending artery supplies the anterior and anterolateral wall of the left ventricle.

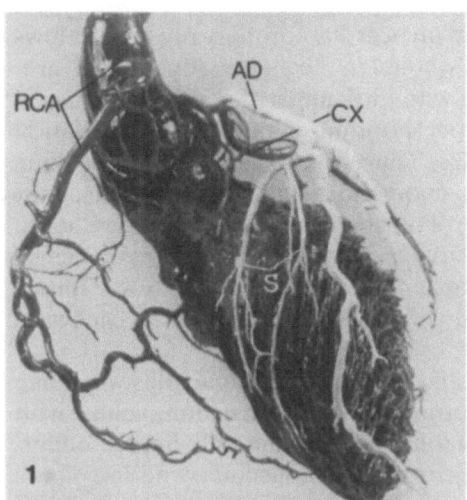

**Fig. 1.** Cast of the left ventricle and the coronary arteries photographed in the right anterior oblique projection. The right coronary artery *(RCA)* originates 1 cm above the anteromedial aortic sinus, but follows a normal course after passing obliquely through the aortic wall. *AD*: anterior descending artery, *CX*: circumflex, *S*: septal arteries. (By permission, from: Allwork SP. A Spectrum of Normal Coronary Artery Distribution in Man. Anatomia Clinica 1:311–319, 1979.)

The circumflex coronary artery, supplies branches to the left atrium and ventricle. The most important branch of the circumflex is the obtuse marginal, which supplies the lateral and posterior walls of the left ventricle.

In a "balanced" circulation, each of the ventricles receives blood from its respective coronary artery. Forty percent of the heart's myocardium is perfused by the right coronary artery and another 40% by the left anterior descending coronary artery, while the left circumflex supplies the remaining 20%. The so-called left dominant situation has the left anterior and left circumflex supplying about 40% each, while the right coronary artery perfuses only 20% of the myocardium. In the case of the right dominant the right cornary artery supplies as much as 45%, the LAD about 40%, and the left circumflex the remaining 15% of the heart. The anterior two-thirds of the septum is usually perfused from the left anterior descending, the posterior third by the right coronary artery.


**Variability Related to Species**

Particular circulation patterns are associated in the human with evolving coronary atherosclerosis, or with multiple periods of myocardial ischemia. The coronary circulations in the dog, baboon, pig, sheep and other animal species, differ from that in the human heart. Historically, much basic knowledge about cardiac physiology and treatments has been derived from animal studies (mostly in dogs), using a variety of models, preparations, and protocols.

In the dog, the right coronary artery assumes a much lesser role than in the human, supplying only the central portion of the free wall of the right ventricle. The dog's collateral circulation tends to be more pronounced and epicardial, as compared to subendocardial collateralization in man. Caution is indicated in applying experimental canine data, but a number of useful dog preparations were designed to study ischemic heart seetings featuring coronary collaterals. The pig and baboon have an epicardial coronary artery pattern similar to the human, with the right coronary dominant and supplying the entire right ventricle as well as the inferior portion of the left ventricle and the posterior ventricular septum. The hearts of baboons and pigs are characteristically poor in collaterals, and may be more suitable to investigate conditions of acute myocardial ischemia. Of the smaller animals, the rat has been extensively employed for exploratory drug screening. Its left anterior descending coronary artery predominates, and supplies most of the left ventricle. Coronary veins, veno-venous shunting, and other communications in various species are discussed in chapter 4 of this book.

*Intramyocardial Coronary System* [9]: This circulation is materially influenced by cardiac contraction, and employs metabolic, hydraulic, mechanical, and neurohumoral controls to provide an adequate coronary blood supply, accommodating oxygen demands over a wide range of physiologic conditions. Following their epicardial branching, coronary arteries ramify into vessels (e.g., .5 mm in diameter) which penetrate into the myocardium at essentially right angles (Fig. 2) and then subdivide in a tree-like manner. Two patterns have been observed and described [9]. One pattern features early division into a fine anastomosing network

3

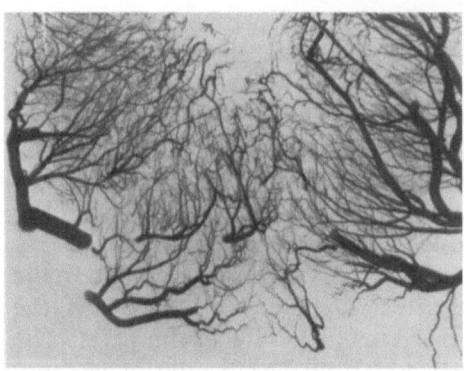

**Fig. 2.** Arterial supply to the left ventricular wall in a normal heart. Two types of vessels can be recognized: 1) Vessels rapidly dividing to give direct supply to various layers of the heart wall; 2) channels which pass with little or no branching directly to the deepest layers and terminate in the subendocardial plexus. (By permission, from: Fulton WFM (1985) Morphology of the myocardial microcirculation. In: Tillmanns H, Kubler W, Zebe H (eds) Microcirculation of the heart, pp 15−25.)

of 70−120 μm vessels that supplies up to three-fourths of the underlying myocardium. Subsidiary networks, paralleling the epicardial surface, lead to pre- and capillary vessels. The other pattern exhibits penetrating intramyocardial coronary arteries of about 0.5 mm diameter, which pass with little or no reduction in vessel size directly to the endocardium. These vessels terminate in an extensive vascular anastomosis (down to 20 μm), providing ample blood supply to the entire subendocardial myocardium. Inter-regional coronary arterial interconnections are frequent in the subendocardial zone. Capillary beds proliferate throughout the ventricular wall, and are particularly prominent subepicardially and subendocardially.

Although less is known about the intramural venous system (see chapter 4 of this book), it seems that there are at least twice as many venules than arterioles in the heart wall. A parallel arrangement between arteries and veins has been noted, with arteries usually accompanied by two veins. Pulsatile pressures within the arterial vessels might further enhance the systolic squeezing-induced peristaltic venous blood drainage through the myocardial veins.

Discussion of myocardial blood flow usually emphasizes two main categories of small coronary arteries: vessels which normally supply the heart muscle, and anastomotic connections which come into play during disease. The normal heart may have evident coronary arterial communications, but it is chronic cardiac disease and regional ischemic injury that promote mobilization of the connecting vessel system, with some epicardial collaterals being as large as 1 mm in diameter. Striking experimental insight into the microcirculatory anatomy and its function has been recently achieved in vivo, by transillumination microscopy of the heart [12−13]. These techniques make use of a small light-transmitting needle inserted into myocardium, along with a floating focus to keep the needle moving in unison with the cardiac surface. Such a method allows quantitative assessment of the behavior of the coronary microcirculation in the left ventricular wall. Flow directions in capillaries and smallest arterioles of corresponding myocardial areas could be determined, and reveal a mixed countercurrent system, apparently well suited for optimal myocardial oxygen supply. In some instances, investigators were able to determine microcirculatory vessel configurations, dimensions, and blood flow velocities.

Intramyocardial arterioles (vessels of a diameter of less than 100 μm) divide and run generally in both directions, parallel to the cardiac muscle fibers, before finally branching and feeding the extensive capillary network. Capillaries also parallel and accompany the muscle fibers, and exhibit many anastomotic loops. Capillary density has been reported to be of the order of 2,500−3,400 vessels/mm. Individual myocardial capillary diameters averaged 5.7 μm, and intercapillary diffusion distance in rats and dogs was found to be about 18 μm. Venules, arising from a confluence of branching capillaries and postcapillary vessels, feature a profusion of intercommunications. Initially, venules parallel the muscle fibers, but then they change directions and join up in larger myocardial coronary veins.

In addition to the above observations, microcirculatory vessel intraluminal pressures have also been recently obtained in the beating myocardium of the cat, rat or rabbit [14]. Thus, average pressure in myocardial coronary veins was 6 mmHg (mean pulse pressure 5.8 mm Hg) and their average diameter was 101.5 μm (range 60−192 μm).

*Direct Heart Cavity Communications:* Thebesius, as early as 1708 [15], and subsequently Wearn [16], demonstrated the existence of myocardial vessels which directly communicate with the heart chamber. Actually, a number of observed or inferred vascular entities have been reported: Thebesians, myocardial sinusoids, arterio-sinusoidals, and arterio-luminal vessels. The myocardial sinusoids are short irregularly shaped branches (diameter 50−350 μm), whose walls have lost their arterial structure and are composed of an endothelial layer. Their efferent arteries, the arterio-sinusoidal vessels, may enter the heart cavity independently, through a common opening, or else communicate with one another via the myocardial capillary network. Arterioluminal vessels are 40−200 μm in diameter and exhibit an arteriolar structure almost as far as their entry into the heart cavity. They are most frequent in the left ventricular wall, communicate with each other and particularly with the capillary network. Thebesian veins (200 and 400 μm diameter) are direct venous communications into the heart cavities, mostly into the right ventricle (see chapter IIIA). The existence of some of the above vessels is accepted, but their significance and descriptive details are still controversial.

*Arteriovenous anastomoses:* Intramyocardial connections between the coronary arteries and veins would bypass the capillaries and thus represent a potentially non-nutritional shunt. Some investigators concluded that such anastomoses are rare or nonexistent, while others report definitive and frequent appearance of such direct connections between the coronary arterial and venous systems. Thus, in one study of human hearts, direct arteriovenous anastomoses were found present in all cases, particularly subendocardially in the anterior left ventricle and the apex (discussed in chapter 3 of this book).

*Primary vs Anastomotic Circulation:* The primary coronary circulation consists of the arterial blood supply to the capillary system, which provides nutrition to the myocardial cells and supports the cardiac contraction and metabolism. The other important system consists of many shunts and communications between the various components of the epicardial and myocardial vessels. These compensatory communications have been of great interest in relation to evolving cardiac disease and its treatment. The intramyocardial coronary veins collect and deliver the blood, along with metabolic products, into the epicardial venous system.

*Epicardial Coronary Venous System* is basically composed of two parts, the coronary sinus system and the anterior cardiac venous system. Before discharging into the right atrium through its ostium, the coronary sinus collects blood from most of the coronary veins serving the left heart. The right heart anterior cardiac veins drain directly into the right atrium. Coronary venous anatomy and its variants (crucial to the retrograde interventions) is the subject of chapters 4 and 5.

## 2. Coronary Supply, Distribution and Drainage [17–19]

In terms of supply, arterial blood from the major coronary arteries flows to the various branching vessels, which perfuse distinct regional myocardial territories. This blood supply is normally very flexible, satisfying myocardial requirements ranging from sedentary conditions through strenuous exercise, and a great many physiological manipulations. The vascular reserve can, however, become greatly limited by disease, for example during severe obstruction of one or more of the major coronary arteries. In such a case, the heart tries to employ its other compensatory supply mechanisms, e.g., via available arterial intercoronary collaterals. When the coronary reserve is exhausted, and blood supply is sharply reduced, there ensue significant ischemic derangements, and potentially regional myocardial necrosis.

It has already been pointed out that attention should be paid not only to the large epicardial coronary vessels, but also to the manner in which perfusion is distributed to all portions of the myocardium. While flow and pressure in the major epicardial supply conduits are indeed a primary determinant of myocardial blood flow, it is equally true that other factors can and do influence the extent and manner in which this supply is distributed to various regions of the heart. Intramyocardial stress and extravascular pressures, for example, can cause an unbalanced perfusion, primarily jeopardizing the endocardial layer. On the other hand, intercoronary collaterals may redistribute arterial blood flow and they help maintain the viability of regionally jeopardized cardiac tissue. Some current treatment modalities may indeed depend on pre-existing anastomoses and shunts, e.g., for delivery of pharmacologic agents to deficient regions.

A number of past investigators have studied the normal antegrade patterns of coronary arterial inflow and drainage. In some studies, the total coronary flow distributed itself to produce approximately a 50%–60% drainage into the right atrium via the coronary sinus and a 20%–25% drainage via the anterior cardiac veins, while only a few percent each entered the right and left ventricular chambers directly. Clearly, the inflow-outflow relationships will depend on both the anatomy of the vasculature and the prevailing physiologic state, including blood pressure and force of contraction.

One characteristic of the heart which will be repeatedly emphasized is that the blood pressure in its small vessels is determined not only by the arterial supply driving pressure, but also by the phasic effects caused by cardiac contracting and relaxation. In systole, extracoronary compression impedes the inflow into intramyocardial vessels and microcirculatory blood volume is largely "squeezed" into the coronary veins. In terms of overall coronary venous drainage, it is well known that the outflow from the coronary sinus occurs predominantly in systole. Arterioles, venules and capillaries have all been observed to constrict during systole, while dilating during diastole.

Effluents must, of course, be properly drained from the capillary system, a function normally well performed by the coronary veins. If a recent study [20] is correct, two microcirculatory venous systems are operative: one always closely approximating the myocardial coronary arteries, and another well removed from the arteries in interstitial spaces. Hindered drainage, due for example to coronary venous obstruction or inadequate ischemic region circulation with poor runoff, tends to exacerbate myocardial injury, because of metabolite accumulations and deleterious myocardial edema. Fortunately, myocardial and epicardial coronary veins are numerous and exhibit a profuse network of anastomosing connections, so that in most circumstances, the venous blood of the heart can be satisfactorily drained, usually into the right atrium. In fact, this anastomosing plexus is in many instances so extensive that satisfactory temporary drainage may be achieved even when a large epicardial coronary vein is fully occluded.

## 3. Coronary Regulation and Hyperemia [1, 2, 4]

The intramyocardial coronary arterioles, with their smooth muscle walls, appear to be the preponderant site for microcirculatory vascular resistance control in the normal heart. Neither collaterals nor venules and capillaries possess the same capacity for expanding when increased myocardial perfusion is required. Regulation of the coronary system is very complex. Among the many facets studied have been physical, myogenic, and metabolic influences. Physical factors regulating the blood flow have already been mentioned, e.g., input pressures and extravascular forces. As intramyocardial force or wall stress increases or decreases, corresponding microvessels such as capillaries and venules are compressed or relax, consequently modifying the blood flow. Myogenic response is also dependent on changes in perfusion pressure and intramural pressures. Neural and neurohumoral factors exert indirect vasodilatory or vasoconstrictive effects on resistance vessels. Among myocardial metabolic substrates observed to have a regulating effect on coronary blood flow are: carbon dioxide, oxygen, lactic acid, hydrogen ions, histamine, potassium ions, increased osmolarity, polypeptides, and adenosine nucleotides. As myocardial activity increases, there is an increase in products of metabolism, some of which may exert a vasodilatory effect on the smooth muscle of the regulatory arterioles, thereby reducing coronary arterial resistance. Conversely, as myocardial activity decreases, arterial resistance increases.

It is worth noting how well a normal myocardial negative feedback operates in various physiological states. For example, a major disturbance of the regional myocardial blood perfusion generally results almost instantly in metabolic derangements. If protracted, these could by themselve be very damaging, yet the consequences may be ameliorated since the very same metabolites serve as a powerful inducement to decrease arteriolar resistance and help delimit the under-perfusion.

In relatively normal states, maximal arteriolar dilatation leads to hyperemia of a type and extent reflecting the state of available coronary reserve. In a reactive hyperemia, following very brief coronary occlusions (lasting from a few seconds to 1 min), the hyperemic flow far exceeds the flow debt incurred during the acute ischemic period. With longer coronary occlusions, the "overpayment" of the flow debt decreases, and with occlusion times beyond a few minutes, the flow debt is actually underpaid. Apparently, some regional reactive hyperemic response can still be elicited even after relatively long occlusions, which lead to irreversible regional myocardial damage. Figure 3 illustrates regional blood flow measurement (using radionuclide microsphere study of a left ventricular slice) at 3 min and 90 min of an experimental left anterior descending coronary occlusion in the dog, followed by repeat measurements at 5, 15, and 30 min reperfusion. A hyperemia is noted in the subepicardial layer after the 90 min occlusion, but not in the subendo-cardial and midmyocardial layer (near transmural hemorrhagic subendocardial infarction). Hyperemia (indicative of coronary reserve) is also seen following phar-macologic interventions, for example, intracoronary injection of dipyridamole, papaverine, or angiographic contrast agents.

## 4. Coronary Blood Flow Dynamics [21–25]

There has been continuous interest and progress in both coronary measurements and improved definition of the coronary blood pressure-flow relationships. The past decade in particular, has not only provided meaningful data and interpretation of the pressure-flow dynamics, but has also evolved new insights into significant components of the coronary circulation and their apparent function. In spite of a dearth of information on physical properties, and still inadequate clinical quantita-tion of transmural perfusion, investigators have proceeded to construct complex (and apparently validated) mathematical models. Such modeling is carried further in chapter 13 of this book and applied to phasic coronary venous interventions, such as the pressure controlled intermittent coronary sinus occlusion or syn-chronized regional diastolic retroperfusion with systolic coronary venous drainage.

One of the important observations which raised questions about conventional use of coronary resistances, was that diastolic coronary inflow ceases at a relatively elevated arterial pressure (e.g., 20 mmHg or higher), higher than the right atrial (coronary outflow) or the diastolic ventricular pressure. Extravascular intramyocardial compressive forces and vasomotor tone were generally implicated in a so-called "waterfall" phenomenon. The zero flow pressure (Pzf) could be derived by plotting the near linear coronary input pressure vs flow, the curve recip-rocal slope representing vascular resistance. Various interventions, such as coro-

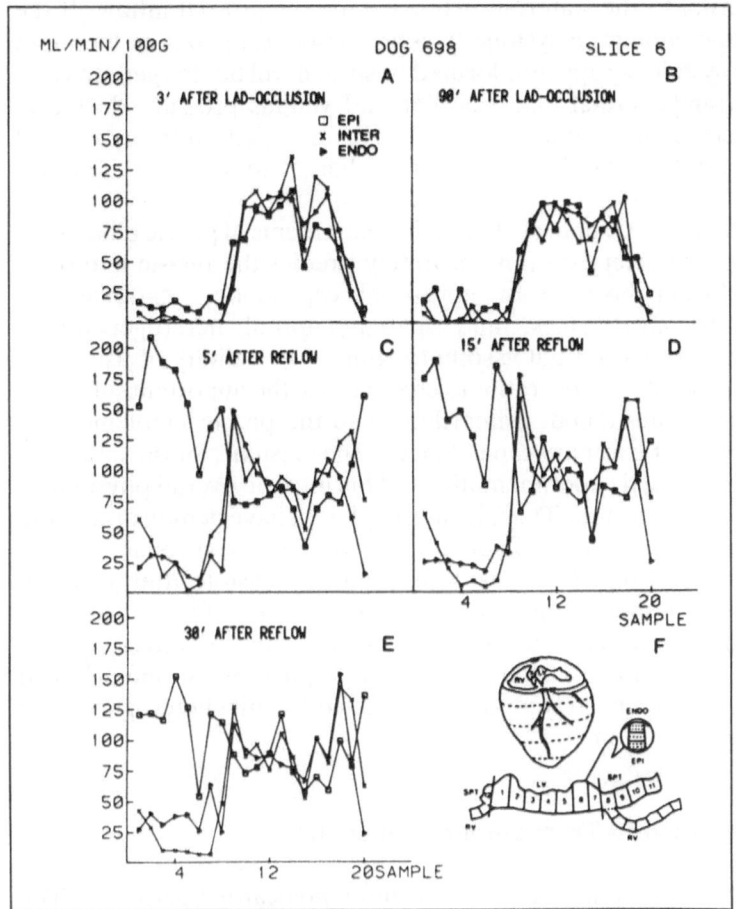

**Fig. 3.** Occlusion and subsequent reperfusion of the proximal left anterior descending artery (LAD). *Panels A and B,* 3 and 90 min after LAD occlusion, *panels C, D and E,* 5, 15 and 30 min after reflow, *panel F,* schematic presentation of preparation of tissue smaples. (By permission, from: Schaper W (1985) Patterns of regional blood flow following reperfusion of ischemic myocardium. In: Tillmanns H, Zubler W, Zebe H (eds) Microcirculation of the heart.)

nary hyperemia or hemodilution, coronary sinus occlusion, rate of change of input pressure, heart rate, etc., were found to decrease or increase Pzf. Evidence has also been presented indicating that Pzf values may vary in various portions of the myocardial circulation.

Consideration of issues related to Pzf led to examination of coronary capacitances (in addition to resistances), which are now understood to play an important role in circulatory dynamics, and are substantial in the myocardial microcirculation. Capacitive flows cause differences in magnitude and phase between coronary flows at different levels of the circulation. Mates et al. [26] recently presented a sophisticated conceptual electrical analog of myocardial coronary pressure-flow relations and described experimental observations. In the normal state, systolic extravascular squeezing causes coronary venous discharge (from a capacitor) and raises the pressure distal to the waterfall, while decreasing arterial inflow. If the arterial pressure is low enough, a systolic backflow may occur, but as the input pressure increases in systole, a capacitor located upstream will be charged. In early diastole, as both intramyocardial extravascular and venous pressure decrease, arterial inflow will increase and venous outflow will diminish as the intramyocardial capacitor recharges and the proximal capacitor discharges to augment the intramyocardial inflow.

As recently emphasized by Chilian et al. [27], measurements of phasic blood flow in an epicardial coronary artery do not accurately predict the phasic nature of intramyocardial perfusion because of epicardial capacitance, extravascular intramyocardial compressive forces, and spatial/temporal heterogeneity of myocardial perfusion, given a normal vasomotor tone. The validity of using systolic-to-diastolic coronary blood flow ratios to characterize the important subendocardial perfusion is also questioned, primarily due to the phasic supplementary charging and discharging of coronary capacitances. The existence of the latter has by now been well documented experimentally, and fits into the overall physiologic model of the myocardial perfusion. Thus, Kajiya et al. [28] have demonstrated the existence of the capacitances by experimental measurements of coronary venous outflow which persisted for some time during an extended diastole after the input arterial flow was totally stopped. Other data show the presence of flow reversal in coronaries under some conditions associated with extravascular compression, vessel tone and other factors. Finally, Chilian et al. [27] applied measurements with tracer microspheres, and demonstrated the temporal and spatial heterogeneity of the normal myocardial perfusion.

## 5. Myocardial Perfusion and Its Transmural Distribution

Hoffman [6] recently reviewed the topic of transmural myocardial perfusion. The tendency and risks of subendocardial ischemia or necrosis are primarily attributed to a regionally decreased myocardial blood flow, although greater subendocardial (vs subepicardial) shortening, a higher regional myocardial wall stress, and the apparently greater subendocardial vascularity should also be kept in mind. Reduction of the coronary flow below its "resting level," or progressively greater coronary stenosis, leads to a fall in subendocardial contractility, along with changes in the endocardial-to-epicardial (endo-epi) blood flow ratio. In normal conscious states, the endo-epi ratio is usually slightly above 1 (1.2–1.4). Anesthesia and other manipulations tend to somewhat reduce this ratio. When the coronary vasculature is maximally dilated, the endo-epi ratio drops with decreasing perfusion pressure, so that there is more perfusion in the subepicardial muscle, and the subendocardium has a lower flow reserve. The endo and epi flow reserves in dogs with impaired autoregulation (lowered perfusion pressure such as with subtotal occlusion) are shown in Fig. 4 [29]. In contrast with dogs in whom perfusion (and viability) tends to persist in the subepicardium, lack of collaterals (e.g., in pigs), results in parallel subendocardial and subepicardial reduction in coronary reserve with successive stenosis, and leads to essentially no perfusion in any of the myocardial layers after sudden coronary occlusion.

10

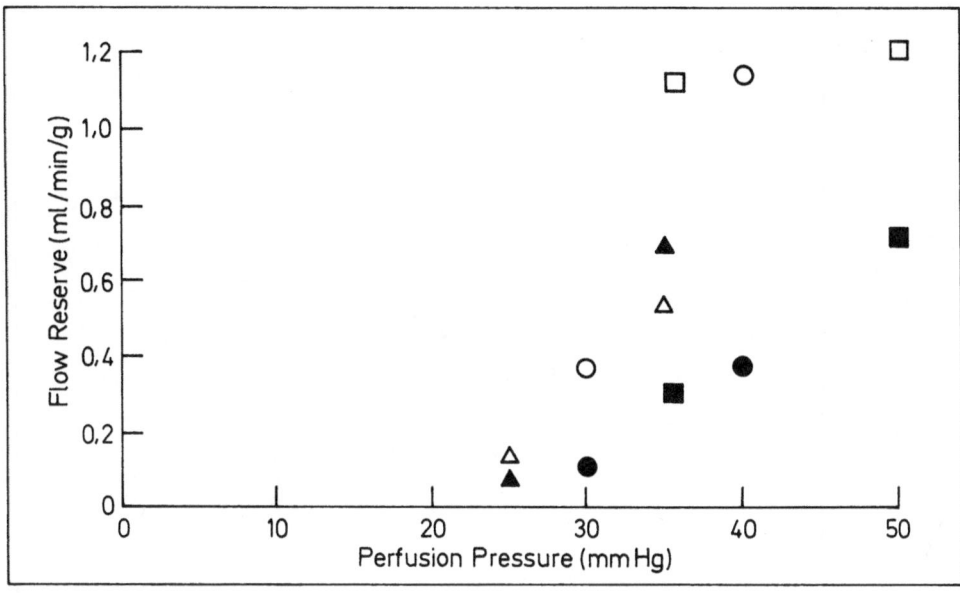

**Fig. 4.** Subendocardial (solid symbols) and subepicardial (open symbols) flow reserve plotted against perfusion pressure in three studies. (■, □) Heart rates 127 and 140 beats/min at perfusion pressures of 50 and 38 mm Hg, (respectively, (▲, △) Heart rate 61 beats/min (●, ○) Heart rate 100 beats/min (By permission, from: Hoffman IIE (1987) Transmural myocardial perfusion. Prog Cardiovasc Dis 28(6):429−464.)

During systolic contraction, the lower subendocardial coronary flow might be logically attributed to the differentially high regional intramyocardial pressure or stress. Thus, the important subendocardial layer will greatly depend on a sufficient blood supply during diastole, and short diastoles and/or low perfusion pressures would imply minimal subendocardial perfusion. To some extent, the subepicardium, too, might not receive an adequate supply in systole, because of the above mentioned systolic charging of coronary capacitances. Actually, Hoffman [6] relates the subendocardial ischemia to an interaction between systole and diastole, with the subendocardium receiving its portion of the flow somewhat later than the (lower resistance) epicardial vasculature. Apart from the heart rate, contractility, afterload, preload and other factors may also influence the endo-epi ratio.

The relationship of the % diastolic coronary inflow (about 75%−85% in control) to the endo-epi distribution is shown in Fig. 5 [30], indicating that in many diseased states, endo-epi ratio tends to be lower, along with an increased percent systolic flow.

It is necessary to re-emphasize that myocardial perfusion may remain "normal" over a wide range of conditions, even down to seemingly pronounced epicardial coronary artery stenosis. The clinical coronary reserve concepts have been very useful in helping to differentiate coronary and perfusion states for vessels with moderate stenosis, larger or less than, say, 50%. Recently developed measurement techniques facilitate clinical assessment of regional perfusion and coronary reserve.

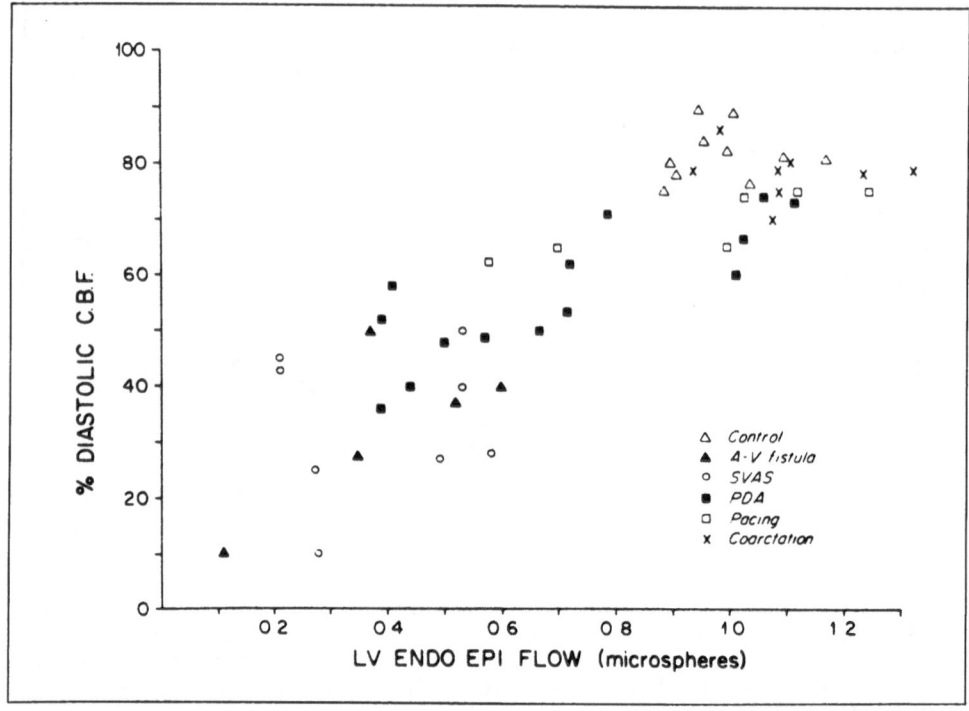

**Fig. 5.** Relationship of inner:outer flow ratio (endo:epi flow) to percent phasic coronary flow in diastole. The measurements are made in a variety of acute animal models of heart disease. (By permission, from: Lea & Febiger, publishers (1976) Progress in Cardiology, vol 5, pp 37–89 (Hoffman, Buckberg).)

## 6. Coronary Venous Modeling

Reverse coronary venous perfusion can be envisaged either in the setting of cardiac arrest or during coronary artery occlusion, and it is usually instituted by coronary vein obstruction, along with retrograde infusion of oxygenated blood. Major modification of myocardial perfusion must be anticipated with such an intervention, and differences will also depend on whether, where, and the extent to which there is a residual antegrade coronary artery flow input. Wong et al. [31] describes a physiologically based model (Fig. 6) which might be employed to analytically study retrograde interventions. Chapter 13 in this book covers some of the progress made in such mathematical/physiologic modeling, in spite of a shortage of pertinent measurements. Certain additional significant factors must be kept in mind. First, high coronary venous and transmitted intravascular blood pressures could elevate interstitial pressure and cause edema, creating a potential safety issue which must be appreciated and addressed, e.g., by delimiting the occluded coronary vein pressures and flows. Secondly, Thebesian connections between the intramyocardial vasculature and the cardiac chambers play a much more important role in providing some of the drainage during coronary vein occlusions, and their potentials as well as limitations need to be assessed. Equally, epicardial coronary

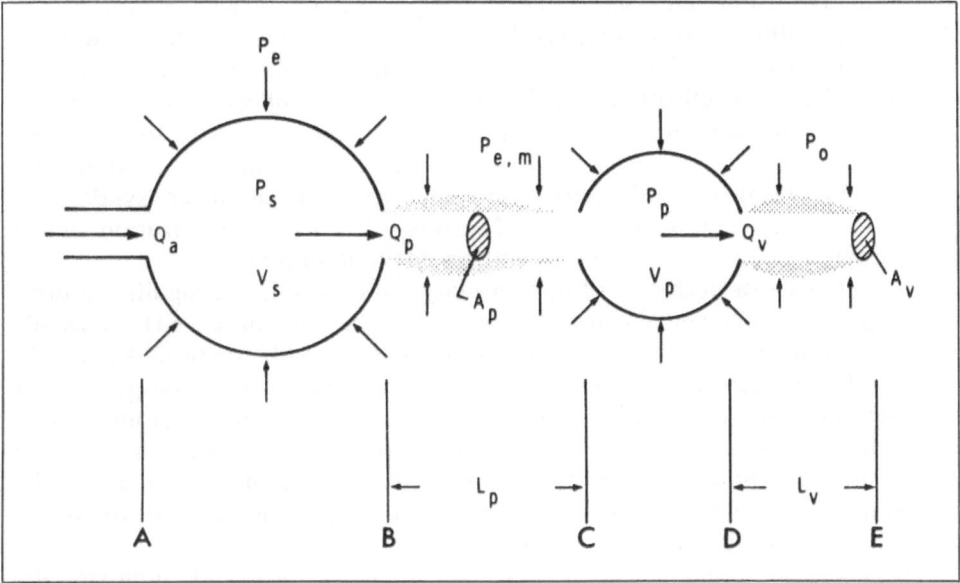

**Fig. 6.** Schematic diagram of the coronary venous system. The arterial blood flows into an intramyocardial storage (AB), whose volume is dependent on the transmural pressure ($P_s - P_e$). From the storage, blood flows into the peripheral veins represented by a collapsible tube BC and a volume segment CD. The cross section area $A_p$ and the effective length $L_p$ are dependent on ($P_p - P_{e.m}$). The volume ($V_p$) is determined by the inflow $Q_p$ and the outflow $Q_v$. The epicardial vein is represented by a compliant segment DE with variable area $A_p$ and effective length $L_v$. (By permission, from: Wong AYK et al. (1984) The dynamics of the coronary venous system in the dog. J Biomechanics 17(3):173–183.)

veno-venous anastomoses (often very sizeable shunts) will be important in determining the reverse coronary pressure-flow relationships, safety, and intervention effectiveness. In studying the basic mechanisms of retrograde myocardial perfusion dynamics, capacitances as well as resistances will have to be evaluated in different segments of the coronary circulation. For the clinically oriented phased coronary venous interventions, it will be necessary to better understand and obtain data on intramyocardial and epicardial vessel volumes, resistances, capacitances, along with effects of such factors as heart rate, contractility, ventricular preloading and afterload.

## 7. Ischemic Alterations of Myocardial Perfusion

It is noteworthy that only after a coronary artery obstruction progresses to 80%–90% of the lumen does the heart experience a significant loss of blood supply. The coronary reserve is however frequently impaired at much less severe stenosis (e.g., 50%), and potentially hazardous conditions may be appreciated by suprimposing an extra stress and thus increasing the myocardial oxygen demands. Once a critical constriction of a major coronary artery is exceeded, the regional myocardial perfusion diminishes rapidly. The perfusion deficiency is exacerbated when there is little or no coronary collateral compensation, e.g., due to multiple coronary vessel disease. The most drastic consequences ensue when a major coronary artery becomes fully occluded, for example, by thrombosis at a stenotic site. Prompt intervention is then required in acute ischemia, e.g., with thrombolytic treatment. Of course, there is a large spectrum of coronary obstructions and myocardial blood flow insufficiencies associated with various clinical syndromes. Some of these may require application of coronary artery revascularization, others can be treated medically or corrected with an angioplasty procedure.

Severe stenosis or occlusion of the coronary artery results in a significant pressure drop between the neighboring territories, i.e., the zone subserved by normally perfused arteries and the ischemic zone distal to the vessel obstruction. As a result, arterial collateral flow may be increased, but if often fails to acutely compensate for the underperfusion, perhaps because of insufficient preexisting anastomotic vessels or their size and tortuosity. The collateral passages do not seem to have a reserve capacity, characteristic of the native circulation. On the other hand, as the collateral circulation develops over time in response to insufficient myocardial perfusion, it can eventually provide major benefits.

Experimentally simulating severe coronary artery stenosis, and employing the previously described methods of intramyocardial microscopic viewing, the initial microvascular alterations observed are dilatation of the small arterioles, while venules and capillaries seem unchanged [32]. In the arterioles, the blood pressure pulsatility becomes damped and the diastolic peak velocity is reduced. All microcirculatory vessels exhibit reduced red cell velocities. Peak flow in capillaries and venules is markedly reduced in systole. Significant systolic pressure reduction in the venules have also been attributed to loss of extravascular compressive force in the dysfunctioning myocardium. The intracapillary distance appears to be increased as a result of myocardial ischemia.

Measurements were carried out in animals during simulated 50%, 70% and 90% coronary artery stenosis, and with coronary supply pressures compatible with either autoregulation or else during maximal adenosine vasodilatation [33]. In the normal autoregulatory setting, 50% and 70% stenosis was associated with an insignificant pressure gradient, but a 90% stenosis increased the gradient significantly, and the regional endo/epi myocardial perfusion ratio decreased relative to that in the normal zone. Yet, blood supply to the myocardium beyond even the severe stenosis was still well compensated and essentially undiminished. During maximal adenosine-induced vasodilatation, the blood flow to the involved post stenotic myocardium was significantly lower at all levels of coronary artery stenosis than flow in the normal perfusion zone. The coronary flow decreased in direct relation to the severity of the stenosis, and perfusion beyond the stenosis was also redistributed away from the subendocardial layer, the endo-epi ratio dropping to about 0.7 (vs 1.2 in the normal zone). It is found during stress-induced elevation of flow and oxygen demands that the subendocardial coronary reserve may be exhausted and ischemia could set in even with a 50% vessel stenosis.

*Effects of Coronary Artery Occlusion:* Many animal research studies have elucidated the consequences of acute myocardial ischemia, following deliberate

14

obstruction of one of the coronary arteries. Thus, in the pig regional blood flow measurements indicate a drop in perfusion down to 1% −5% of normal, while perfusion in the dog (with its variable and numerous collaterals) may drop to 20% of normal. The latter level is apparently not adequate to permanently support muscle viability, although it is, of course, a less severe condition than absence of collateral circulation.

The consequences of acute ischemia are almost immediate and can be profound. Thus, regional contraction often ceases, and the locally ischemic zone of the heart may expand paradoxically during systole. Depending on the size of the endangered zone of the heart (risk zone), global hemodynamic and functional derangements can ensue, (further decreasing perfusion), including increased end-diastolic left ventricular pressure and a compensatory dilatation of the heart. Along with loss of regional contraction, the involved myocardium will exhibit a rapid reduction of its ATP and CP content, depletion of endogenous glycogen stores, appearance of lactic acid, edema, and reduction of cell and tissue pH.

Systematic mapping during the ischemia will also reveal significant ECG changes, e.g., of ST elevations, which have been used in assessing the extent of ischemic injury. Other effects of acute coronary occlusion may include increased heart rate, appearance of arrhythmias, increased systemic vascular resistance, decreased blood pressure, and other alterations which are − to some degree − also associated with neural and/or hormonal reflexes.

Detailed microcirculatory studies [34] during the earliest phase of acute ischemia indicated limited alterations in the capillary structure, mostly endothelial swelling. With experimental coronary occlusions lasting less than 40 min duration, ischemia did not appear to greatly change the capillaries. On the other hand, longer durations of severe ischemia, are associated with some myocyte death, and an accompanying significant capillary encroachment by endothelial blebs was noted. It is generally accepted that microvascular injury is a later event, as compared to irreversible myocardial damage.

Unless reversed, more than 20−40 min acute ischemia (without extensive collaterals) leads to profound morphologic changes in the dog, involving progressive cellular alterations from reversible to irreversible damage. Many investigations of physiologic mechanisms and treatment addressed the process of ischemic injury and infarction. Recently, oxygen free radicals have been implicated in both ischemic and reperfusion injury. Treatments are aimed at salvaging jeopardized myocardium, and also at preventing the so-called "stunning" of reperfused cardiac tissue function.

## References

1. Berne RM, Rubio R (1979) Coronary circulation. In: Handbook of physiology, the cardiovascular system. The heart. In: Berne RM (ed) Bethesda, MD, American Physiologic Society, sec 2, vol I, ch 25
2. Feigl EO (1983) Coronary physiology, Physiol Rev 63(1):1−205
3. Schaper W (ed) 1979 The pathophysiology of myocardial perfusion. Elsevier Biomedical, Amsterdam
4. Marcus ML (1983) The coronary circulation in health and disease. New York, McGraw-Hill

5. Klocke FJ, Mates RE, Canty JM Jr., et al. (1985) Coronary pressureflow relationships – controversial issues and probable implications, Circ Res 56:310−323
6. Hoffman JIE (1987) Transmural myocardial perfusion. Prog Cardiovasc Dis 28(6):429−464
7. James T (1961) Anatomy of coronary arteries. Hober, New York
8. Fulton WFM (1965) The coronary arteries. Charles C Thomas Publishers, Springfield IL
9. Fulton WFM (1985) Morphology of the myocardial microcirculation. In: Tillmanns H, Kubler W, Zebe H (eds) Microcirculation of the heart, pp 15−25
10. Schaper W, Schaper J (1977) The coronary microcirculation. Am J Cardiol 40:1008−1012
11. Grayson J, Davidson JW, Fitzgerald-Finch A, Scott G (1974) The functional morphology of the coronary microcirculation in the dog. Microvasc Res 8:20−43
12. Steinhausen M, Tillmanns H, Theredan H, Leinberger H (1985) Methods for direct evaluation of the terminal vascular bed of the ventricular myocardium. In: Tillmanns H, Kubler W, Zebe H (eds) Microcirculation of the heart, pp 56−60
13. Steinhausen M, Tillmanns H, Theredan H (1978) I a method for in vivo observation of the microcirculation of ventricular myocardium of heart and capillary flow pattern under normal and hypoxy conditions. Pflugers Arch 378:9−14
14. Nellis SH, Liedtke AJ (1985) Pressure and dimensions in the terminal vascular bed of the myocardium determined by a new freemotion technique. In: Tillmans H, Kubler W, Zebe H (eds) Microcirculation of the heart, pp 61−74
15. Thebessius AC (1708) Dissertatio medica de circulo sanguinis in corde. Lagdumi Batavorium Amsterdam, Elsevier
16. Wearn J (1928) The role of the thebessian vessels in the circulation of the heart. J Exp Med 47:293−304
17. Anrep GV, Blalock A, Hammonda M (1929) The distribution of the blood flow in the coronary blood vessels. J Physiol 67:87−96
18. Driscol TE, Eckstein RW (1967) Coronary inflow and outflow responses to coronary artery occlusion. Circ Res 20:485−495
19. Hammond GL, Austen WG (1967) Drainage patterns of coronary arterial flow as determined from the isolated heart. Am J Physiol 212(6):1435−1440
20. Hutchins GM, Moore GW, Hatton EV (1986) Arterial-venous relationships in the human left ventricular myocardium: Anatomic basis for countercurrent regulation of blood flow. Circulation 74(6):1195−1202
21. Bellamy RF (1978) Diastolic coronary artery pressure-flow relation in the dog. Circ Res 43:92−101
22. Downey JM, Kirk ES (1975) Inhibition of coronary blood flow by a vascular waterfall mechanism. Circ Res 36:753−760
23. Ellis AK, Klocke FJ (1980) Effects of preload on the transmural distribution of perfusion and pressure-flow relationships in the canine coronary vascular bed. Circ Res 46:68−77
24. Uhlig PN, Baer RW, Vlalickes GJ, Hanley FL, Messina LM, Hoffman JIE (1984) Arterial and venous coronary pressure-flow relations in anesthetized dogs: Evidence for a vascular waterfall in epicardial coronary veins. Circ Res 55:238−248
25. Klocke FJ, Canty JM Jr., Mates RE (1984) Evolving concepts of coronary pressure-flow relationships. In: Able FL, Neuman WH (eds) Functional aspects of the normal, hypertrophied and failing heart. Nijhoff, Boston, pp 40−56
26. Mates RE, Klocke FJ, Canty JM (1988) Coronary capacitance. Progress in cardiovascular diseases 31(1):1−15
27. Chilian WM, Eastham CL, Layne SM, Marcus ML (1988) Small vessel phenomena in the coronary microcirculation: Phasic intramyocardial perfusion and coronary microvascular dynamics. Prog Cardiovasc Dis 31(1):171−238
28. Kagiya F, Tsujoka K, Goto M et al. (1986) Functional characteristic of intramyocardial capacitance vessels during diastole in the dog. Circ Res 58:476−485
29. Gratten MT, Hanley FL, Stevens MB et al. (1986) Transmural coronary flow reserve patterns in dogs. Am J Physiol 250:H276−H283
30. Hoffman JIE, Buckberg G (1976) Progress in cardiology. Lea Febiger Publ., 5:37−89
31. Wong AYK, Armour JA, Klassen GA, Lee B (1984) The dynamics of the coronary venous system in the dog. J Biomechanics 17(3):173−183

32. Leinberger H, Tillmans H, Hoppe S, Kuber W (1982) Microcirculatory impairment following transient myocardial ischemia. In: Tillmanns H, Kubler W, Zebe H (eds) Microcirculation of the heart. pp 98–103
33. Klocke FJ (1987) Measurement of coronary flow reserve defining the pathophysiology vs. making decisions about patient care. Circulation 76(6):1183–1189
34. Jennings RB, Kloner RA, Ganote CE, Hawkins HK, Reimer KA (1982) Changes in capillary fine structure and function in acute myocardial ischemic injury. In: Tillmanns H, Kubler W, Zebe H (eds) In microcirculation of the heart

17

# Methods of coronary and perfusion measurement

S. Meerbaum

There obviously is a need to measure the coronary blood supply and to quantitate myocardial perfusion. Such measurements are more difficult in the human, but a number of useful techniques are being introduced for diagnosis and evaluation of treatments.

**Quantitation of epicardial coronary supply and reserve**

Mean as well as phasic epicardial coronary artery (and in some cases coronary venous) blood flow or velocity can be very accurately determined in most animal preparations, and also in the open chest during cardiac surgery, for example, by means of electromagnetic [1] or ultrasonic [2] probes along with suitable flowmeter techniques. Measurement of epicardial vessel blood flow at a specific site of a coronary artery provides information on the supply to a corresponding subserved myocardial region, even if it will not indicate the detailed intramyocardial transport and distribution of the perfusion. Extensive experimental and some intraoperative application have already yielded quantitative data on resting or potentiated coronary artery flow rate or velocity. These probes are attached to the exterior surface of the coronary vessel, and thus interfere minimally with the intraluminal hydrodynamics. Calibration methods had to be perfected and firm contact between probe and vessel ensured (particularly with electromagnetic probes) to facilitate reliable acute as well as chronic repeat measurement of phasic and mean flow (velocity in the case of the ultrasound probe). One of the more recently reported and validated techniques is a suction-attached single crystal 20 MHz pulsed Doppler flow probe, which is readily attached (without need for vessel dissection) and has been used to intraoperatively measure blood flow velocity and coronary reserve in individual vessels of patients [3]. Another attractive recent technological development, based on earlier research by Cole and Hartley [4], is an intracoronary steerable selective catheter with a miniaturized Doppler flow probe at its tip [5]. Further improvement, demonstration and validation should make these catheter-borne instruments very useful tools for measurement of coronary blood flow and reserve in the human. Using intracoronary papaverine for vasodilatation, substantial coronary reserve data have already been collected in patients undergoing catheterization [6].

Currently, changes in coronary blood flow are also being assessed indirectly in the clinical setting. Thus, coronary sinus outflow measurement with thermodilution can characterize changes in mean coronary blood supply of the left ventricle [7,

8]. Thermodilution is a simple and widely applied technique, which basically depends on thermal measurements within the coronary sinus and/or the great cardiac vein. It uses multiple brief saline injections of a predetermined rate through special coronary venous catheters, and depends on excellent saline-blood mixing. Full quantitation should not be expected over a wide range of flows, primarily because of the nature of the vascular system and presence of coronary shunts. Another problem is precise and stable positioning of the catheter tip within the coronary sinus. It is not practically feasible to measure the total coronary venous outflow, because of major branching vessels issuing flow at the very ostium of the coronary sinus. Also, while measurement of coronary flow is generally satisfactory in normal states, errors must be anticipated during pronounced disease conditions. Nonetheless, thermodilution is very convenient and continues to be employed to semi-quantitatively record coronary flow changes in various physiologic states, and with interventions.

Digital subtraction angiography and videodensitometric methods using radiographic contrast materials have also been employed, and some of the current computerized systems show promise for clinical definition of the coronary supply [9, 10]. When contrast is infused into a coronary artery, changes in mean blood flow can be determined by videodensitometric measurement of contrast transit time, while angiograms also provide assessment of vessel dimensions. Apparently, rapid changes in flow cannot be quantitated. Small and complex vessels present problems, and absolute measurements depend on improved control of contrast injections and advanced computerized digital subtraction methods.

Coronary reserve, defined as the ratio of maximally potentiated to resting coronary artery flow at a given supply pressure, serves as an index of available myocardial perfusion. Coronary artery flow is inherently or deliberately augmented, e.g., during hypoxia or ischemia, with pacing or exercise, pharmacologically using specific vasodilators, or during intracoronary contrast administration. Safe application of dipyridamole or papaverine has been demonstrated.

Maximal vasodilation substantially increases the coronary artery flow in the normal heart, (by a factor of $4-5$), whereas in severe coronary artery disease this factor is generally low (e.g., close to 1). Important description of regional coronary reserve (e.g., in subepicardial and subendocardial layers), requires accurate measurement of local myocardial perfusion, the ultimate objective of sophisticated techniques now being developed.

## Myocardial perfusion and its distribution

Many techniques have been considered to accurately determine regional myocardial perfusion, or conversely, to measure the degree of myocardial underperfusion. Most of these techniques are quite useful for study of flow changes but cannot as yet be considered fully quantitative for study of absolute flows. The one accepted "gold standard" for quantitative regional myocardial perfusion measurements, i.e., the radionuclide microsphere technique, features excellent spatial resolution, does not alter the circulation or function, and can be applied repeatedly to measure global and regional myocardial blood flows. However, it is limited to experimental studies and does not provide information on phasic flows. Only a few semiquantitative methods are available for clinical use, and intensive current development is still aimed at quantitative assessment of regional perfusion.

*The radionuclide microsphere technique* [11, 12] uses uniformly sized radionuclide tagged microspheres, $9-15$ μm in diameter. These microspheres are to mix uniformly with the flowing blood, and are delivered into the myocardium where they become entrapped in capillary or precapillary vessels. The number of microspheres which enter the myocardium during the first (and only) transit is proportional to myocardial blood flow. The numnber of the isotope-labelled spheres in myocardial regions is readily measured, since it is proportional to the counting rate. Microsphere injection is usually into the left atrium, and a withdrawal rate of about 25cc per minute from the aorta is used to provide a reference sample for count measurement. Assuming no arterio-venous shunting (apparently a reasonable but not totally correct assumption), the operative equation states that total coronary inflow (ml/min) is equal to a calibration factor multiplied by the sum of the measured activity of individual tissue samples (counts/min) times the weight of the tissue samples. Among the radionuclides applied for computerized measurements by gamma counters (up to 6 or even 9 have been found distinctly resolvable) are: $^{125}$I, $^{111}$Ce, $^{51}$Cr, $^{85}$Sr, $^{95}$Nb, $^{46}$Sc.

After sacrificing the animal, the heart is cut transversely, and the resulting short axis slabs are further subdivided into subepicardial, midmyocardial and subendocardial layers, each of which is cut into anatomically regional tissue samples (e.g., myocardium subserved by the left anterior descending or left circumflex coronary artery. It has been shown that for validity and accuracy, individual tissue sample size should be larger than 1 gm and contain about 400 microspheres/gm. To minimize errors, about 2 million microspheres or more are often used per 100 gm myocardium. There are further constraints for special physiologic states. Thus, one must consider and correct potential errors due to gradual loss of spheres from capillaries, and excessive microsphere dilution in the presence of myocardial edema. Numerous reproducibility and validation studies have now been performed, using modern computerized radionuclide microsphere systems. It was concluded that this experimental measurement technique offers sufficiently quantitative results to serve as a "gold standard" for proper evaluation of various new clinically oriented approaches to assessment of myocardial perfusion.

*Xenon clearance assessment of myocardial perfusion:* This radioactive tracer technique was introduced by Cannon [13] and is a clinical myocardial blood flow measurement method which is similar to a number of other inert gas clearance methods. $^{133}$Xe dissolved in saline is injected into a coronary artery, and activity curves are obtained periodically to estimate regional blood flow. The radioactive inert gas diffuses into myocardial tissue subserved from the injection site in the coronary artery, and is then washed out from the tissue as a direct function of blood flow. Washout curve analysis uses a monoexponential curve and specifically its initial slope of "washout" which is proportional to blood-flow. The applicable Kety-Schmidt equation indicates myocardial blood flow (ml/min/gm of myocardium) being equal to the above slope computed from the semi logarithmic washout curve, times the myocardium-to-blood partition coefficient, divided by the specific gravity of the myocardial tissue. Constant flow, instantaneous indicator-myocardium equilibration and spatial uniformity of myocardial partition coefficient are assumed.

Generally, 10 mCi of $^{133}$Xe are injected into either the left or right coronary artery, following which gamma ray emission is registered on a gamma scintillation camera. Computerized analysis is then performed in recorded consecutive images and data processing yields washout curves which characterize myocardial blood flow, and its spatial as well as temporal distribution. Reproducibility and validity (vs radionuclide microsphere measurements) have been reported, indicating potential applicability of this methodology. Safe application of Xenon perfusion measurements is feasible in the catheterization laboratory, and perfusion has indeed been studied both at rest and during exercise or drug potentiation to evaluate the significance of obstructions [14]. Sequential studies (intervals as short as 10 min) can be carried out because of the short half-life of Xenon.

There are problems with the Xenon technique. One relates to Compton scatter, which reduces measurement accuracy. Another is Xenon solubility in cardiac fat limiting the number of measurements, and rather different characteristic partition coefficients of myocardial muscle, scar or fat. For high blood flows, Xenon may not be adequately diffusible. Recirculation and contamination of washout curves by non-myocardial counts necessitates limiting curve analysis to the initial (30–40 s) washout period. The difficulty due to assumption of uniform gas concentration may be reduced when studying individual regions, e.g., normal or underperfused myocardium. Finally, a practical hindrance to Xenon perfusion measurements applications is the elaborate angiography-nuclear setup needed to perform controlled intracoronary tracer injections.

*Thalium myocardial perfusion scintigraphy:* This qualitative radionuclide imaging technique is perhaps the most commonly clinically applied approach to spatial assessment of myocardial perfusion, by labeling normal tissue and slowing abnormal zones as voids [5]. External imaging is carried out after intravenous injection of the agent 201 Thallium, a radiopharmaceutical which is efficiently extracted from the blood by the cardiac muscle. The myocardial uptake is primarily affected by the regional blood flow, although it also depends on myocardial metabolism. Thus, regionally reduced uptake ("cold" spot) generally signifies an inadequate myocardial blood flow. Noninvasive and safe Thallium assessment of myocardial perfusion is widely used during exercise testing for evaluation of patients with coronary artery disease. After 201 Thallium is injected intravenously at an appropriate exercise load, gamma imaging is performed in various patient positions. The entire procedure is generally completed within 40 min.

In normal states, the heart is imaged by Thallium as a horseshoe-shaped myocardial structure, exhibiting almost homogenous radionuclide uptake. Perfusion defects appear as regions of reduced radionuclide uptake. Exercise images showing 15% of more of the horseshoe circumference as ischemic are usually considered clinically significant. The Thallium technique has been enhanced by computer processing and analysis of data. Although certainly successfully applied in assessment of perfusion defects, further studies of tracer kinetics have yet to prove that the method could also provide a quantitate assessment of myocardial perfusion.

*Positron emission computerized tomography:* This is a diffusible indicator technique, using an intravenously injected positron emitting radionuclide $^{13}$N-ammonia agent, for measurement of myocardial perfusion and metabolism [16]. Employing appropriate corrections and computerized analysis, the absolute regional myocardial radionuclide content is measured with a gated fast positron camera. Experimental validations against radionuclide microspheres indicate very good agreement. It is intended to measure subendocardial and subepicardial perfusion, and number of impressive study applications have been reported, including PET in conjunction with dipyridamole, but there are some questions about the temporal resolution of the technique. Positron emission computerized tomography appears to be headed for extensive clinical application, particularly if it be further improved and the cost of the system can be reduced. Thus, use of Rubidium 82 appears to be advantageous since it does not require an on-site cyclotron [17].

*Digital angiographic procedures [18]* with intravenous contrast injection and video-densitometric measurements have been applied to study the regional myocardium during normal contrast filling, or in the presence of localized underperfusion. Similarly, recent research indicates that *myocardial contrast two dimensional echocardiography* is capaple of outlining regions of myorcardial perfusion or perfusion defects, and possibly also assess the level of myocardial perfusion [19]. Most recently, clinically oriented myocardial perfusion measurement was found to be feasible and satisfactory with a *rapid acquisition computed tomography* technique, along with intravenous injection of a contrast agent. Indicator dilution type analysis of contrast clearance in images of regional myocardium yielded several indices which correlated well with radionuclide measurements of regional myocardial perfusion. Finally, *magnetic resonance imaging* may also be applied to myocardial blood flow measurements, but this, as well as the preceding methods, are still under development [21].

## 3. Anatomy and physiologic significance of coronary stenosis

The understanding and evaluation of the functional significance of coronary artery stenosis was discussed by Gould in August, 1988 [22]. Anatomic and physiological quantification of coronary artery stenosis is clearly desirable, e.g., for decisions relative to pharmacologic interventions, PTCA, thrombolysis or bypass surgery.

Utilizing coronary arteriography, percent narrowing, absolute diameter and length of stenosis can be determined, and such dimensional data can then be applied in fluid dynamic analysis yielding estimates of stenotic vessel resistance, its pressure-flow characteristics, and of its coronary flow reserve. But, the severity and importance of a coronary stenosis is also affected by concommittant physiological characteristics of the vascular system, particularly the aortic input pressure, the prevailing coronary vasomotor tone, and baseline flow levels which reflect the perfused tissue's metabolic demands. Hence, quantitative assessment of a particular vessel's stenosis requires not only measurement of the anatomic lesion geometry, but also knowledge of the normal vessel's flow reserve, determined after normalization for an assumed standardized aortic pressure.

Actually, the commonly employed percent vessel narrowing, and even the absolute cross sectional area of the coronary lumen, are no longer considered adequate descriptors. Percent narrowing clearly does not account for such other geometric characteristics as length, shape, eccentricity, and multiple lesions. The absolute lumen area appears to usually correlate with directly measured coronary flow reserve in the case of the LAD vessel [23], but apparently not for other coronary artery stenosis. An appropriate flow reserve assessment, under standardized hemodynamic conditions, is said to be obtained when all the geometric dimensions of a stenosis are considered, i.e., narrowing, absolute diameter and length [24].

Assessment of the physiological functional effects of a coronary stenosis requires measurement of myocardial perfusion in resting and maximum coronary flow states. The previously mentioned invasive methods such as coronary sinus thermodilution, Doppler-tipped coronary artery catheters and digital subtraction angiography, can provide some data, but it is now hoped to develop noninvasive semiquantitative and quantitative myocardial perfusion imaging techniques. The coronary blood flow is increased pharmacologically, usually with intravenous dipyridamole for the noninvasive (vs intracoronary papaverine for the invasive) studies. Coronary flow reserve is then defined as the maximal potentiated flow divided by the resting flow. Since this computed flow reserve varies with factors such as changing afterload, it is more appropriate to define a relative maximal flow, i.e., the stenotic coronary region's peak potentiated flow related to an equivalent peak flow from a corresponding nonstenotic vessel. This relative regional perfusion reserve can be assessed by radionuclide imaging, contrast echocardiography, nuclear magnetic resonance imaging, or computed tomographic scanning.

According to Gould, all the prospective methods of perfusion measurement are inadequate because their output is not proportional to flow at the high levels induced by flow potentiation. He points out that radionuclide methods, fast computed tomography, microbubble contrast echocardiography, and nuclear magnetic resonance imaging all have signals which lead to plateaus as coronary flow rises at these high levels. The image reconstruction techniques in positron emission tomography (PET) are said to be clinically reliable and, according to Gould, the PET signal is currently significantly better than with other imaging modalities. When used with dipyridamole stress testing, positron imaging of the heart is presently deemed optimal for accurate noninvasive diagnostic assessment of myocardial perfusion and of the physiological significance of coronary stenosis severity. Since clinical methods of perfusion measurement are undergoing intensive development, a definitive comparative assessment of the potential of techniques may be premature.

### References

1. Marsdon EL, Barefoot CA, Spencer MP (1959), Non-cannulating measurement of coronary blood flow. Surg Forum 10:636−639
2. Reneman RS (ed) (1974) Cardiovascular applications of ultrasound. Proceedings of an international symposium. Janssen Pharmaceutica, Beerse, Belgium, May 29−30, 1973. American Elsevier, New York

3. Marcus ML, Wright C, Dotey D et al. (1981) Measurements of coronary velocity and reactive hyperemia in the coronary circulation of humans. Circ Res 49:877–891
4. Cole JS, Hartley CJ (1977) The pulsed doppler coronary artery catheter; preliminary report of a new technique for measuring rapid changes in coronary artery flow velocity in man. Circulation 56:18–25
5. Wilson RF, Laughlin DE, Ackell PH et al. (1985) Transluminal subselective measurement of coronary artery blood flow velocity and vasodilator reserve in man. Circulation 72:82
6. Wilson RF, White CW (1986) Intracoronary papaverine, and ideal coronary vasodilator for studies of the coronary circulation in conscious humans. Circulation 73:444–451
7. Ganz W, Tamura K, Wallace JC et al. (1968) Measurement of great cardiac vein and coronary sinus flow in man by continuous local thermodilution. Circulation [Suppl 6] 37:6
8. Hayward R, White J, Ead H et al. (1983) A microcomputer based catheter laboratory system for analysis of coronary blood flow in man. J Biomed Eng 5:248
9. Foerster JM, Lantz BM, Holcroft JW et al. (1981) Angiographic measurement of coronary blood flow by video dilution technique. Acta Radiol (Diagn) 22:121
10. Vogel R, Lefee M, Bates E, O'Neill W, Foster R, Kirlin P, Smith D, Pitt B (1984) Application of digital techniques to selective coronary arteriography: use of myocardial contrast appearance time to measure coronary flow reserve. Am Heart J 107:153–164
11. Domenech RJ, Hoffman JIE, Noble MIM et al. (1969) Total and regional coronary blood flow measured by radioactive microspheres in conscious and anesthetized dogs. Circ Res 25:581–596
12. Baer RW, Payne BD, Verrier ED et al. (1984) Increased number of myocardial blood flow measurements with radionuclide labelled microspheres. Am J Physiol 246:H418–434
13. Cannon PJ, Dell RB, Dwyer EM Jr (1972) Measurement of regional myocardial perfusion in man with 133-Xenon and Scintillation camera. J Clin Invest 51:964–977
14. Maseri A, L'Abbate A, Michelassi C, Pesola A, Pisani P, Marzilli M, Denes M, Mancini P (1977) Possibilities, limitations and techniques for the study of regional myocardial perfusion in man by xenon 133. Cardiovasc Res 11:277–290
15. Strauss HW, Lebowitz E, Pitt B (1974) Myocardial perfusion scanning with thalium 201. Circulation [Suppl III] 50:26
16. Schelbert HR, Phelps ME, Hoffman EJ et al. (1979) Regional myocardial perfusion assessed with N-13 labelled ammonia and positron emission computerized axial tomography. Am J Cardiol 43:209–218
17. Gould KL, Goldstein RA, Mulliani NA et al. (1986) Noninvasive assessment of cornary stenoses by myocardial perfusion imaging during pharmacologic coronary vasodilatation: VIII. Clinical feasibility of positron cardiac imaging without cyclotron using generator-produced rubidium-82. J Am Col Cardiol 7:775–789
18. Mancini GBJ, Higgins CB (1985) Digital subtraction angiography: a review of cardiac applications. Prog Cardiovasc Dis 18:111
19. Kaul S, Glasheen W, Ruddy TD, Pandian NG, Weyman AE, Okada RD (1987) The importance of defining the left ventricular "Area at Risk" in vivo during acute myocardial infarction and experimental evaluation with myocardial contrast 2D-echocardiography. Circulation 75:1249–1260
20. Rumberger JA, Feirin AJ, Higgins CR et al. (1987) Use of ultrafast CT to quantitate myocardial perfusion: A preliminary report. J Am Col Cardiol 9:59–69
21. Singer JR (1986) Nuclear magnetic resonance blood flow measurements. Cardiovasc Interven Radiol 8:251–259
22. Gould KL (1988) Identifying and measuring severity of coronary artery stenosis. Circulation 78(2):237–245
23. Harrison DG, White CW, Hiratzka LF, Doty DB, Barnes DH, Eastman CL, Marcus ML (1984) The value of lesion cross-sectional area determined by quantitative angiography in assessing the physiological significance of poximal left anterior descending coronary artery stenoses. Circulation 69:1111–1119
24. Kirkeeide RL, Gould KL, Parsel L (1986) Assessment of coronary stenoses by myocardial perfusion imaging during pharmacologic coronary vasolilatation. VII. Validation of coronary flow reserve as a single integrated functional measure of stenosis severity reflecting all its geometric dimensions. J Am Col Cardiol 7:775–789

# Reperfusion

S. Meerbaum

Although coronary artery reperfusion is a frequently reviewed and discussed subject [1–17], this chapter will address a limited number of aspects which could have a bearing on the subsequently discussed coronary venous interventions. It is necessary to first briefly consider the rationales, mechanisms, applications, experience and evident benefits of coronary artery reperfusion or revascularization, and the clinical or surgical modalities for treatment of certain manifestations of coronary artery disease, including an acutely evolving myocardial infarction. Secondly, one should be aware that the potentially very effective reperfusion process is not always successful, and is frequently associated with myocardial injury and functional derangements, which may require supplemental treatment or temporary support. Along with other possible interventions, an adjunct retrograde delivery and restoration of nutritive perfusion (via coronary veins) might well be considered to protect the jeopardized myocardium during the primary procedure. The objectives of a complementary treatment would be to 1) extend cardiac tissue viability pending successful revascularization, 2) to counteract reflow injury and prevent infarct extension, and 3) to accelerate recovery from the frequently encountered postreperfusion dysfunction (stunning).

## 1. General rationales of coronary artery reperfusion

The past three decades have seen major conceptual changes in relation to acute myocardial ischemia and infarction. The prevailing older view was that, once a major coronary artery is acutely obstructed, the subserved myocardium is immediately and inevitably doomed to become necrotic. The most appropriate myocardial treatment in these circumstances was deemed to be an amelioration of the consequences, primarily by lowering the heart's oxygen demands, e.g., by reducing the heart rate, contractility and left ventricular loading. Apart from rest and control of arrhythmias, supplemental pharmacologic treatments might include systemic vasodilators and beta blockers. An intraaortic balloon counterpulsation method was also developed to provide effective mechanical unloading of the ischemic left ventricle.

On the basis of experimental observations in the 1960s and 1970s, e.g., Jennings and Reimer [1, 6], it was postulated that 1) acutely ischemic tissue (even in the absence of coronary collaterals) is capable of fully surviving for a certain limited period (say, 20 min), 2) there ensues over the next several hours a wave front propagation of necrosis in the underperfused "risk" zone, from endocardium to epicardium, and to a lesser extent laterally, and that 3) reversibly injured ischemic myocardium may be effectively treated to prevent or minimize necrosis. There ensued in the 1970s a massive amount of research in which many indirect (e.g., systemic) or direct myocardial modes of treatment were studied in animals, usually following an acute coronary artery occlusion. Evaluation of the newly proposed interventions was aided by significantly improved methods of assessing cardiac function, perfusion, metabolism, along with measurement of ischemic risk areas as well as the extent of irreversible or eventual myocardial damage. It was demonstrated in dogs, pigs and other animal species that myocardial viability and function could be maintained with some of the treatments by modifying the oxygen demand-to-supply ratio, and that infarct size could be significantly reduced when compared to untreated coronary occlusions. However, many of the reported favorable results were predicated on extremely early institution of the treatment, and also, presumably, on sufficient collateral communications allowing passage of directly acting myocardial drugs into the region distal to the coronary occlusion.

Unfortunately, extensive clinical trials and followup failed to corroborate the generally sweeping and optimistic experimental conclusions regarding various pharmacologic or mechanical assist treatments of acute myocardial ischemia. It was then concluded that only one clinical intervention seemed strikingly effective: prompt coronary artery reperfusion. Definitive salvage of acutely ischemic myocardium requires early reestablishment of its blood flow, at least to a level compatible with maintained tissue viability and survival. This minimal level is occasionally assumed as between 30% and 50% of the normal myocardial blood flow, but perfusion thresholds vary and are poorly defined. Continued research is providing improved understanding and guidelines needed for refinement of reperfusion methods.

In noncollateralized acutely ischemic dog myocardium (e.g., the posterior papillary muscle) necrosis might normally begin to appear as early as 20 min following a coronary occlusion (30–40 min in the variably collateralized canine ventricle free wall), and, as already indicated, the necrosis then increases with time by primarily encompassing a growing portion of the wall thickness (from endocardium toward the epicardium). Because of its characteristically profuse coronary artery collaterals, epicardial sparing is often observed in the dog, but near transmural infarction generally develops within 3 to 6 h from the time of coronary artery occlusion. A conceptualization of the time propagation of necrosis during a coronary occlusion, and the alterations due to subsequent reperfusion, is shown in Fig. 1, from one of the review articles [15]. Note the differences in infarct transmurality, and the extent, as well as distribution of hemorrhages caused by microvascular damage.

It should be noted that salvage is influenced by both the severity of the blood flow deficit and oxygen demands imposed upon the myocardium. The extent of collaterals as well as the physiologic state play a significant role. Generally, reestablishment of the dog's coronary flow after 2 h and even 3 h major coronary artery occlusion still often resulted in significant myocardial wall salvage [18], whereas reflow after 4 h or more frequently failed to offer permanent benefits [19]. Actually, a slower advance of necrosis was noted in conscious animals (lower heart rate), as exemplified by reperfusion after 4-h occlusion still providing a significant 39% myocardial salvage [20]. Much earlier infarction is found in the essen-

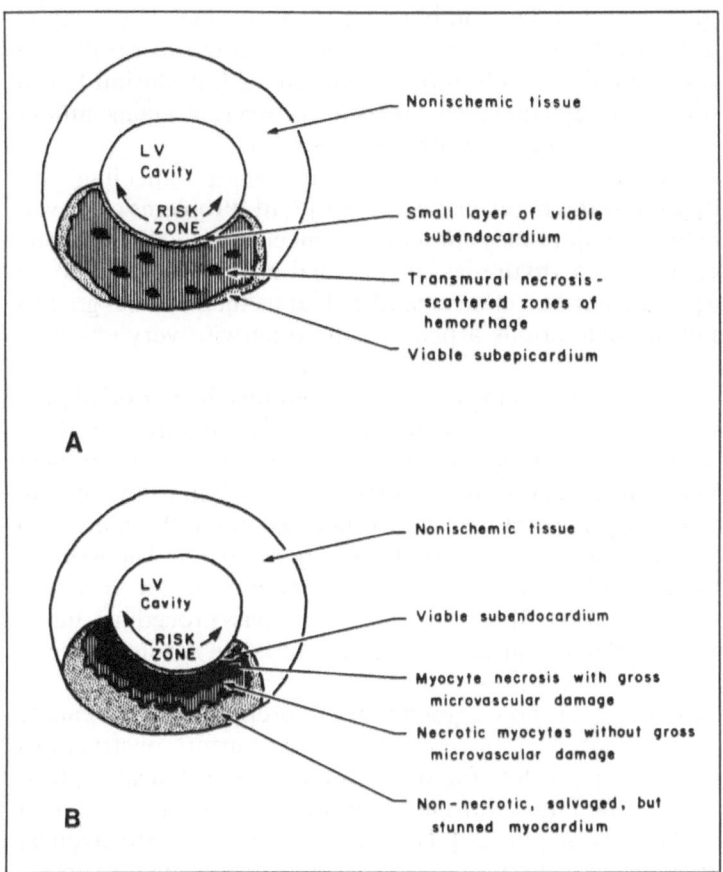

Fig. 1. *(A)* Schematic diagram showing a transverse section through a canine left ventricle subjected to a permanent coronary occlusion without reperfusion. The white area represents nonischemic myocardium supplied by the nonoccluded vessel. The infarct (hatched area) is transmural or neartransmural. There are scattered zones of hemorrhage (solid black). A small layer of viable subendocardium is present which derives its oxygen directly from the ventricular cavity. Where collateral flow is high, there may be a small rim of surviving subepicardium (stippled areas). *(B)* Schematic diagram showing a transverse section through a canine left ventricle subjected to coronary occlusion followed within 1 or 2 h by coronary reperfusion. The hatched and solid black areas represent the infarct which is confined to the inner half of the myocardium. The solid black area represents the zone of gross microvascular damage including the zones of no-reflow and hemorrhage. It is smaller than the total infarct. The remainder of the infarct without severe microvascular damage is represented by the hatched area and is located in the mid-myocardium. The epicardial portion of the ischemic zone (stippled area) has been salvaged by coronary reperfusion. It is nonnecrotic but stunned (postischemic ventricular dysfunction) for hours to days following coronary reperfusion. (By permission, from: Braunwald E, Kloner R (1985) Myocardial reperfusion: a double-edged sword? J Clin Invest 76:1713–1719.)

tially collateral free pig, in the rat and in non-human primates. Thus, Fig. 2 shows that reperfusion after a 30-min main coronary artery occlusion in rats resulted in contained nontransmural infarction, while reperfusion after 2-h occlusion led to a full transmural infarction, although there was still some protection against infarct extension when compared to persisting (2-week) occlusion [21].

Research also revealed characteristic functional derangements coinciding with or even being directly caused by the reperfusion process [7, 8]. Problems observed in spite of the reflow included major delays in return of mechanical and metabolic function, early post reperfusion arrhythmias, myocardial edema and occasional hemorrhages. Many experimental and clinical studies had as their special goal to devise means for countering deleterious aspects of the otherwise very effective reperfusion treatment.

In the surgical arena, emphasis was being placed on extending the period of protected global ischemia, to provide more time and a quiescent operative field for delicate multiple coronary vessel revascularization. Very extensive surgical research studies of myocardial preservation, recently reviewed [22], culminated in use of sophisticated cardioplegic, solutions, and otherwise carefully controlled reperfusion protocols, for cardiac surgery periods as long as 4−6 h. The issue of nonuniform distribution of hypothermic cardioplegia in the presence of coronary artery occlusion has been addressed [23]. Coronary artery bypass procedures have, in some instances, provided effective surgical treatment of acute coronary occlusions [24].

After protracted controversy, it is now agreed that the preponderance of acute myocardial infarctions are initiated by thrombotic coronary artery obstructions [25]. This conclusion was the basis for the rapid development and clinical application of intracoronary and intravenous thrombolytic methods for treatment of evolving acute myocardial infarction [26−27]. Having established the effectiveness of several thrombolytic agents, attention is now centered on increasing the success rate and on overcoming the early reperfusion derangements (e.g., stunned contraction), which are also observed during human reperfusion.

Just as exciting are the percutaneous transluminal coronary artery angioplasty (PTCA) techniques [28], now widely employed to electively balloon dilate severely stenosed vessels, and − to a lesser extent − for acute mechanical reopening of thrombosed coronary arteries. While dilatation of a stenosis is not generally analogous to the reperfusion most often investigated in the laboratory, PTCA does involve a series of brief vessel occlusions with intervening reflow periods [29]. Primary PTCA treatment, or post-thrombolysis dilation of residual stenosis, results in an increased coronary flow reserve and improved myocardial perfusion, yielding enhanced cardiac performance.

Following years of study and debate, it is now agreed that the perhaps only truly effective approach to reversing consequences of acute ischemia, before tissue damage becomes irreversible, is to promptly reestablish the myocardial perfusion. It is currently recognized that thrombolysis (and possibly aggressive angioplasty) constitutes a choice treatment in the emergency setting of evolving acute myocardial infarction, provided it is administered before substantial ventricular wall necrosis has supervened. It is also recognized that modern surgical revascularization [30] remains the most comprehensive approach in multivessel disease and constitutes

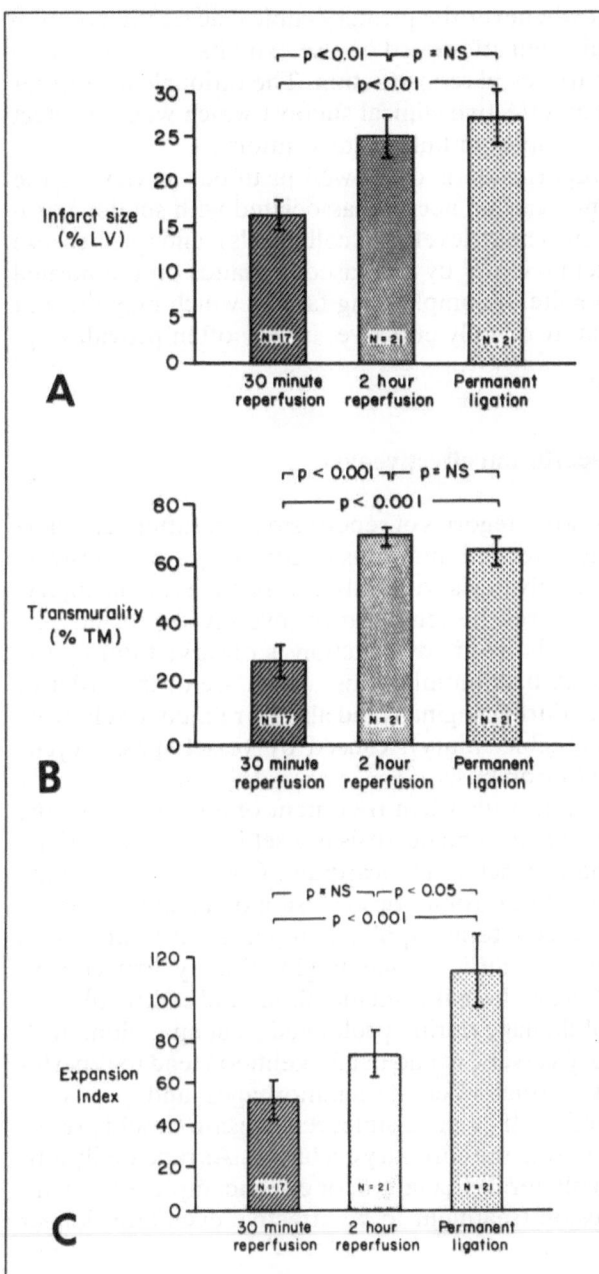

**Fig. 2.** Infarct size *(A)*, extent of transmurality *(B)*, and expansion index *(C)* for 30 min reperfused, 2 h reperfused, and permanently ligated hearts (mean ± SEM). (By permission, from: Hochman JS, Choo H (1987) Limitation of myocardial infarct expansion by reperfusion independent of myocardial salvage. Circulation 75(1):299–306.)

31

the necessary standby treatment whenever the primary clinical acute intervention is counterindicated, unsuccessful when attempted or else exhibits only temporary partial reflow followed by coronary vessel reobstruction. The rationale for further intensive research is to provide an effective clinical support which would protect myocardial viability during such complex or failed interventions.

The aggressive approach of reperfusion is often well justified, in view of the otherwise rapid progression of myocardial necrosis associated with sudden coronary artery obstruction (without sufficient developed collaterals), and also because of possible exacerbation of ischemic injury by increased demands on the normal segments of the heart. Thus, in spite of complicating factors which may prevent reperfusion from being immediately or fully effective, it very often provides significant benefits.

## 2. Primary factors influencing reperfusion effectiveness

It is perhaps useful to distinguish two categories of reperfusion, one after very short durations and the other following extended durations of coronary artery obstruction. In the former case, one generally deals with fully reversible ischemic injury, although the cardiac function may well be temporarily severely depressed as a result of the coronary obstruction. Ischemic dysfunction is observed during the fully occlusive intracoronary PTCA balloon inflations, which are of the order of 1 min duration. Ischemic episodes during angina could also be related to relatively short coronary obstructions (reversible injury), caused by vessel spasm, cyclic platelet deposition or intermittent thrombosis.

A somewhat different condition prevails when treatment of an acute ischemic injury is delayed beyond the time when some necrosis has set in, perhaps initially in the subendocardial layer of the ventricular myocardium. Consider an evolving acute myocardial infarction caused by a thrombotic occlusion of a major coronary artery in the absence of adequate collateral supply. If reperfusion treatment is started 3−4 h after the coronary occlusion, one might already expect substantial cardiac tissue necrosis, along with significant metabolic and contractile dysfunction. Irreversible myocardial damage during prolonged ischemia, along with potential injury to microcirculatory vessels, would in all likelihood lead to reperfusion complications such as myocardial edema, hemmorhages and persisting metabolic and contractile dysfunction. If at all feasible, reperfusion should preferably be applied within one or two h from the coronary occlusion. As repeatedly intimated, presence of a substantial collateral circulation, or else incomplete coronary occlusion, should permit reperfusion treatment to be effective even after longer periods of occlusion.

*Experimental evidence:* The following will illustrate the type of effects that may be anticipated with reperfusion. Figures 3 and 4 and Tables 1 and 2 indicate hemodynamics, percent histologic infarction and radionuclide microsphere measurements of regional myocardial perfusion (as well as its epicardial-endocardial distribution) in awake dogs, during 2 h or 6 h circumflex coronary artery occlusion, followed by reperfusion, vs permanent (7−10 day) occlusion [30]. Note the

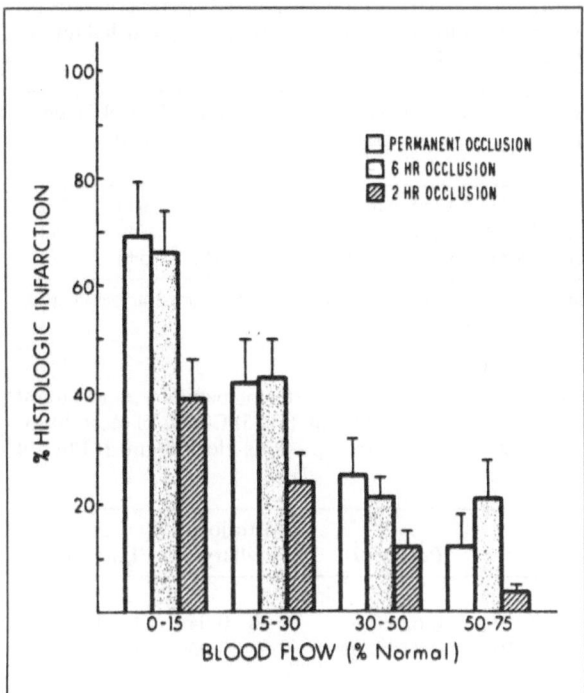

**Fig. 3.** Extent of histological infarction in myocardial samples grouped according to normalized ischemic blood flow ranges in epicardial layers of dogs subjected to permanent occlusion or occlusion for 6 or 2 h followed by reperfusion. Values are means ± SE. (By permission, from: Murdock RH et al. (1985) Effects of reestablishing blood flow on the extent of myocardial infarction in conscious dogs. Am J Physiol 249(18):H783–H791.)

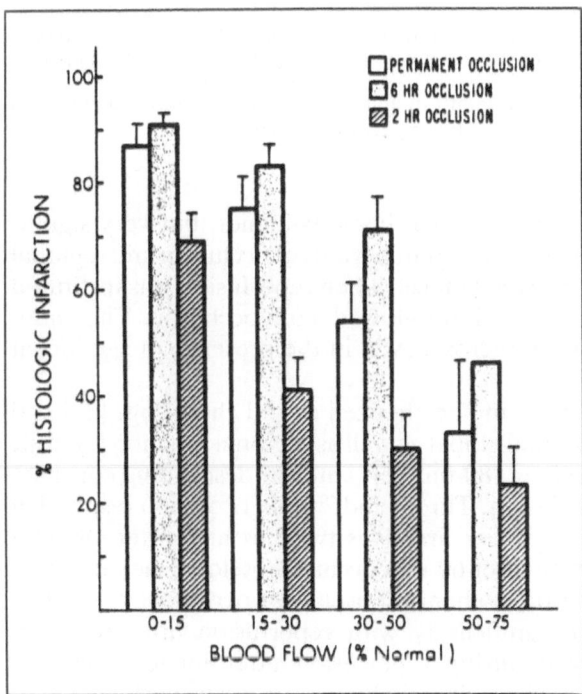

**Fig. 4.** Extent of histological infarction in myocardial samples grouped according to normalized blood flow ranges in endocardial layers of dogs subjected to permanent occlusion and occlusion for 6 or 2 h followed by reperfusion. Values are means ± SE. (By permission, from: Murdock RH et al. (1985) Effects of reestablishing blood flow on the extent of myocardial infarction in conscious dogs. Am J Physiol 249(18):H783–H791.)

**Table 1.** Hemodynamic measurements before and 2 and 6 h after occlusion. (By permission, from: Murdock RH et al. (1985) Effects of reestablishing blood flow on the extent of myocardial infarction in conscious dogs. Am J Physiol 249(18):H783–H791.)

| | Heart Rate, beats/min | | | Mean Aortic Pressure, mmHg | | | Mean Left Atrial Pressure, mmHg | | |
|---|---|---|---|---|---|---|---|---|---|
| | Control | 2 h | 6 h | Control | 2 h | 6 h | Control | 2 h | 6 h |
| Permanent occlusion | 84±7 | 113±6 | 122±6 | 97±5 | 102±4 | 92±3 | 4±1 | 6±1 | 7±2 |
| 6-h Occlusion + reperfusion | 102±11 | 110±8 | 122±8 | 98±6 | 106±4 | 90±4 | 2±1 | 4±1 | 3±1 |
| 2-h Occlusion + reperfusion | 94±6 | 101±4 | 120±8 | 105±2 | 108±4 | 95±3 | 5±1 | 5±1 | 5±1 |

Value are means ± SE. Hemodynamic measurements between the 3 groups were not significantly different at each interval.

**Table 2.** Blood flow to nonoccluded anterior and to posterior, posterior papillary, and lateral regions 2 h after occlusion. (By permission, from: Murdock RH et al. (1985) Effects of reestablishing blood flow on the extent of myocardial infarction in conscious dogs. Am J Physiol 249(18):H783–791.)

| | Anterior | Posterior | Posterior Papillary | Lateral |
|---|---|---|---|---|
| **Permanent occlusion** | | | | |
| Epi | 1.08±0.10 | 0.62±0.08 | 0.33±0.04 | 0.50±0.07 |
| Endo | 1.11±0.10 | 0.33±0.09 | 0.07±0.03 | 0.30±0.07 |
| **6-h Occlusion + reperfusion** | | | | |
| Epi | 0.88±0.06 | 0.52±0.06 | 0.26±0.04 | 0.33±0.04 |
| Endo | 0.95±0.08 | 0.24±0.04 | 0.04±0.03 | 0.16±0.03 |
| **2-h Occlusion + reperfusion** | | | | |
| Epi | 1.19±0.16 | 0.64±0.07 | 0.17±0.05 | 0.36±0.08 |
| Endo | 1.23±0.17 | 0.44±0.11 | 0.06±0.03 | 0.33±0.10 |

Values are means ± SE for blood flow measurements in epicardial (epi) and endocardial (endo) layers 2 h after occlusion in the permanent, 2-, and 6-h reperfusion groups.

nonsignificant reperfusion differences in global hemodynamics, but very significant differences in percent infarction for different periods of occlusion and regional blood flow levels. Again, infarct size was reduced when reperfusion was instituted 2 h postocclusion, but not when reperfusion followed a 6 h occlusion. The endocardium reflected a greater extent of necrosis even in the presence of significant residual myocardial perfusion.

Table 3 and Fig. 5 are from a study in anesthetized closed chest dogs [31] and show the changes in hemodynamics and global as well as regional function (systolic wall thickening) caused by 3, 20, 60 or 180 min left anterior descending coronary artery occlusion, followed by reperfusion. Three- and 20-min occlusion resulted in no necrosis. Infarct size as percent of risk area was twice as high after 180 min occlusion as compared to reperfusion after 60-min occlusion. Global ejection fraction did improve with reperfusion only when the occlusion period was less than 3 h. Regional function returned significantly with reperfusion after the very short occlusion, but return lagged during reperfusion after longer occlusion periods. An interesting feature was the strikingly high early postreperfusion end-

**Table 3.** Effects of coronary artery occlusion on hemodynamic variables in the four experimental groups (By permission, from: Haendchen RV et al. (1984) Increased regional end-diastolic wall thickness early after reperfusion: a sign of irreversibly damaged myocardium. J Am Col Cardiol 3(6):1444−1453.)

| | Group I (3 minutes' occlusion) | Group II (20 minutes' occlusion) | Group III (60 minutes' occlusion) | Group IV (180 minutes' occlusion) |
|---|---|---|---|---|
| Heart rate (beats/min) | | | | |
| Preocclusion | 84 ± 6.6 | 88 ± 7.3 | 78 ± 3.8 | 76 ± 7.3 ‡ |
| Postocclusion | 94 ± 9.8 | 94 ± 2.6 | 93 ± 4.3 | 97 ± 5.5 |
| Mean aortic pressure (mm Hg) | | | | |
| Preocclusion | 98 ± 2.4 | 90 ± 3.7 | 95 ± 3.6 | 97 ± 1.9 ¶ |
| Postocclusion | 93 ± 2.3 | 95 ± 2.8 | 95 ± 1.8 | 108 ± 4.3 |
| Left ventricular end-diastolic volume index (cc/m²) | | | | |
| Preocclusion | 68 ± 1.5 * | 71 ± 2.7 * | | |
| 71 ± 2.8 * | 74 ± 3.2 † | | | † |
| Postocclusion | 80 ± 5.5 | 89 ± 5.1 | 88 ± 5.1 | 85 ± 4.4 |
| Ejection fraction (%) | | | | |
| Preocclusion | 58 ± 2.8 † | 59 ± 0.5 † | 58 ± 1.0 † | 60 ± 1.1 † |
| Postocclusion | 36 ± 4.1 | 39 ± 2.0 | 35 ± 1.9 | 36 ± 1.4 |

*, †, ‡ p (probability) < 0.01(*), < 0.001(†), < 0.05(‡).

diastolic wall thickness noted in the ischemic zone after the longer occlusion, presumably reflecting edema, and also correlating with evidence of reflow hemorrhages (Figs. 6, 7).

It has been conclusively demonstrated that, even with occlusions compatible with reversible ischemic injury, the heart experiences a temporary yet very significant metabolic and contractile "stunning" (see below). The general pattern noted by many investigators is for regional function to be immediately and severely diminished during a coronary occlusion. With reperfusion following relatively short periods (e.g., 15 min) of acute ischemia, function tends to increase at first, but is subsequently stunned, to only gradually return toward normal levels. Most evidence also points to myocardium generally remaining viable after the short coronary occlusions. Reestablishment of the coronary artery flow reverses (or eventually normalizes) most of the common signs of significant ischemia, such as the ECG-ST segment elevations, elevated left ventricular end-diastolic pressure and increased chamber volume. Reperfusion following more extended periods of coronary occlusion (beyond, say, 2 h) results in partial but incomplete myocardial salvage. The functional and hemodynamic benefits of reperfusion may still be significant, yet normal function is seldom fully restored.

A more favorable (yet, admittedly, more difficult to analyze) situation prevails when the coronary occlusion is associated with substantial collateral blood supply to the ischemic myocardium. This is somewhat analogous to conditions prevailing with significant coronary stenosis, as compared to a total interruption of the coronary flow. As already pointed out, depending on the level of such residual myocardial perfusion, reperfusion (e.g., reversal of thrombotic occlusion by lytic or

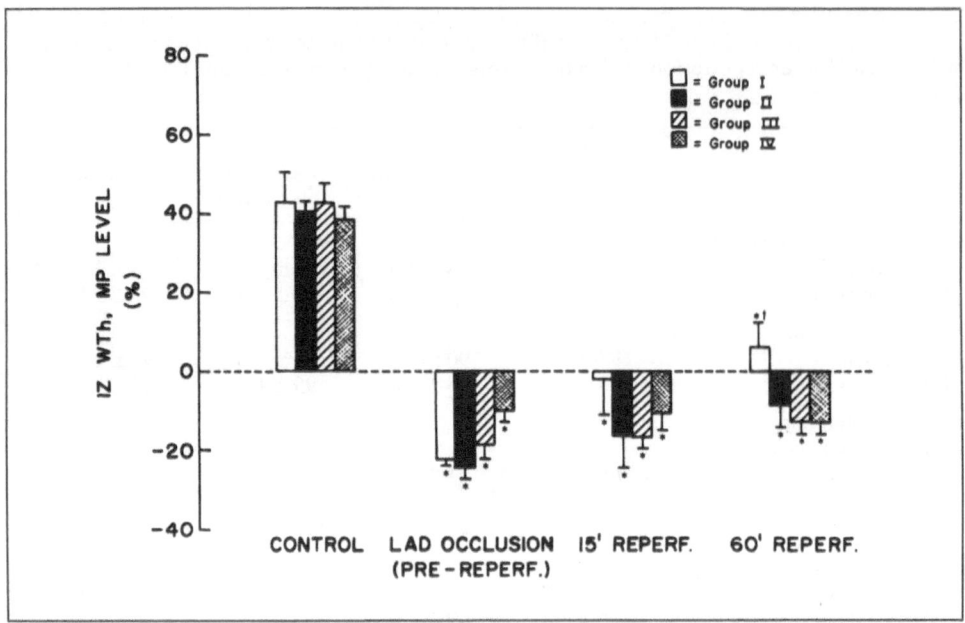

**Fig. 5.** Segmental systolic wall thickening (WTh) measured in the center of the ischemic zone (IZ) at the midpapillary (MP) short-axis level of the left ventricle during the control period, immediately before reperfusion and 15 and 60 minutes after reperfusion. Group I = 4 dogs with 3 minutes of occlusion of the left anterior descending coronary artery. Group II = 4 dogs with 20 minutes of occlusion. Group III = 5 dogs with 60 minutes of occlusion and Group IV = 12 dogs with 180 minutes of occlusion, followed (in all groups) by reperfusion. * = p < 0.05 relative to the control period, † = p < 0.05 relative to the period immediately before reperfusion (PRE-REPERF). (By permission, from: Haendchen RV et al. (1984) Increased regional end-diastolic wall thickness early after reperfusion: a sign of irreversibly damaged myocardium. J Am Col Cardiol 3(6):1444–1453.)

mechanical means) may prove effective even after longer periods of acute ischemia, yielding long-term myocardial salvage as well as enhanced cardiac function and reserve. Some of the major controversy surrounding animal experiments can indeed be traced to the highly variable yet substantial compensatory coronary collateral circulation in coronary occluded dogs, resulting in an almost 0% to 100% range of necrosis of the ischemic risk region. A wide range might also be expected when extensive chronic ischemic history leads to variably substantial coronary collateral development. On the other hand, the pig, an experimental animal whose coronary system resembles the human coronary anatomy, features very limited collaterals. Such a species may therefore be considered a reasonable model for the noncollateralized human heart.

## 3. Reperfusion mechanisms, injury, and treatment

Without attempting to cover the complex and repeatedly argued reperfusion mechanisms, it is necessary to be aware of terms including the "calcium and oxygen

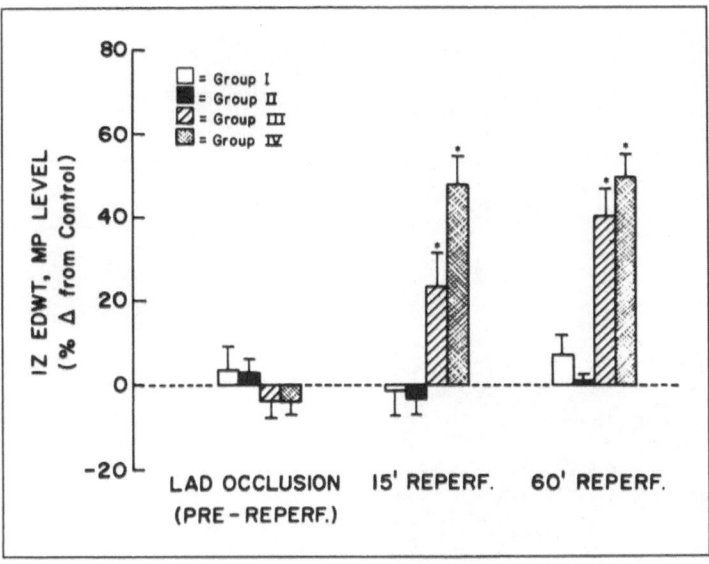

**Fig. 6.** Ischemic zone (IZ) end-diastolic wall thickness (EDWT) expressed as percent change ($\Delta$) from control in the four groups of dogs, measured at the midpapillary (MP) level of the left ventricle. There were minor changes in regional end-diastolic wall thickness from control to the pre-reperfusion period in all the groups; at 15 minutes after reperfusion there was a marked increase in this index in dogs with 60 (Group III) and 180 (Group IV) minutes of occlusion. * = $p < 0.05$ relative to the pre-reperfusion period. (By permission, from: Haendchen RV et al. (1984) Increased regional end-diastolic wall thickness early after reperfusion: a sign of irreversibly damaged myo-cardium. J Am Col Cardiol 3(6):1444–1453.)

paradox", the "no-reflow phenomenon", "post-reperfusion arrhythmias," and "temporary stunning of cardiac function". Recently, attention is centered on myocardial injury apparently induced by oxygen free radicals, both during the period of acute ischemia and particularly at the time of reperfusion. Deleterious effects characteristic of reperfusion after longer ischemia can be appreciated from striking morphologic changes shown in Fig. 8 taken from a review article by Becker and Ambrosio [16], which also contrasts reperfused vs nonreperfused infarcts. One notes myocyte disintegration, explosive cell swelling, sarcolemal disruption, mitochondrial fragmentation, sarcomere hypercontraction and contraction band necrosis.

Calcium overload relates to a sudden reperfusion in-rush of calcium (after ischemic calcium denudation), which could overwhelm the injured but potentially viable myocyte calcium balance, and thus cause contracture, swelling and cell necrosis. Actually, such a mechanism relates more properly to surgical calcium-free solution cardioplegia, whereas clinical myocardial ischemia does not involve a calcium-free environment. Reintroduction of oxygen to anoxic mycardium can also be detrimental, leading up to excess liberation of enzymes and contracture. This seems to contrast with anoxic fluid reperfusion of an ischemic heart. The "no reflow" phenomenon involves persistence of some regional myocardial underperfusion even after reopening of a previously occluded coronary artery. This can

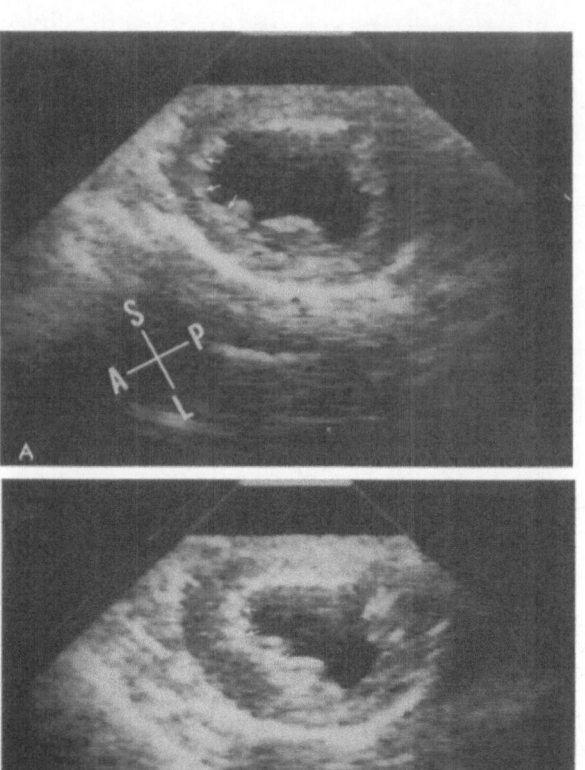
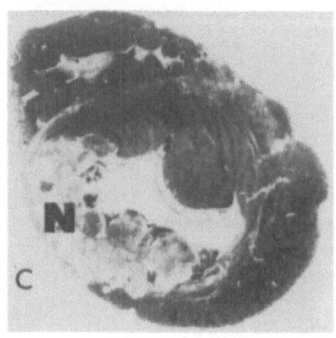

**Fig. 7.** Two-dimensional echocardiographic short-axis sections at the midpapillary muscle level of the left ventricle and pathologic study in a dog with 3 hours of occlusion followed by 7 days of reperfusion, presenting with nearly transmural myocardial infarction. **A,** Two-dimensional echocardiographic end-diastolic frame obtained at 3 hours of coronary occlusion, immediately before reperfusion (**arrows** indicate epicardial and endocardial interface in the ischemic zone). A = anterior; L = lateral; P = posterior; S = septum. **B,** End-diastolic frame obtained 15 minutes after reperfusion, showing marked increase in wall thickness in the ischemic anterolateral region. **C,** Transverse myocardial slice in an equivalent level of the left ventricle stained with triphenyltetrazolium chloride at 7 days after reperfusion. Note the necrotic area (N) in the anterolateral wall, which corresponds to the region of myocardium with increased end-diastolic wall thickness early after reperfusion shown in **B.** (By permission, from: Haendchen RV et al. (1984) Increased regional end-diastolic wall thickness early after reperfusion: a sign of irreversibly damaged myocardium. J Am Col Cardiol 3(6):1444–1453.)

→

**Fig. 8.** Histologic and ultrastructural appearance of necrotic myocardium (A and B) that has been reperfused (contraction band necrosis) or (B and C) not reperfused (coagulation necrosis). (A) Intracellular transverse eosinophilic bands alternating with cleared cytoplasmic spaces, representing clumps of sarcomeres separated by myofibrillar rupture (original magnification ×400). (B) Grossly disrupted cellular architecture, with contraction bands (left side of fig.) and cleared spaces (upper center and right), along with swollen mitochondria containing fragmented cristae and dense bodies representing precipitated calcium (original magnification ×8,000). (C) Characteristic changes of

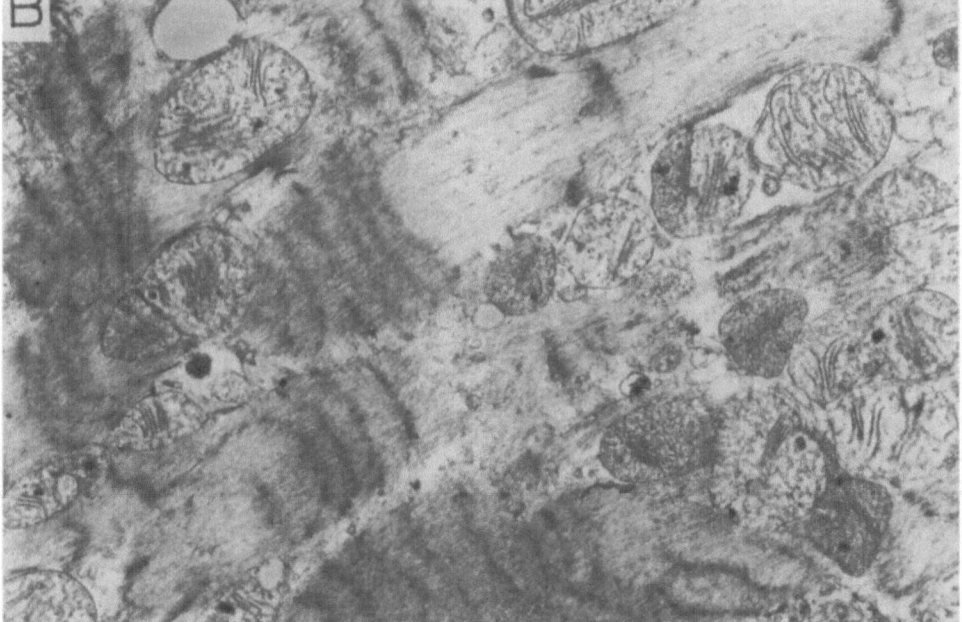

coagulation necrosis with nearly normal looking myofibrils, loss of nuclei, leukocyte infiltration, and moderate edema (original magnification ×125). (D) Marginated and clumped nuclear chromatin (center) without disruption of myofibrils or significant swelling of mitochondria (original magnification ×4,000). (By permission, from: Becker LC, Ambrosio G (1987) Myocardial consequences of reperfusion. Progr Cardiovasc Dis 30(1):23−44.)

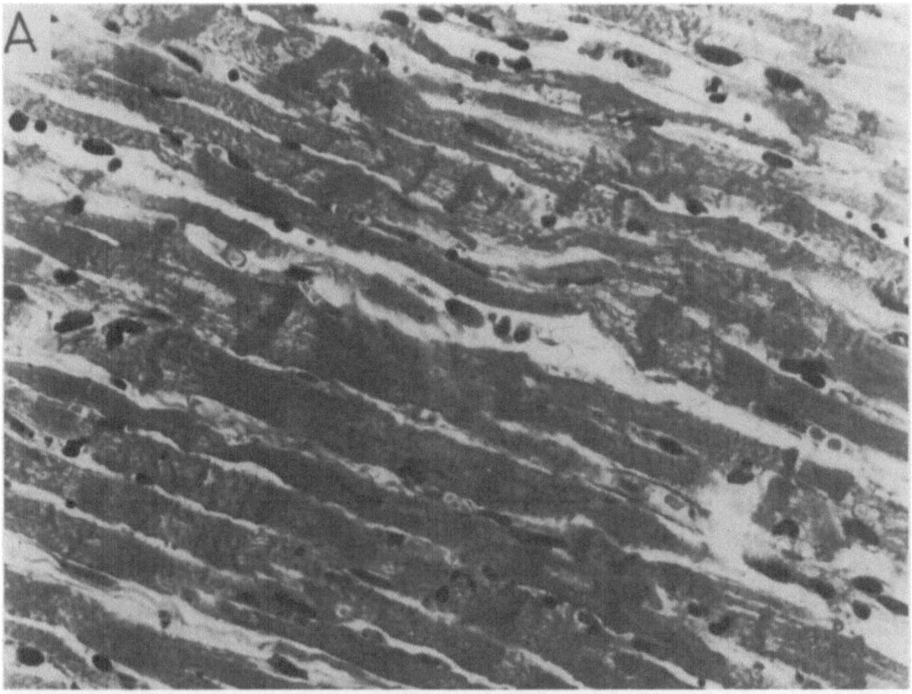

39

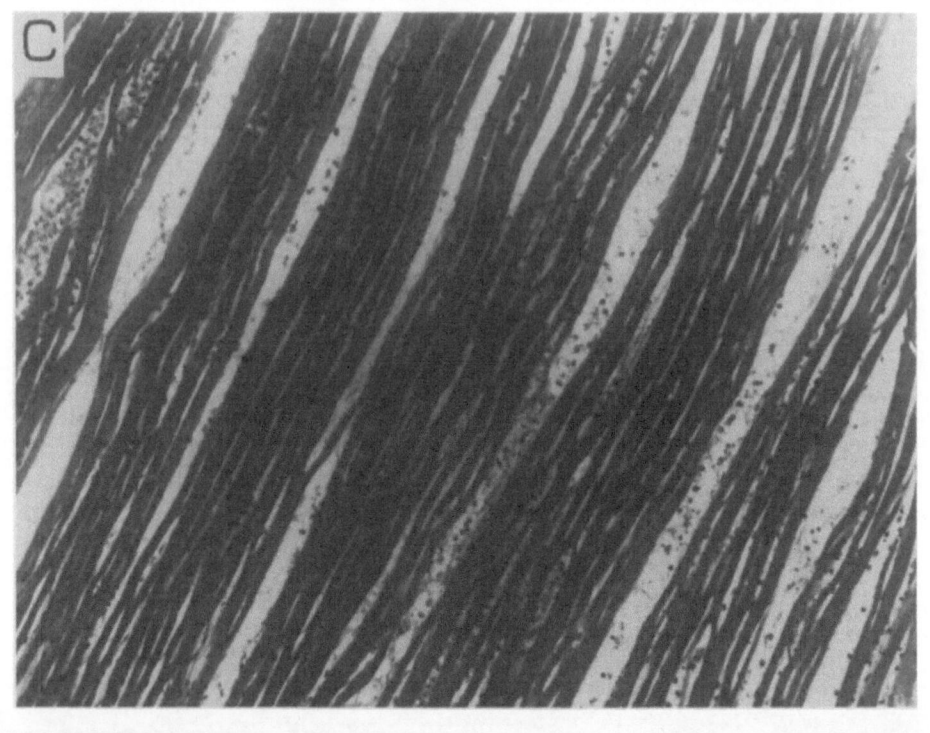

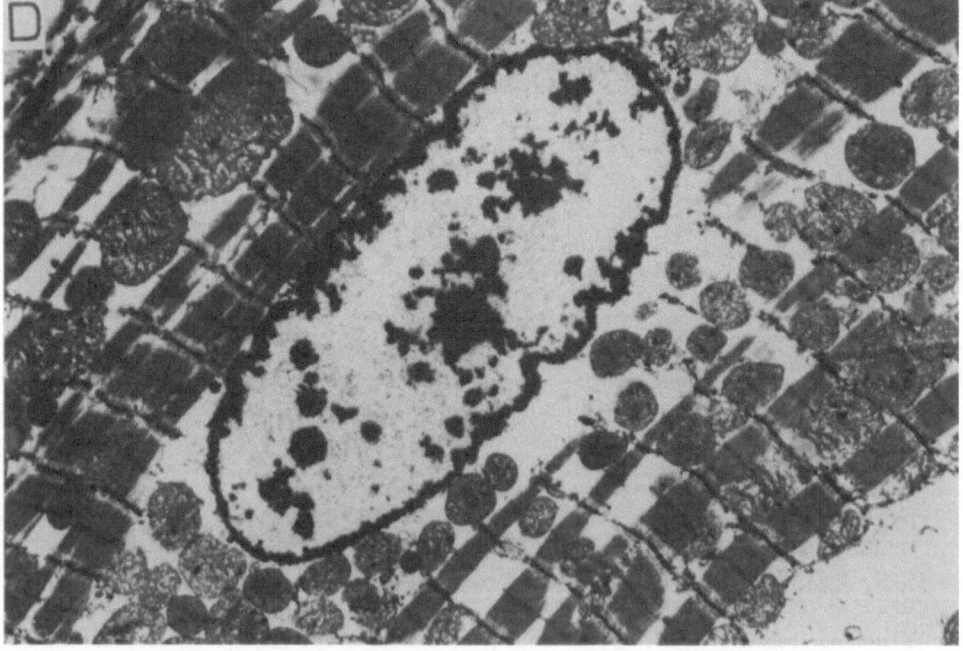

40

extend ischemic injury, and is believed to be associated with cellular and endothelial swelling which could obstruct microvascular flow, or else vessel plugging due to leukocyte deposition or thrombosis. It appears though that these mechanisms apply primarily in the case of irreversible ischemic injury. In the latter setting, i.e., after longer coronary occlusion causing pronounced ischemic injury, there is also evidence of significant myocardial edema and even microvascular injury with hemorrhages. These are generally found to be contained within the zone of infarction. Various measurements, such as those of diastolic ventricular wall thickness with two-dimensional echocardiography [31], have indicated distinct effects of reperfusion on myocardial edema.

Reperfusion can also result in rapid and abrupt onset of malignant ventricular arrhythmias and ventricular fibrillation [32]. These arrhythmias are more prevalent when reperfusion is instituted after about 30-min occlusion, and appear refractive to standard antiarrhythmic treatment. Increased automaticity is implicated, and alpha adrenergic blockade or depletion of norepinephrine are found to reduce the incidence of reperfusion ventricular fibrillations. Increase in intracellular calcium, washout of accumulated extra cellular potassium, and oxygen derived free radicals, are thought to play a role in arrhythmias during reperfusion after coronary artery occlusion [33, 34].

Much reperfusion study has been devoted to often significant persisting myocardial underperfusion, contractile dysfunction and metabolic derangements. The operative mechanisms are uncertain, but, in addition to the cellular effects of reperfusion, myocardial edema and its resorption may play a role in temporarily restricting perfusion. Figures 9 and 10 are from an early study by Ellis et al. in a 1983 Reperfusion Symposium [34], and clearly show the delay in return of function (% systolic wall thickening), and subepicardial as well as subendocardial ATP levels with reperfusion in dogs after 120 min occlusion. Other studies have indicated substantial reperfusion "stunning" even after occlusions of 5 or 15 min. Thus, after 5 min of canine coronary occlusion in the writer's laboratory (unpublished), regional contraction remained significantly depressed for more than 1 h. Following a 15 min dyskinesis, dysfunction clearly persists over longer periods, and after 2 h or 2-h coronary occlusion recovery may take several days or weeks. As demonstrated, there are delays in return of ATP from its depressed level caused by an acute coronary occlusion, while creatine phosphate seems to recover more rapidly after reperfusion. Viability of the stunned reperfused myocardium can be demonstrated, for example, by pharmacologic potentiation using dobutamine. The temporarily stunned reversible injured myocardium will generally recover within days and at most weeks. Ultimate infarct extension and lack of return to normal function are reasons for concern when the reperfusion is begun after periods of coronary occlusion exceeding 4–6 h.

Although not fully resolved, it appears that the deleterious effects noted following reperfusion result both from the underlying acute ischemic injury, and also from the process of reperfusion per se. Figure 11 shows a possible schema for myocardial cell death after coronary occlusion and reperfusion, as discussed in the Becker and Ambrosio review [16]. Many investigators now consider reperfusion injury to be due to generation of toxic free radicals upon oxygenated reperfusion. Conversely, investigators have shown that treatment with free radical scavengers

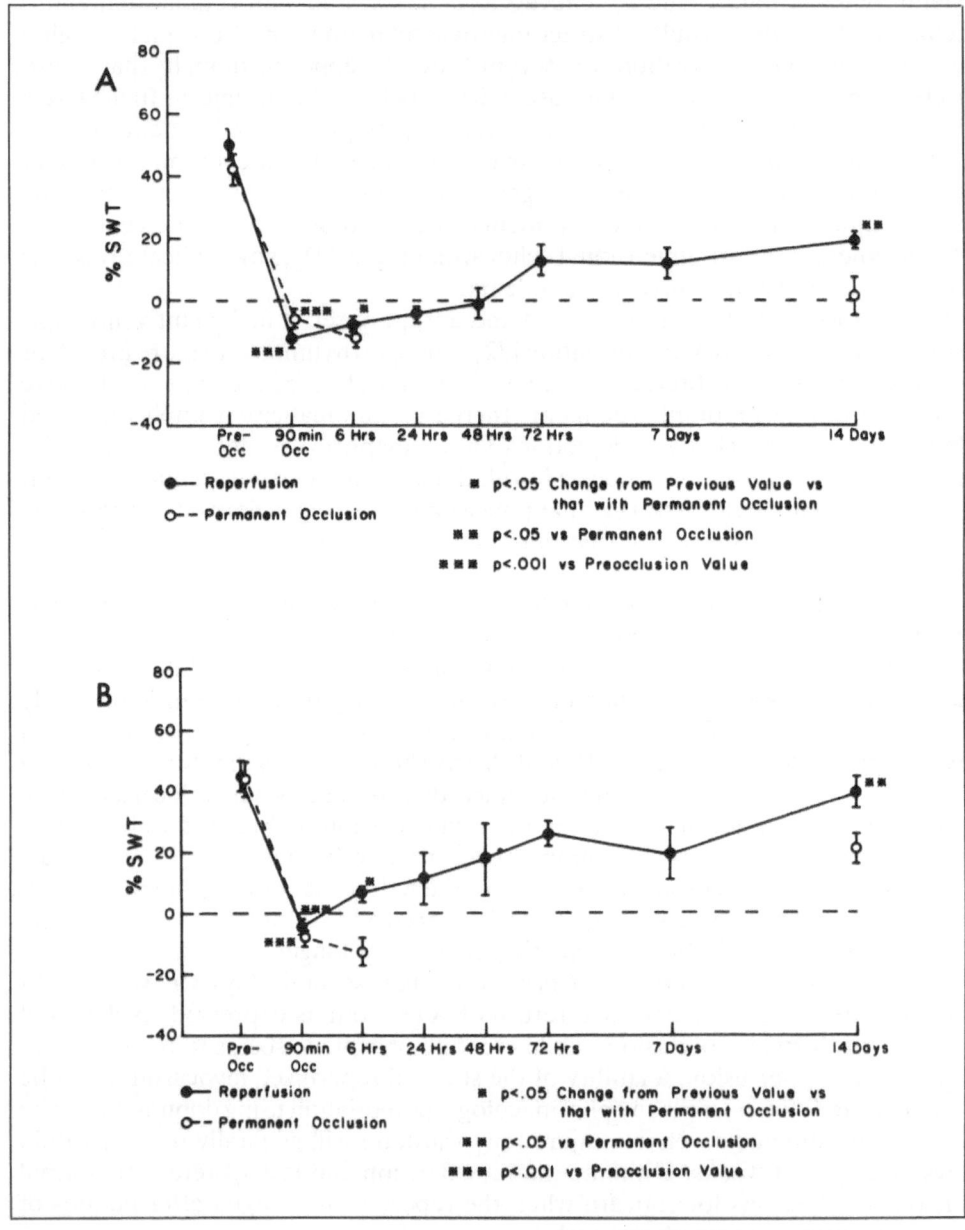

**Fig. 9.** Contractile function in the central ischemic zone **(A)** and peripheral ischemic zone **(B).** Shown on the **abscissa** is time after coronary occlusion and on the **ordinate,** percent systolic wall thickening (% SWT). No values were obtained for dogs with permanent occlusion between 6 hours and 14 days. Occ = occlusion; p = probability; Pre-Occ = preocclusion measurement. (By permission, from: Ellis SG et al. (1983) Time course of functional and biochemical recovery of myocardium salvaged by reperfusion. J Am Col Cardiol 1(4):1047–1055.)

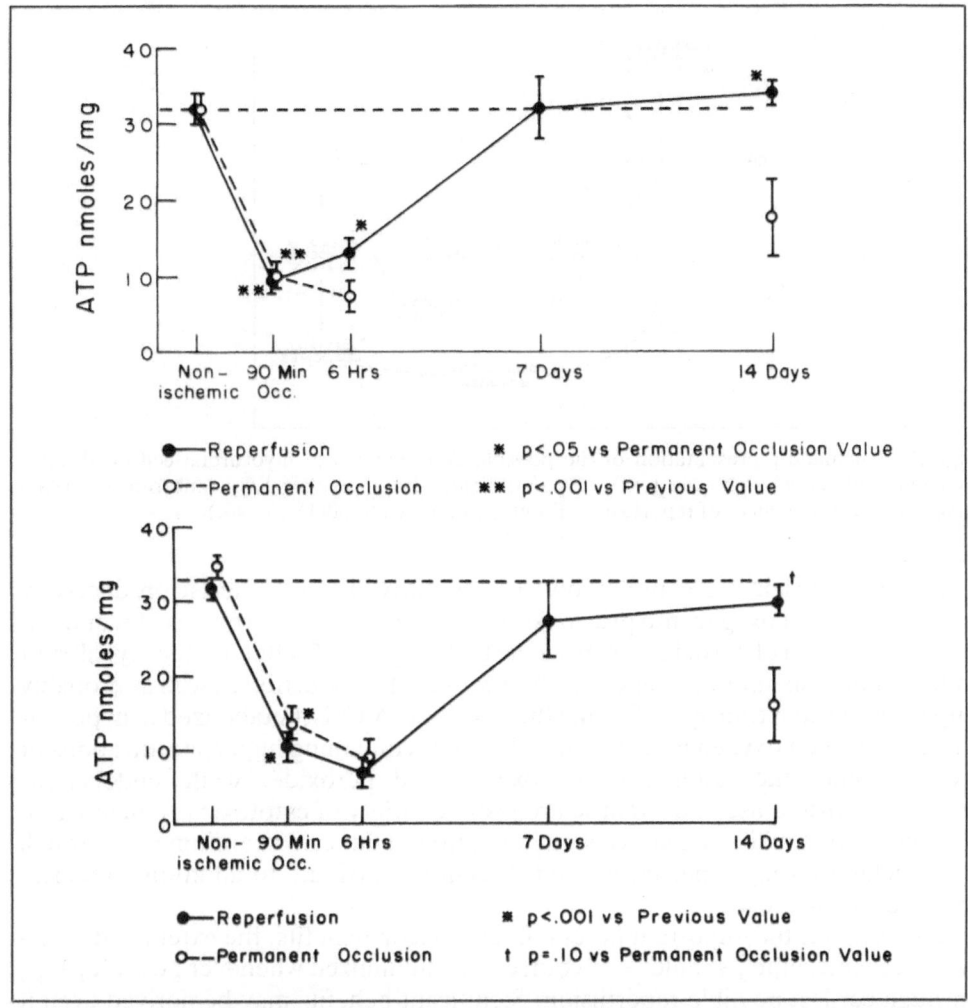

**Fig. 10.** Recovery of subepicardial **(top)** and subendocardial **(bottom)** adenosine triphosphate (ATP) with reperfusion. Shown on **abscissa** is time after occlusion and on the **ordinate.** ATP in nmol/mg cardiac protein. No values were obtained for dogs with permanent occlusion between 6 h and 14 days. (By permission, from: Ellis SG et al. (1983) Time course of functional and biochemical recovery of myocardium salvaged by reperfusion. J Am Col Cardiol 1(4):1047–1055.)

can benefit the reperfusion and results in eventual infarct size reduction. Thus, superoxide dismutase (SOD) plus catalase treatment begun 15 min before reperfusion following a 90 min circumflex coronary artery occlusion in the dog reduced infarct size from 44% to 22% of the occluded coronary risk area [35]. There was no salvage when the treatment began before the coronary occlusion. Even when recombinant human superoxide dismutase was applied at the moment of reperfusion, infarct size could be reduced by 40% in intact dogs [36]. Figure 12 indicates this result, and other data showed a better creatine phosphate and left ventricular function recovery when SOD was administered at the time of reperfusion.

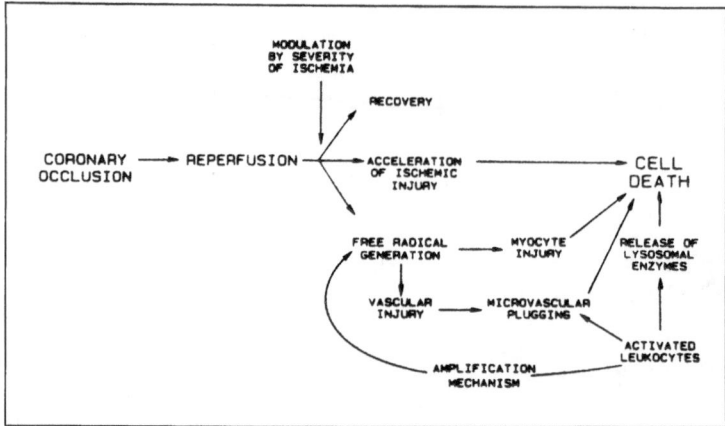

**Fig. 11.** Schematic representation of the possible determinants of myocardial cell death after ischemia and reperfusion (see text). (By permission, from: Becker LC, Ambrosio G (1987) Myocardial consequences of reperfusion. Progr Cardiovasc Dis 30(1):23−44.)

The mechanisms are complex but oxygen-derived free radicals, produced by reintroduction of oxygen into previously ischemic tissue, cause severe cellular damage by formation of peroxides, which lead to disruption of cellular integrity, disrupt mitochondrial membranes, and interfere with ATP production as well as probably impairing calcium transport. During the ischemia, ATP is metabolized to hypoxanthine, and when oxygen is reintroduced to cells containing high concentrations of hypoxanthine, the latter releases oxygen and peroxide, while endogenous superoxide dismutase and catalase are lost. A variety of cytotoxic products cause vascular permeability, leukocytes migrate into ischemic myocardium and through endothelium during reperfusion, and evidently contribute to additional reperfusion induced necrosis.

To maximize the important potential reperfusion benefits, the extent and severity of ischemic injury should − of course − be minimized whenever possible, e.g., through earliest possible reperfusion. Significant benefits may be derived even if only a portion of the myocardial wall (e.g., subepicardial layer) can be kept viable, and certainly presence of substantial collaterals should allow effective reperfusion even after longer periods of coronary obstruction. Since reperfusion is associated with some deleterious effects and can present hazards, potential supplementary treatment prior to and during the reperfusion should be considered. Any practicable means to reduce the ischemic injury before reperfusion can be instituted is clearly desirable, and this may perhaps be accomplished pending the intended primary intervention, e.g., thrombolysis. Thus, beta-adrenergic blocking agents, calcium channel antagonists, or coronary venous retroperfusion, might all help maintain myodardial viability up to the time of reperfusion. On the other hand, it is logistically advantageous and of great interest to have modes of treatment which can also be effective when applied during reperfusion. Hypertonic mannitol, fluorocarbons or scavengers of free radicals have been shown to have promise, preferably by direct application into the zone of the acute ischemia [15]. Any effective intravenous treatment at the time of reperfusion, would be desirable.

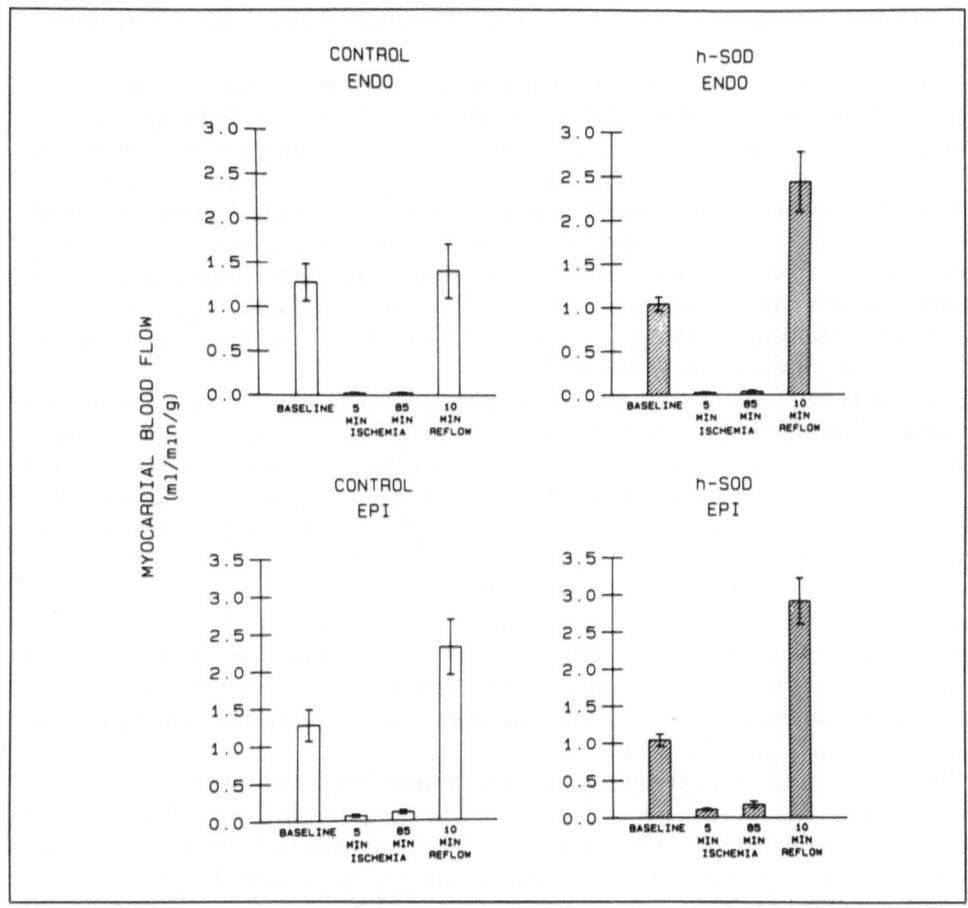

**Fig. 12.** Myocardial blood flow at different times during the study in the control *(left)* and the h-SOD Group *(right)*. (By permission, from: Ambrosio G et al. (1986) Reduction in experimental infarct size by recombinant human superoxide dismutase. Insights into the pathophysiology of reperfusion injury. Circulation 74:1424–1433.)

Among special methods also proposed to overcome the postreperfusion derangements have been a gradual or staged reflow, intraaortic balloon pumping support, and intermittent coronary sinus occlusion. All new methods must first be validated in experimental studies, encompassing blood flow determination, measurement of regional and global ventricular function, assessment of biochemical or metabolic alterations, observation of myocardial edema and any hemorrhages, and monitoring of postreperfusion arrhythmias. We may conclude that some evidence points to reperfusion per se causing myocardial injury, in addition to the significant damage that can be generated by the preceding ischemia, obviously depending on its severity and duration. At the same time, there is a growing indication that special interventions or combined treatments may eventually overcome or at least minimize the reperfusion injury.

## 4. Surgical coronary revascularization and intraoperative myocardial protection

Surgical graft bypass of severely obstructed coronary arteries plays a major role in the treatment of ischemic heart disease. The large number of bypass operations being performed, and the extremely low associated mortality and morbidity, are clear evidence of the contribution made to cardiac patient survival and quality life. No attempt is made to summarize this important and well documented field, beyond simply pointing to special reperfusion circumstances during coronary artery graft bypass revascularization, performed in the globally ischemic and non-contracting heart [37]. Initial research and applications highlighted specific types of early post-reperfusion derangements (e.g., "stone heart"), and there was and still is occasional perioperative infarction.

Recent developments in cardioplegia have played a major role in improving myocardial protection, even during protracted and complex surgical procedures. The aim has been to rapidly institute a total cardiac arrest, maintain the arrested heart in a still state and assure myocardial protection in spite of extended global ischemia, and then cause a fast and eventually full return to normal metabolic and mechanical cardiac function. The surgical procedure must meet practical requirements of a static, readily observable setting for the often delicate coronary revascularization procedure. These requirements were not satisfactorily met in the beating heart, or even when the heart's metabolism was decreased through topical hypothermia or induction of ventricular fibrillation, procedures which also caused regional dysfunction and maldistribution of myocardial perfusion (away from the jeopardized endocardial layer).

Following an interim, decade-long, compromise approach to avoiding reflow derangements by intermittent 15−20 min aortic clamping periods producing global ischemia, the now prevalent intraoperative technique is to apply cold cardioplegic solutions as intracoronary perfusate, aimed at myocardial protection and prevention of ischemic damage during protracted ascending aorta cross clamping [22]. The new methods are based on rapid induction of complete electromechanical arrest and adequate hypothermia, with buffering and avoidance of excessive myocardial edema. This allows lengthy anastomotic coronary reconstruction in a flaccid bloodless arrested heart, followed by rapid resumption of normal cardiac function. Membrane stabilizing agents and various metabolic manipulations are being considered as part of optimizing the cardioplegic procedure. Ideal markers of ischemic injury are still being sought.

There are several practical difficulties and unresolved issues concerning myocardial protection during the revascularization and extended periods of extracorporeal circulation. Either blood or crystalloid cardioplegia have been used in various studies. Delivery distal to significant coronary artery obstructions may not be adequate [23], causing temperature gradients and occasional perioperative infarction. Methods such as distal anastomosis delivery of the solution and higher aortic root perfusion pressures have been used [37]. Retrograde coronary sinus cardioplegia perfusion (following the initial antegrade cardioplegic arrest) appears to have special merit, and is currently being explored (see chapter 9).

Surgical coronary reperfusion treatment within about 5 h of the onset of transmural myocardial infarction was reported to result in significantly improved cardiac function [24], whereas periods beyond 6 h were not associated with benefits, unless adequate collateral circulation was present. Acceptable low mortality was reported by some, but results of acute surgical intervention are not from control randomized studies. The method has been used in only in a few centers, and would have to be further demonstrated.

## 5. Thrombolytic treatment of evolving acute myocardial infarction

In recent years we have seen a dramatic rise in interventional cardiology, primarily due to thrombolytic and angioplastic techniques. As is evident from numerous reports, both intracoronary and intravenous thrombolytic treatments of evolving acute myocardial infarction are now well established. Streptokinase, tissue plasminogen activator, and other agents are being applied, without or with subsequent PTCA for dilation of significantly stenosed culprit coronary arteries. Complicating factors include multivessel disease, collaterals or subtotal occlusion. Thrombolytic therapy is effective, particularly when applied within the first few hours from a thrombotic occlusion event, and can dramatically improve the infarcting patient's clinical history. Sometimes it fails to lyse a clot, possibly due to the site or conditions of the thrombus, and it may also be ineffective when the coronary occlusion is associated with spasm or plaque hemorrhage. Investigations point to possible further treatment to prevent any rethrombosis and reinfarction, e.g., by using platelet anti-aggregating agents and beta blockers.

Important national and international trials have now provided evidence of significantly reduced mortality, along with reopening of occluded coronary arteries and improved function, particularly when reperfusion is instituted promptly. Recanalization with intracoronary streptokinase thrombolysis after a mean 4 h of coronary occlusion was reported as 68% − 75%, with higher rates of success when the reperfusion was instituted more rapidly, and high global as well as regional function were also reported. Tissue plasminogen activator appeared in NIH studies to provide a higher success rate than was the case with streptokinase [27]. The intravenous thrombolytic treatment has several advangates, including logistics and saving crucial time which is otherwise needed for catheterization laboratory procedures.

## 6. Angioplasty and coronary reperfusion

PTCA may be applied, not only to dilate tight coronary stenoses, but even for dislodging a coronary thrombus, followed by dilatation of the vessel. Indeed, there have been major trials of such emergency PTCA in the setting of evolving acute myocardial infarction, after a thrombotic occlusion. Thus, based on their experience, O'Neill et al. [38] considered emergency PTCA superior to intracoronary streptokinase, in terms of recanalization and cardiac function. However, such a method for clinical reperfusion is primarily applied in special circumstances, e.g., when the thrombolytic treatment is found ineffective. For total coronary occlusions, PTCA represents a technique to be applied in conjunction with thrombolysis. PTCA is of particular significance whenever severe stenosis of a coronary artery underlies the clot.

In terms of the elective PTCA dilatation of stenotic vessels, and apart from the brief occlusion-reperfusion conditions during the dilatation process itself, a change from a very tight to mild stenosis represents quasi reperfusion or improved perfusion. In the case of dilatation from moderate to mild stenosis, the effectiveness may have to be judged in terms of significantly improved coronary reserve, as determined by angiography coupled with papaverine potentiation. PTCA dilatation of coronary vessels may also be associated with a gradual return of function, but systematic evidence on such quasi "stunning" is not available. Similar considerations of improved coronary flow reserve and regional/global cardiac function should apply when the severe coronary artery stenosis is treated using any one of a number of newer research techniques, such as laser or mechanical ablation of the plaque.

## 7. Enhancement of myocardial reperfusion with coronary venous interventions

Briefly, without dwelling on retroperfusion which will be dealt with in subsequent chapters, there are three approaches to applications of coronary venous assist, aimed at improved retrograde reperfusion of ischemic territory following a coronary artery occlusion. Retroperfusion may be effectively applied prior to reperfusion, and indeed as early as is practicable during the evolving myocardial infarction, to assure significant myocardial salvage. This approach is exemplified in a 7-day experimental reperfusion study with 3 h coronary occlusion, in which untreated occlusion-reperfusion was compared with hypothermic synchronized retroperfusion support initiated 30 min post occlusion and applied for 2.5 h [39]. Results demonstrated enhanced myocardial salvage, improved function, and reduced post reperfusion derangements, when the acute ischemia was treated by the coronary venous technique. In another approach, the institution of the retrograde treatment is timed to coincide with the reperfusion, in the hope of reducing reflow damage. Pressure controlled intermittent coronary sinus occlusion was studied in this manner (see chapter 6). Indications were that retrogradely enhanced ischemic zone circulation and effective metabolite washout contributed to improvement in function and myocardial salvage. The third coronary venous method is actually an alternate thrombolytic treatment, in which streptokinase or another lytic agent is injected or retroinfused into a regional coronary vein. It was demonstrated in animals that it is indeed possible in this manner to lyse a coronary artery thrombus (see chapter 8).

One might wonder how the coronary venous interventions can provide sufficient benefits in terms of ischemic zone reperfusion. Based on prior experimental studies, one might hypothesize one or more reasons. Thus, even a limited retrogradely delivered reperfusion of the ischemic myocardium (as compared to the often more effective antegrade reperfusion) could support the jeopardized tissue viability, and do so during the coronary artery occlusion and/or during its reopening. Furthermore, there is evidence that most of the coronary venous interventions potentiate circulation within and toxic metabolite washout from the ischemic zone, thus improving function and metabolism. The more gradual character of the retrogradely induced ischemic zone reperfusion might also be beneficial, e.g., in avoiding postreperfusion arrhythmias and edema. It is anticipated that ongoing research and clinical studies will clarify the extent and mechanisms of benefits when coronary venous interventions are applied in conjunction with current reperfusion techniques.

## 9. Summary

Reperfusion has now moved from extensive experimentation to widespread clinical application. The beneficial results are already sufficiently dramatic to have changed the entire scene of interventional cardiology. There still are difficult problems to be overcome, and some controversy persists about the most effective clinical approach in specific patient subsets. These differences are not surprising, considering the complex issue of categorizing a highly diverse population, the varying practical difficulties of promptly initiating reperfusion procedures, some uncertainty regarding occasional deleterious effects of acute interventions, and the obvious need for more effective measurement of treatment effectiveness and patient followup. The impressive success of the new techniques should stimulate further efforts to insure an even higher rate of success, and to address cases where presently available reperfusion proves inadequate. It is believed there is ample rationale for supplemental support treatment during the thrombolytic, emergency bypass and PTCA reperfusion procedures.

## References

1. Jennings RB, Sommers H, Smyth GA, Flock HA, Linn H (1960) Myocardial necrosis induced by temporary occlusion of a coronary artery in the dog. Arch Pathol Lab Med 70:68–78
2. Maroko PK, Libby P, Ginks WR, Bloor CN, Shell WF, Sobel BE, Ross J (1972) Coronary reperfusion. J Clin Invest 51:2710–2716
3. Kloner RA, Ganote CE, Jennings RB (1974) The "no reflow" phenomenon after temporary coronary occlusion in the dog. J Clin Invest 54:1496–1508
4. Meerbaum S, Corday E (1975) Symposium on reperfusion during acute myocardial infarction. Parts 1 and 2. Am J Cardiol 36:211–261, 368–406
5. Hearse DJ (1977) Reperfusion of the ischemic myocardium. J Mol Cell Cardiol 9:8
6. Reimer KA, Jennings RB (1979) The "wavefront phenomenon" of myocardial ischemic cell death. II. Transmural progression of necrosis within the framework of ischemic bed size (myocardium at risk) and collateral flow. Lab Invest 40:633–644
7. Kloner RA, Braunwald E (1980) Observations on experimental myocardial ischemia. Cardiovasc Res 14:371–395
8. Braunwald E, Kloner RA (1982) The stunned myocardium: Prolonged postischemic ventricular dysfunction. Circulation 66(6):1146–1149
9. Gould KL (1982) Coronary reperfusion: medical, surgical or not at all. JAMA 248(11): 1362–1363
10. Corday E, Meerbaum S (1983) Part 1. Symposium on the present state of reperfusion of the acutely ischemic myocardium. J Am Col Cardiol 1(4):1031–1036
11. Meerbaum S, Corday E (1983) Part 2. Symposium on the present state of reperfusion of the acutely ischemic myocardium. J Am Col Cardiol 1(5):1223–1234
12. Jennings RB, Reimer KA (1983) Factors involved in salvaging ischemic myocardium: effect of reperfusion of arterial blood. Circulation 68 [Suppl I]:1–25
13. McDonagh PF (1983) The role of the coronary microcirculation in myocardial recovery from ischemia. Yale J Bio Med 56:303–311
14. Kloner RA, Ellis SG, Large R, Braunwald E (1983) Studies of experimental coronary artery reperfusion: effects on infarct size, myocardial function, biochemistry, ultrastructure and microvascular damage. Circulation 68 [Suppl]:1–8

15. Braunwald E, Kloner R (1985) Myocardial reperfusion: a double edged sword? J Clin Invest 76:1713—1719
16. Becker LC, Ambrosio G (1987) Myocardial consequences of reperfusion. Progr Cardiovasc Dis 30(1):23—44
17. Weisfeld ML (1987) Reperfusion and reperfusion injury. Clin Res 35(1):13—20
18. Constantini C, Corday E, Lang TW et al. (1975) Revascularization after 3 h of coronary arterial occlusion: Effects on regional cardiac metabolic function and infarct size. Am J Cardiol 36:368—384
19. Mathur VS, Gink GA, Burris WH (1975) Maximal revascularization (reperfusion) in intact conscious dogs after 2—5 h of coronary occlusion. Am J Cardiol 36:252—261
20. Khalafbeigni F, Becker LC, Hutchins CM et al. (1981) Time course of necrosis after coronary occlusion in the conscious dog (Abstract) Circulation 64 [Suppl 4]:IV-98
21. Hochman JS, Choo H (1987) Limitation of myocardial infarct expansion by reperfusion independent of myocardial salvage. Circulation 75(1):299—306
22. Silverman NAL, Levitsky S (1987) Intraoperative myocardial protection in the context of coronary revascularization. Prog Cardiovasc Dis 29(6):413—438
23. Hilton CJ, Teubel W, Acker M et al. (1979) Inadequate cardioplegic protection with obstructed coronary arteries. Ann Thorac Surg 28:323—334
24. De Wood MA, Heit J, Spores J, Berg R, Selinger SL, Rudy LW, Hendes G, Shields JP (1983) Anterior transmural myocardial infarction: effects of surgical coronary reperfusion in global and regional left ventricular function. J Am Col Cardiol 1(5):1223—1234
25. De Wood MA, Spores J, Notske R et al. (1980) Prevalence of total coronary occlusion during the early hours of transmural myocardial infarction. N Engl J Med 303:897—902
26. Rentrop KP (1985) Thrombolytic therapy in patients with acute myocardial infarction. Circulation 71:627—231
27. TIMI Studygroup (1985) The thrombolysis in myocardial infarction (TIMI) trial. Phase 1 findings. N Engl J Med 312:932—936
28. Gruntzig AR, Senning A, Siegenthaler WE (1979) Nonoperative dilatation of coronary artery stenosis: N Engl J Med 301:61—68
29. Williams DO, Reilly RS, Singh AK, Most AS (1983) Coronary circulatory dynamics before and after successful coronary angioplasty. J Am Col Cardiol 1(5):1267—1272
30. Murdock RH, Chu A, Grubb M, Cobb FR (1985) Effects of reestablishing blood flow on the extent of myocardial infarction in conscious dogs. Am J Physiol 249(18):H783—H791
31. Haendchen RV, Corday E, Torres M, Maurer G, Fishbein MC, Meerbaum S (1984) Increased regional end-diastolic wall thickness early after reperfusion: a sign of irreversibly damaged myocardium. J Am Col Cardiol 3(6):1444—1453
32. Balke CW, Kaplinsky E, Michaelson EL et al. (1981) Reperfusion ventricular tachyarrhythmias: correlation with acute coronary artery occlusion tachyarrhythmias and duration of myocardial ischemia. Am Heart J 101:449—455
33. Corbalan R, Verrier RL, Lown B (1966) Differing mechanisms for ventricular vulnerability during coronary artery occlusion and release. Am Heart J 92:223—230
34. Ellis SG, Haenchke CI, Sandor T, Wynne J, Braunwald R, Kloner RA (1983) Time course of functional and biochemical recovery of myocardium salvaged by reperfusion. J Am Col Cardiol 1(4):1047—1055
35. Jolly SR, Kane WJ, Bailie MB et al. (1984) Canine myocardial reperfusion injury. Its reduction by the combined administration of superoxyde dismutase and catalase. Circ Res 54:277—285
36. Ambrosio G, Becker LC, Hutchins GM et al. (1986) Reduction in experimental infarct size by recombinant human superoxide dismutase. Insights into the pathophysiology of reperfusion injury. Circulation 74:1424—1433
37. Rosenkranz ER, Buckberg GD (1983) Myocardial protection during surgical reperfusion. J Am Col Cardiol 1(5):1235—1246
38. O'Neill W, Timmis GC, Bourdilla PD et al. (1986) A prospective randomized clinical trial of intracoronary streptokinase vs coronary angioplasty for acute myocardial infarction. N Engl J Med 314:812—818
39. Haendchen RV, Corday E, Meerbaum S, Povzhitkov M, Ritt J, Fishbein MC (1983) Prevention of ischemic injury and early reperfusion derangements by hypothermic retroperfusion. J Am Col Cardiol 1(4):1067—1080

50

# The coronary venous anatomy

E. Ratajczyk-Pakalska

## Introduction

The venous component of the coronary circulation has usually been ignored by anatomists and the scientific community in general. Yet, knowledge of the venous vascular system is particularly important, not only in relation to surgical retroperfusion using aorto-venous bypass anastomoses (e.g., the historic Beck II procedure), but also for the more recent applications of retrograde coronary sinus cardioplegia, synchronized coronary venous retroperfusion (SRP), and pressure control intermittent coronary sinus occlusion (PICSO) [2, 4, 5, 7, 13, 23, 29, 38, 43, 45, 46, 48, 49].

In the normal anatomical circumstances, myocardial venous drainage takes place along three pathways: 1) through anterior right ventricular veins into the right atrium, 2) through the coronary sinus, also emptying into the right atrium, and 3) through the Thebesian veins (ThV) into all the chambers of the heart (Schema 1).

Apart from the larger superficially and subepicardially situated veins, which empty into the right atrium, either directly or indirectly via the coronary sinus, the following significant vascular interconnections have been ascertained at various levels of the myocardial circulation: veno-venous anastomoses, arterial-venous anastomoses, and the Thebesian veins.

Physiologic studies [27–31] indicate that about 60% of the coronary venous blood returns to the right atrium through the coronary sinus. The remaining 40% of the drainage is distributed through anterior right ventricular veins, the ThV, coronary veno-venous anastomoses, and perhaps via arterio-venous connections. The latter may become apparent in the presence of abnormal pressure gradients, such as when a coronary artery is occluded or narrowed. Postmortem studies suggest that the distribution of the venous return via these alternative routes may depend on the duration and the extent of arterial occlusive disease. Thus, the special ThV vascular interconnections appear to be more numerous in coronary artery disease, when the arterial stenosis is severe and has persisted for some time [28, 37, 55, 69, 70, 72, 76, 80].

## A. Cardiac veins

The nomenclature of the cardiac veins, used in this chapter, is based on results of a meeting of the Nomenclature Commission [40], which met during the first Interna-

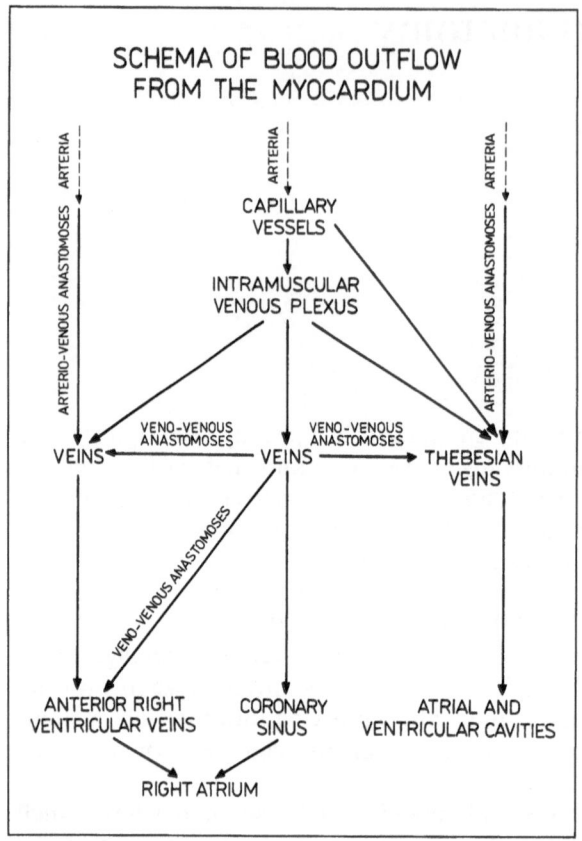

SCHEMA OF BLOOD OUTFLOW
FROM THE MYOCARDIUM

**Schema 1**

tional Symposium on Myocardial Protection Via the Coronary Sinus, in Vienna, 1984 (Schema 2).

As is the case with all venous systems in the human body, the course of the cardiac veins, their tributaries, orifices and extent of veno-venous anastomoses are characterized by a great number of individual variations. The common feature of all superficial cardiac veins is their subepicardial configuration. The major cardiac veins lie parallel (and somewhat superficially) to corresponding branches of the coronary arteries. In the normal heart, there are a large number of cardiac veins as compared to coronary arteries [61, 76]. With advanced age, the mass of an atherosclerotic heart is generally reduced; the cardiac veins become elongated and matted, exhibiting bulges and varice-like changes [61, 76].

With regard to muscular bridges (in human and animal hearts), the posterior wall of the coronary sinus is covered by a thin layer of muscularis from the left atrium. In the pig heart, subepicardial veins are often (40%) covered by muscular bridges [41, 56, 61]. This applies to the left coronary vein in its course in the anterior interventricular sulcus (10%), and to the posterior interventricular veins (30%). In 7% of cases, the muscular bridge over the posterior interventricular vein also covers the right coronary artery. In man, muscular bridges were observed by Polacek

52

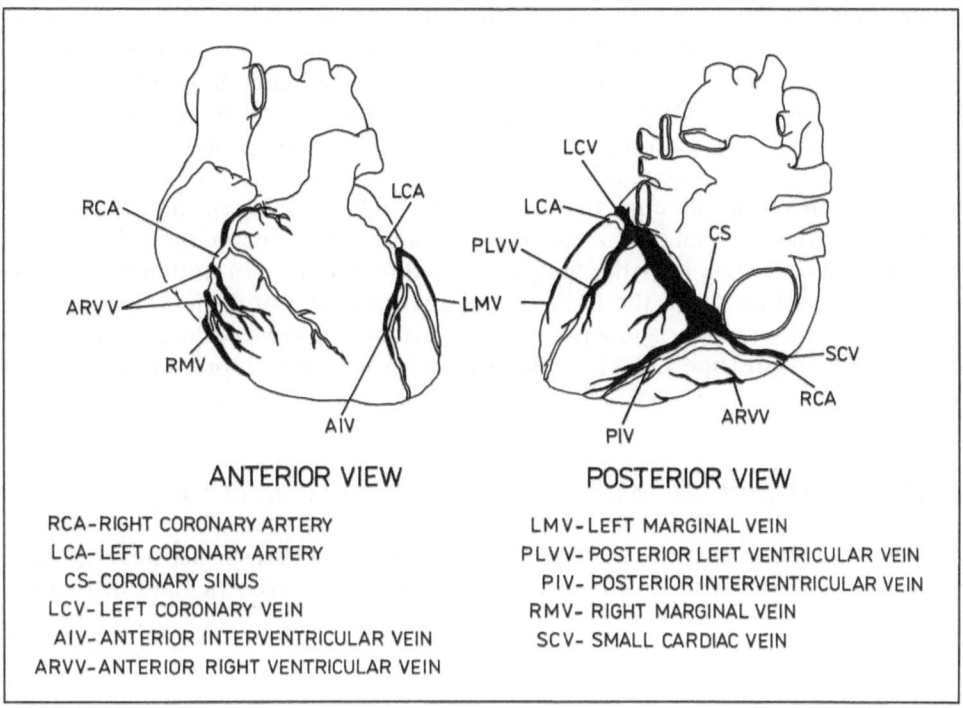

ANTERIOR VIEW

RCA–RIGHT CORONARY ARTERY
LCA–LEFT CORONARY ARTERY
CS–CORONARY SINUS
LCV–LEFT CORONARY VEIN
AIV–ANTERIOR INTERVENTRICULAR VEIN
ARVV–ANTERIOR RIGHT VENTRICULAR VEIN

POSTERIOR VIEW

LMV–LEFT MARGINAL VEIN
PLVV–POSTERIOR LEFT VENTRICULAR VEIN
PIV–POSTERIOR INTERVENTRICULAR VEIN
RMV–RIGHT MARGINAL VEIN
SCV–SMALL CARDIAC VEIN

**Schema 2**

56]. Chen-Jen-Lien and Liao-Rei found muscular bridges over the left coronary vein in 3.1%, and over the posterior interventricular vein in 1% of cases (in 100 human hearts). Recently, muscular bridges have been described by v. Luedinghausen ([41], also see chapter 5).

The individual cardiac veins will now be discussed:

*1. Coronary sinus (CS)*

The CS drains most of the left ventricular venous blood (Schema 2). As previously indicated, 60% of the coronary venous blood returns to the CS [31].

Based on Marshall's embryogenetic division, Tandler [81] considered that the origin of the CS was delimited by the valve of the left coronary vein (LCV) or great cardiac vein, and its other extremity by the valve of the CS. Adachi [1] observed in hearts that the LCV valve was not always present and hence postulated the origin of the CS at the site where it is joined by the left atrial oblique vein. In embryonic life, the CS originates at the terminal segment of the left superior vena cava [72]. Therefore, it is logical to place the beginning of the sinus at the left atrial oblique vein (which is part of the embryonic left superior vena cava), and the end of the coronary sinus at the site of the orifice to the right atrium.

53

The CS is a short, wide vein, lying in the "atrial ventricular sulcus", on the diaphragmatic surface of the heart, between the left atrium and the left ventricle (Fig. 1). The posterior wall of the sinus is covered by a delicate, thin wall of muscle, descending from the left atrium. The muscular coat has no definitive feature, because it is generally not fully circular and covers only the posterior wall of the sinus. Generally, it has been maintained that this muscular coat can be circular and then constitutes a certain constriction of the CS. The opening of the CS into the lumen of the right atrium is situated on the posterior wall, below the valve of the inferior vena cava, just near the margin formed by the inferior and medial atrial walls. The transition from the LCV to the CS is mostly not distinct; sometimes (15%) there is a characteristic separating constriction, corresponding to the valve of LCV. The major superficial branches of the CS will now be presented in more detail.

## 2. Anterior interventricular vein (AIV) and left coronary vein (LCV)

Most of the authors describe the origin of the AIV on the heart surface near the "incissura apicis" of the heart, from the side of the anterior interventricular sulcus. The

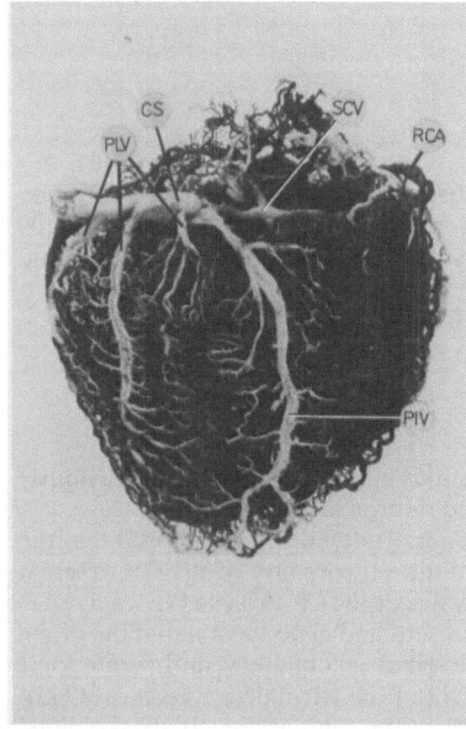 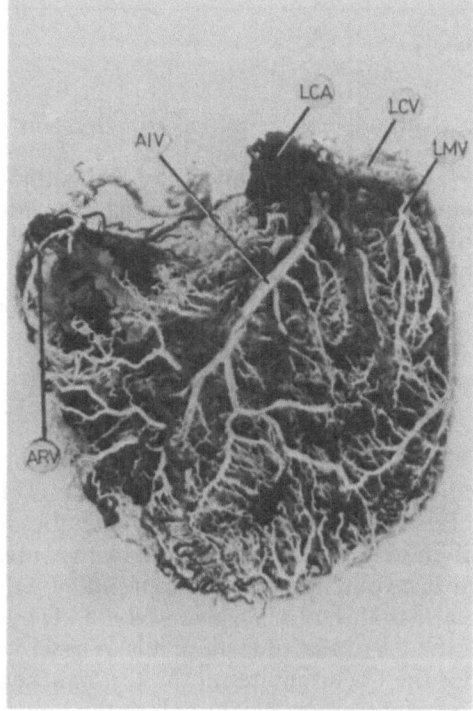

**Fig. 1.** Diaphragmatic surface of the heart; three PLVV.

**Fig. 2.** Corrosion preparation of the coronary vessels.

author of this chapter found such an origin in 70% of cases [61]. Only Adachi [1] found the beginning of this vein to be in the anterior interventricular sulcus above the apex, at a half of the height of this sulcus (Ratajczyk-Pakalska in only 8% of cases; [61]). The LCV opens into the CS [61, 74]. Cases in which the LCV communicated directly into the left atrium reflect a rare modification of the mouth of the LCV. Adachi [1] described a small enlargement of the LCV (1% 160 case) which presumably corresponded to that presented by Gruber as "sinus venae coronariae magnae proprius".

The longest portion of the cardiac vein ascends in the "anterior interventricular sulcus", then turns down to reach the "coronary sulcus" as LCV. At this site, the AIV forms a right or acute angle. From the anatomical point of view, these angles may have a particular effect in determining the blood flow resistance in the AIV and LCV (Fig. 2).

Among the superficial branches of the AIV, emptying into its main trunk at different angles, the following venous vessels were discerned: a) vessels of the anterior wall of the right and left ventricle, b) oblique vein of the left ventricle, c) left marginal vein, d) posterior veins of the left ventricle (also tributary of the CS), 3) venous vessels of the interventricular septum, f) vessels of the left atrial walls (empty into the LCV). One frequent vein (90% of cases), which empties into the AIV, has a diameter equal to that of the AIV; it runs on the anterior wall of the left ventricle and enters the trunk of the AIV at an angle of 45°−90°, either in the middle of the anterior interventricular sulcus (81%; Figs. 2, 3) or else in the upper one-third of this sulcus (9%; [61]). This is the oblique vein of the left ventricle (LVOV). The remaining venous vessels emptying into the AIV feature specific characteristics to be described subsequently.

## 3. The left marginal vein (LMV)

The LMV belongs to branches of LCV. Various authors present its occurrence in different degrees. Adachi [1] found it in 20.5%, Parsonet [54] in 75%, Gruber [81] in 80%, Tschabitscher [86] in 60%, Ratajczyk-Pakalska [61] in 100% of cases. Most of these authors consider it to be a single vessel, but some [60, 61, 74] cite the possibility of double veins. According to the opinion of most investigators, the LMV feeds into the LCV (Figs. 3, 4). Cases, as seen by Gruber [81], in which the LMV opens into the CS (81%) constitute a rare modification.

## 4. Posterior left ventricular veins (PLVV)

The PLVVs are not stable veins, and their incidence is variously reported. According to Parsonnet [54], they are present only in 63% of cases, and incidence reported by Adachi [1] is 89%. Authors also differ as to the number of PLVV vessels. They were described as two veins [72], two or three veins [72], and two to seven veins [1]; reports [1, 62, 86] describe the PLVV as numerous veins. Ratajczyk-Pakalska [61] found three PLVV in 66% of cases (Fig. 1), two PLVV in 24%, and one PLVV in 10% (Table 1). In no case were more than three PLVV observed (Fig. 1).

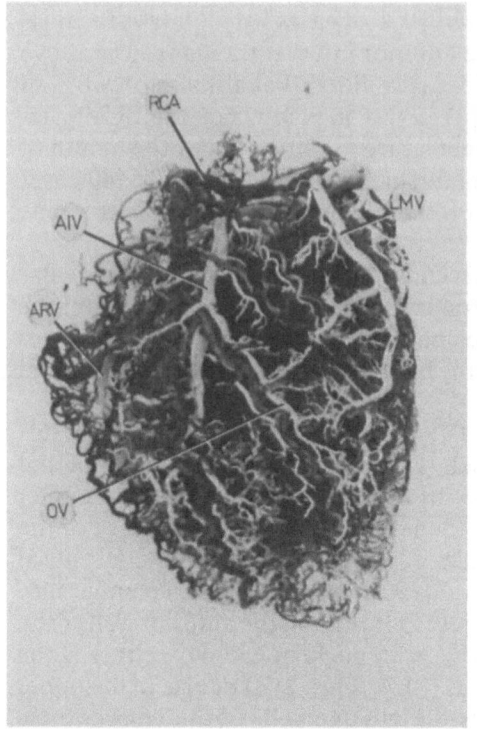 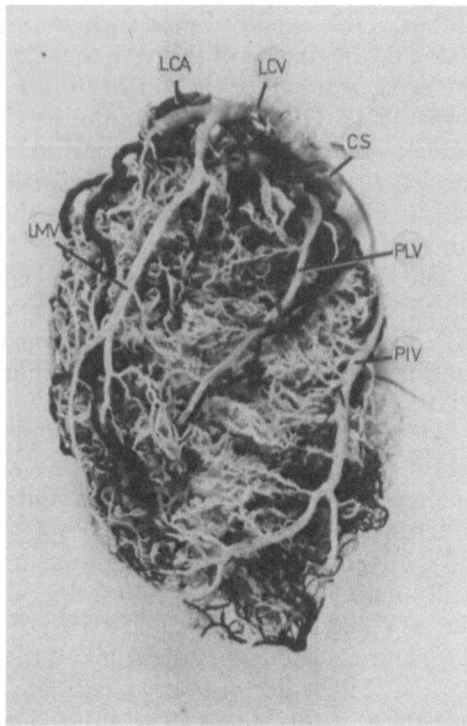

**Fig. 3.** Corrosion preparation of the coronary vessels; compression of AIV by sclerotic LCA (frontal view).

**Fig. 4.** Corrosion preparation of the coronary vessels; one LMV.

Depending on their diameter, the PLVV were of various lengths. The veins with the larger diameter were also longer, originating nearer to the apex, whereas the veins with smaller diameters were shorter, originating farther from the apex. Observing the branches of the PLVV, it appeared that each PLVV had two until six branches, which opened either into the main trunk along the entire course, or else two to three branches appeared below the orifice, entering either into the LCV or the CS. The PLVV always discharges either into the CS or into the LCV.

**Table 1.**

| PLVV | One | Two | Three |
|---|---|---|---|
| Frequency per Adachi | 0 | 6.6 | 32 |
| Frequency per Ratajczyk-Pakalska | 10 | 24 | 66 |

## 5. Posterior interventricular vein (PIV)

Most authors describe the superficial origin of the PIV in the posterior interventricular sulcus, near the apex of the heart. Ratajczyk-Pakalska [61] found such an origin in 78% of cases. Only Adachi [1] placed the onset of this vein in the anterior interventricular sulcus near the apex.

A rectilinear course of the PIV was found in 71% (Fig. 1), and in those cases, the venous branches from the posterior surface of both ventricles ran in a parallel configuration as three to eight vessels, entering perpendicularly into the PIV [61]. Such a course of the PIV with its branches is considered typical (Fig. 1). In less common cases (29%), the PIV ran meanderously, and its branches were present in dendrific forms, three to eight in number on each side [61].

The PIV opens mostly into the final segment of the CS near its ostium valve (89%, [61]), and occasionally directly into the lumen of the right atrium (10%, [61]). In only one case, was the orifice of the PIV found to be at the initial segment of the CS on the level of the left atrial oblique vein [61]. There seems to be a distinct interdependence between the diameters of the PLVV and PIV. In the case of a well developed PIV, the PLVV vessels are shorter and have a smaller diameter, and vice versa.

## 6. Left atrial oblique vein (LAOV)

The LAOV, also called the Marshall vein, develops from the third left ductus of Cuvier, which disappears in early embryonic life [27], and (similarly to the CS) from the left embryonic form of the venous sinus [72, 81]. In rare anomalies, instead of the residual LAOV, a vena cava superior sinistra [72, 81] occurs persistently. The LAOV is not always present; frequently, at this site, a band of connective tissue is present, called the ligament of the vena cava superior sinistra [81]. According to Parsonnet [54], the LAOV was present in 97% of 30 cases. Adachi [1] reported 13% in 160, and Ratajczyk-Pakalska [61] 84% in 100 cases. The LAOV is a short and thin vein, which originates on the lateral surface of the left atrium, below the inferior left pulmonary vein. It slants to the right side and downward, opening from above into the initial segment of the CS, thus determining its origin. In 71% of cases, there was a valve in the orifice of the LCV just above the opening of the LAOV into the CS. From this observation it appears that in 71% of cases, the presence of the valve of the LCV was associated with the incidence of a LAOV, while in the remaining 29% of cases, the occurrence of LAOV was connected with the valve of LCV [61].

## 7. Small cardiac vein (SCV)

The SCV is usually described as an inconstant vein. According to Parsonnet [54], it is present in 37% of cases. Adachi [1] saw it in 46%, v. Luedinghausen [41] in about 33%, Ratajczyk-Pakalska in 94% of cases. There are also divergencies in the description of its origin. Some authors report that the SCV originates at the right cardiac margin [61, 72, 80], whereas others, among them Adachi [1], observed the onset on the diaphragmatic heart surface, in the right segment of the coronary sulcus. According to most [72, 80, 86], the SCV opens into the final segment of the CS. Adachi [1] described this vein opening either into the CS or the LCV or directly into the right atrium. The authors of this chapter [61] observed that the SCV discharged most frequently into the final segement of the CS (74%; Fig. 1). Rarely did the SCV issue into the right atrium. The following branches of the SCV were noted on the heart surface:

a) venous vessels from the anterior and posterior walls of the right ventricle, numbering 6–11 and running almost parallel to each other before entering almost perpendicularly into the trunk of the SCV, coursing on the right margin of the heart;

b) very small venous vessels from the posterior wall of the right atrium, numbering 1 or 2, entering into the SCV in its course within the coronary sulcus;

c) anterior right ventricular veins presented branches of the SCV in all those cases where the SCV opened into the CS (see below).

## 8. Anterior right ventricular veins (ARVV)

The literature nomenclature of the ARVV is very divergent. They were variously called the accessory, minor, small coronary, Galen or Viessens veins. In the Bazylian nomenclature, the ARVV and SCV actually had a common name (Adachi[1]).. The ARVV veins have an inconstant incidence and appear in various numbers. Adachi [1] found them in 90% of cases, Ratajczyk-Pakalska in 85% [61]. The number varies from two to six, and according to Roberts [72], the number can be higher. It was stated that most frequently, there are two ARVV (in 75% cases, the incidence was 88%, Fig. 7), more rarely one ARVV [61]. The ARVV opened into the right atrium, either with a common orifice (per Parsonnet in 27%, [54], or with separate ostia [1, 54]; the ARVV can also enter into the SCV [1]; one of the ARVV, with a characteristic course, is also called the right marginal vein (RMV) or Galen's vein (Figs. 5, 6). Adachi [1] found it in 33%. Along its course, the ARVV collects numerous branches, characterized by dentric forms. After crossing the coronary sulcus, they open either into the lumen of the right atrium with separate orifices, located on its flank (76% of 65 cases), or else into the SCV (24% of 20 cases).

The ARVV enter into right atrium separately or by forming a trunk consisting of two until three veins [41, 44, 61]. Around the point of entrance of the ARVV there is a pectinate muscle, which acts as a valve [44]. In 60% of cases [41], the veins empty into an intramural tunnel or subendocardial venous lake (the sinus coronarius atrii dextri or right atrial coronary sinus − cit. [41]), which terminates in the anterior and posterior aspect of the atrium. This subendocardial venous lake has been described previously by Parsonnet [54] as subendocardial right ventricular vein.

## B. Valves of the cardiac veins

Such valves are present only sometimes in the ostia of larger cardiac veins [54]. v. Luedinghausen [41] found them in 30–48%. These valves are functionally inefficient, since during retrograde injection of the cardiac veins from the CS, little or

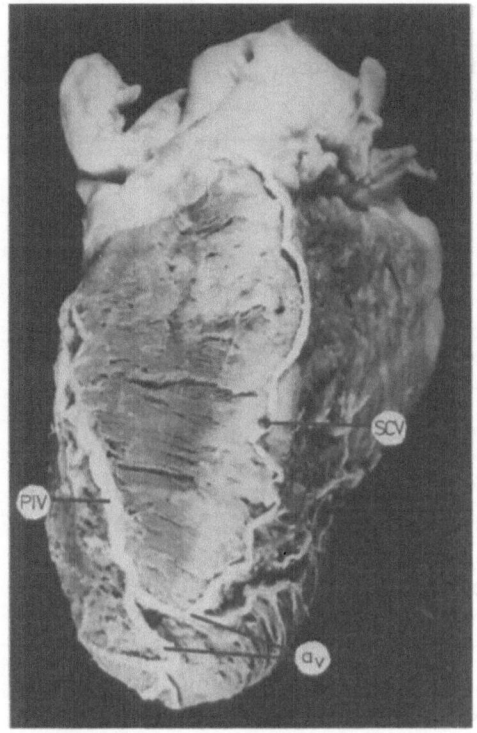

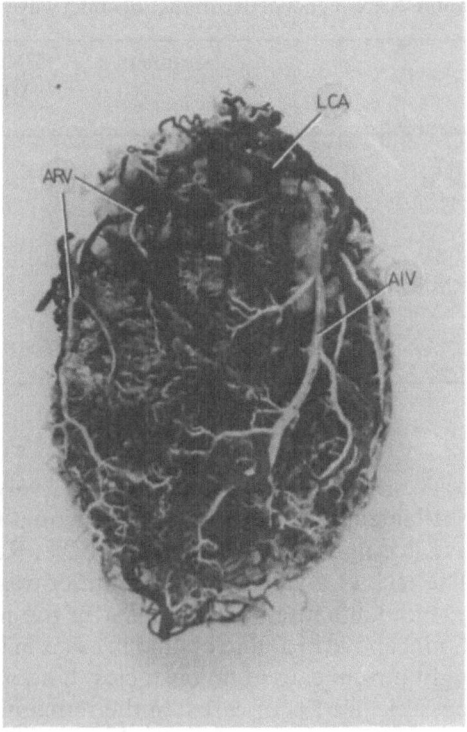

**Fig. 5.** Right margin of the heart; large v-v anastomoses in the region of the apex.

**Fig. 6.** Corrosion preparation of the coronary vessels; compression of AIV by sclerotic LCA (right-frontal view).

no flow resistance was found. They are unilobar valves of semilunar shape and ovally cut margin. The incidence of outlet valves of the cardiac veins, according to the various investigations and this chapter's author [61], is shown in Table 2. Relative to the CS valve, Hellerstein and Orbison [14] found that the anatomical character of the CS valve depends on its structure, cardiac size and also on cardiac anomalies. They discerned six basic orifices of the CS, and studied these orifices from the point of view of catheterizing the sinus from the right atrium. They concluded that satisfactory access to the CS was possible in 75% of cases. Gorlin [25] met difficulties of probing in only about 10%.

With regard to shape and structure, three variations of the CS valves were discerned [61].

a) semi lunar valve, with one cusp in the shape of a crescent; the valvular margin is directed towards the interatrial septum (46%).

b) membranous valve, very thin semi-translucent membrane covering almost the whole orifice, nearest to the interatrial septum. Numerous filiform bands, connecting it with the outlet margin, were found (33%).

c) Trabecular valve, dividing the orifice of CS either with trabaculae or bands connecting the superior and inferior margins of the ostium (5%).

**Table 2.** Frequency of occurrence of outlet valves of the cardiac veins according to various authors

| | Parsonnet % | Hellerstein Orbison % | Adachi % | Ratajczyk-Pakalska % |
|---|---|---|---|---|
| CS | | 85.3 | 71 | 79 |
| LCV | 80 | | | 76 |
| PIV | 87 | | | 80 |
| SCV | | | | 17 |
| LMV | | | | 12 |
| RMV | | | | 10 |
| PLVV right | | | | 10 |
| PLVV middle | | | | 20 |
| PLVV left | | | | 14.5 |

d) A rare Chiari-network-semilunar valves; from its free margins, the network of filaments was outgoing, connecting it with the valve of vena cava inferior; v. Luedinghausen described it in 2%, Ratajczyk-Pakalska in 1%.

The lack of a valve at the CS outlet was found in 15% of cases [61], according to v. Luedinghausen [41] in 38%, in the hearts with valvular shape of CS, as mentional above in a) and c), and in cases missing valve, the introduction of probe from right atrium caused no obstacles. Under clinical conditions, probing of the CS was possible in 61% of cases; in the remaining cases, probing was either impossible or very difficult [61]. v. Luedinghausen [41] stated that there are some other possible anatomic hindrances (as mentioned above) to successful catheterization of the right atrium and its afferent vessels, namely venous valves, intramural courses of AIV of LCV, and occlusion of the CS, atrial hindrances to catheterization.

## C. Venous vascularization of the heart

### 1. Vascularization of the left atrium
Two groups of veins take part in the venous vascularization:
a) Thebesian veins, to be discussed in subsequent sections of this chapter, and
b) Venous vessels of the left atrium (three to five in number), being the branches of either the LCV or CS, originating on the inferior half of the interior and posterior walls, and collecting in their course the small venous branches of tangled root appearance (Fig. 8 [50]). One of the anterior atrial veins vascularizes the auricle, running on its medial surface, adjacent to the pulmonary trunk. The LVOV also takes part in the venous vascularization of the left atrium.

### 2. Vascularization of the right atrium

The blood from atrial walls flows via the ThV into the cavity of this atrium, and from the posterior right atrial wall blood flows directly into the CS (Fig. 38). The blood from the myocardium of the sinus node area flows through the right anterior

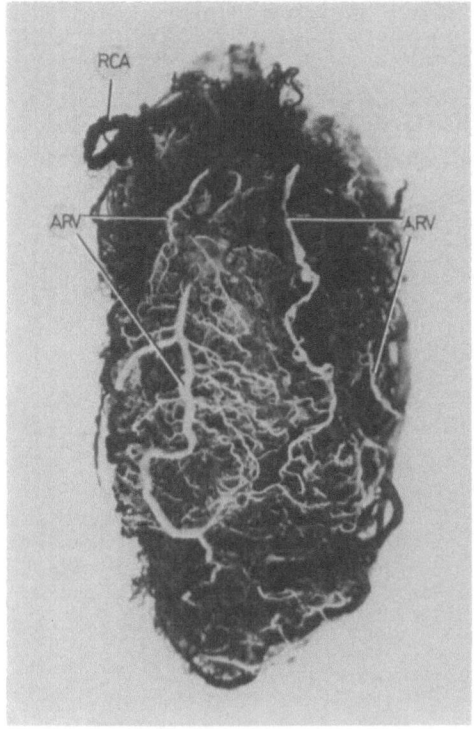 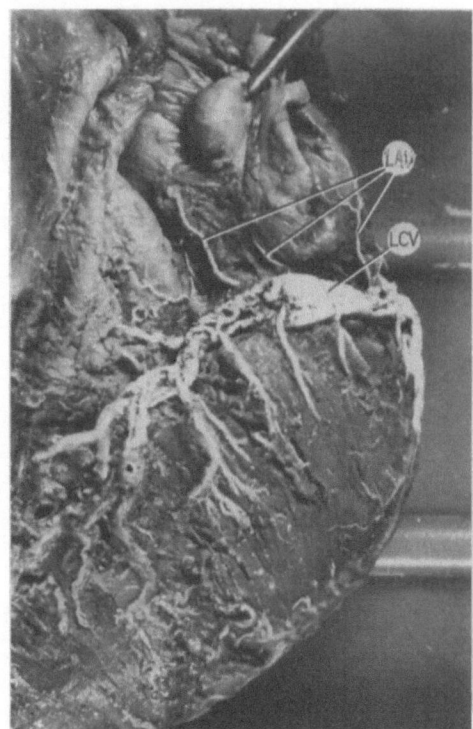

**Fig. 7.** Corrosion preparation of the coronary vessels; triple ARVV.

**Fig. 8.** Anterior surface of the left auricle.

atrial veins [41, 61]. Parsonnet [54] described the subendocardial vein of the right atrium, running either into the CS or directly into the right atrium; this vein had to take part in the vascularization of the right atrium, and also the right ventricle.

### 3. Vascularization of interatrial septum

According to common opinion [51, 72], the blood from the interatrial septum flows via ThV into atrial cavities, and only small venous vessels from interatrial septum open into the LCV.

### 4. Vascularization of the left ventricle

Most authors [54, 61, 72, 80], describing the venous vascularization of the left ventricle, emphasize that blood from this portion of the heart flows either directly into the LCV, LMV and PLVV, or else into the lumen of the cardiac cavities via ThV [55, 61, 72, 82]. Within the wall of the left ventricle there is a multilayer intramuscular venous plexus [61, 72, 85], formed from the tributaries of the above mentioned superficial cardiac veins (Figs. 9, 10).

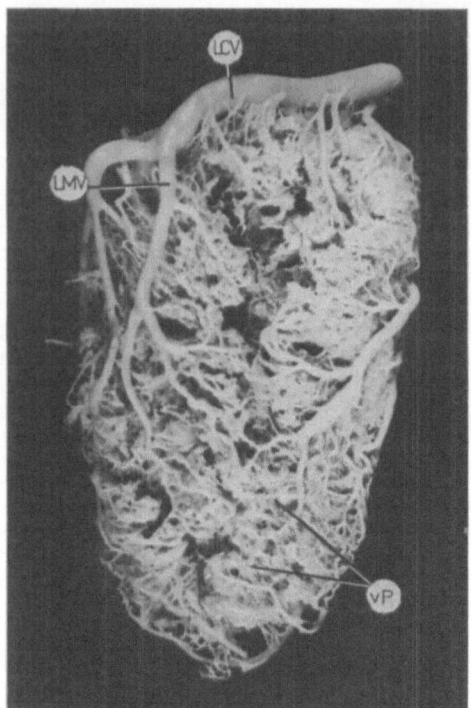 

**Fig. 9.** Corrosion preparation of the cardiac veins; single LMV (lateral frontal view).

**Fig. 10.** Corrosion preparation of the coronary vessels in sheep; intramuscular venous plexus (frontal view).

## 5. Vascularization of the right ventricle

Blood from the wall of the right ventricle flows mainly via the ARV veins [54, 80]. According to most authors, not only the ARV veins, but also the LCV, PIV, and SCV take part in the venous vascularization of the right ventricular walls. Also, as described by Parsonnet [54], the subendocardial right ventricular vein (CSRA) which opens to the ARV veins, takes part in this vascularization. Branches of the above mentioned veins form a single layered intramuscular venous plexus of the right ventricle; it is less developed than the corresponding plexus of the left ventricular wall [55, 72, 61, 82]. The venous plexus of the right ventricle is interconnected (like the left ventricular plexus) with ThV and also ARV, SCV, LCV, PIV, and with their superficial inflows. Venous branches, connecting the intramuscular venous plexus with aforementioned branches, and with the main trunks of superficial veins, enter the large vessels under an acute angle (Fig. 10).

## 6. Vascularization of ventricular septum

In the accessible literature, no detailed description of venous vascularization of the ventricular septum could be found. James [62] was especially interested in

62

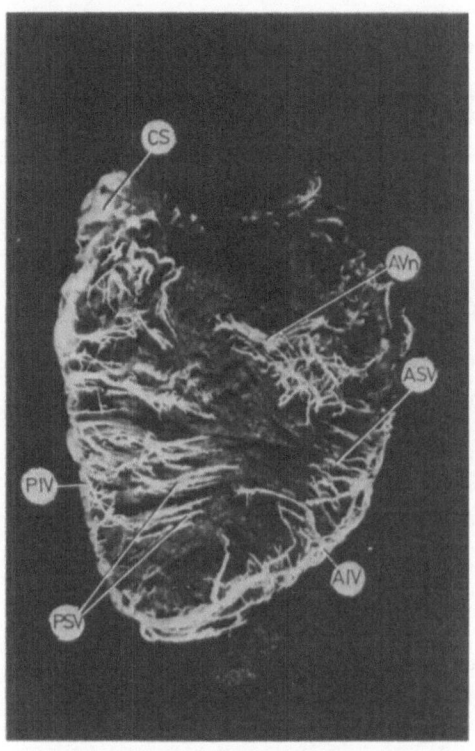

**Fig. 11.** Corrosion preparation of the veins of the ventricular septum; atrio-ventricular node vein (view from the side of the right ventricle).

vascularization of the ventricular septum, and described the arteries at this site without paying much attention to veins. These form an intermuscular plexus which is as well developed as the one in the left ventricular wall. This plexus connects with the cavities of both ventricles and especially with the cavity of the right ventricle via ThV, and with superficial cardiac veins into which veins of the ventricular septum discharge. In the current author's investigation [62], the anterior and posterior veins of the ventricular septum opened in all cases into the LCV and PIV, and also into PLVV − 21% of 21 cases, and LMV − 9% of nine cases (Fig. 11). In 32% of 32 cases, among the ARVV, a vein with a characteristic course was observed [61, 62]: the superior, anterior vein of the ventricular septum − atrial ventricular node vein − AVnV. It is a vein with a diameter larger than any of the other ventricular veins located in the anterior superior part of the septum; this vein slants slightly from down and behind, upward and forward, and after a short course, it opens into the LCV, in the segment located in the superior part of the ventricular anterior sulcus. The superior anterior vein of the ventricular septum collects numerous branches from superior and anterior parts of the ventricular septum, and also additionally, from the myocardium of the area of the atrial-ventricular node and bundle. The anterior veins of the ventricular septum are connected with the posterior of the ventricular septum. The apical part of the ventricular septum is vascularized by anterior and posterior veins of the ventricular septum, being branches of LCV and PIV, and also by the ThV. The medial part of the ventricular septum is vascularized also by ThV.

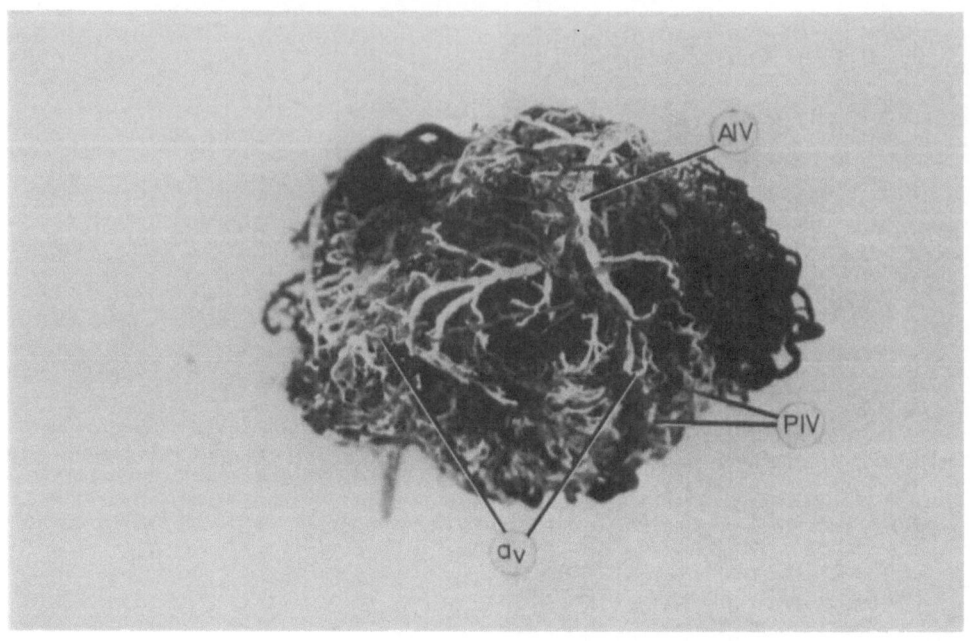

**Fig. 12.** Corrosion preparation of the coronary vessels; large v-v anastomoses (view from the apex).

## D. Veno-venous anastomoses

Veno-venous anastomoses (v-v anastomoses) may occur at various levels of the cardiac veins. According to their diameter, the v-v anastomoses have been classified as three types, viz.: a) large, b) medium, c) small.

a) Large v-v anastomoses are common between the epicardial veins, most often in the region of the cardiac apex. The diameters of their vascular casts (measured with an ocular micrometer) were larger than 1 mm; the amount of v-v anastomoses was characterized by substantial individual variation (Fig. 5). Large v-v anastomoses, situated superficially and connecting the larger cardiac veins, were described by Parsonnet [54], who distinguished two constant venous anastomotic rings in human hearts. The first ring connects the beginning of the posterior interventricular vein and the anterior interventricular vein. He found this in 90% of the observed hearts. Later (1980), Pakalska and Kolff [51] found this type v-v anastomosis only in 40% of the observed human hearts. A second venous anastomotic ring was also described by Parsonnet [54] and was found in 70%. It connects the posterior left ventricular vein to the posterior interventricular vein and left marginal vein. Pakalska and Kolff [51] have found this v-v anastomosis only in 3% in the human hearts, although they described two more groups of v-v anastomoses located in the region of the cardiac apex (Table 3) (Figs. 5, 12).

These types of v-v anastomoses are quite important, from at least two points of view; 1) they may shunt away the blood from the myocardium during retrograde perfusion of the cardiac veins, thus decreasing retroperfusion effectiveness, and 2)

**Table 3.** Occurrence of medium veno-venous anastomoses.

| | Great cardiac vein % of hearts | | | | Position of anastomoses | Middle cardiac vein % of hearts | | | | Position of anastomoses |
|---|---|---|---|---|---|---|---|---|---|---|
| | man | sheep | pig | dog | | man | sheep | pig | dog | |
| Middle cardiac vein | 27 | 30 | 42 | 17 | A, S, VSa, Vsp | | | | | |
| Posterior vein of the left ventricle | 48 | 32 | 30 | 8 | A, VSp | 20 | 10 | 18 | 17 | VSp, A |
| Anterior cardiac veins | 53 | 22 | 40 | 40 | VDa | 7 | 0 | 3 | 0 | A, VDa, VDp |

Explantations:

| | | | | |
|---|---|---|---|---|
| VSa | – Ventricular Sinister anterior | | VDp | – Ventricular Dexter posterior |
| VSp | – Ventricular Sinister posterior | | S | – Interventricular Septum |
| VDa | – Ventricular Dexter anterior | | A | – Cardiac apex |

Great cardiac vein and middle cardiac vein are listed on the top.

Middle cardiac vein, posterior cardiac vein of the ventricle and anterior cardiac veins are listed on the left side of the Table. (Remember that the anterior cardiac veins are on the right side of the heart)

The Table lists the recurrence of medium size veno-venous anastomoses in percentage of hearts counted for each species. There are substantial differences.

In the diagram of Figure 1 only one medium sized veno-venous anastomoses is drawn. Actually, there are many and their importance lies in the fact that they may increase in size.

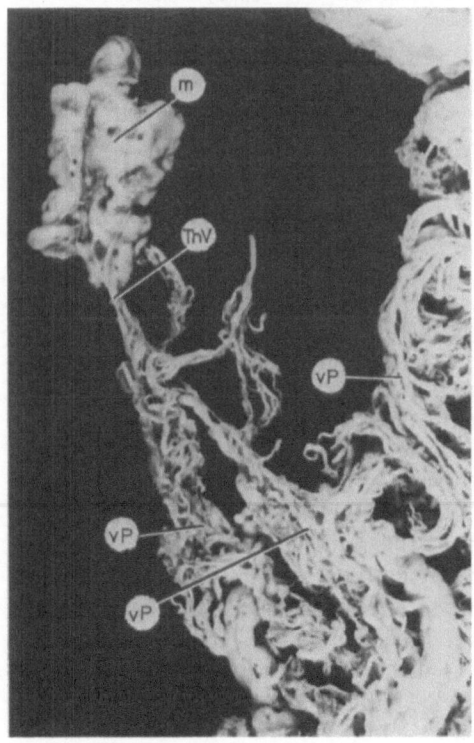

**Fig. 13.** Corrosion preparation of the cardiac veins of the left ventricular wall.

**Table 4.** Localization of ischemia in different areas, veins for perfusion, and possible anastomoses

| Localization of ischemia | Coronary arterial obstructions leading to infarct | Choice of vein for retrovenous perfusion | Possibility of outflow through v-v anastomoses (type A) |
|---|---|---|---|
| antero – lateral – figure a | anterior interventricular branch of left coronary artery | anterior interventricular vein | middle cardiac vein, posterior ventricular vein, anterior cardiac veins |
| antero – spetal – figure b | right division of anterior interventricular branch of left coronary artery | anterior interventricular vein | middle cardiac vein, posterior ventricular vein, anterior cardiac veins |
| apical – figure c | terminal portion of anterior interventricular branch of left coronary artery | anterior interventricular vein in the beginning of its course | middle cardiac vein, posterior ventricular vein, anterior cardiac veins |
| antero – basal – figure d | [a] circumflex branch of left coronary artery | left marginal vein or great cardiac vein on the left margin of heart | middle cardiac vein, posterior ventricular vein, anterior cardiac veins |
| postero – inferior – figure e | posterior interventricular branch of right coronary artery | middle cardiac vein ½ part of its course | great cardiac vein, posterior ventricular vein |
| postero – septal – figure f | right coronary artery or middle cardiac vein its posterior interventricular branch | great cardiac vein, | posterior ventricular vein |
| postero – lateral – figure g | [a] circumflex branch of left coronary artery | posterior ventricular vein (left) | middle cardiac vein, great cardiac vein |
| postero – basal – figure h | [a] circumflex branch of left | posterior ventricular vein (left) | middle cardiac vein, great cardiac vein |
| diffuse ischemia multivessel disease figure 1 | disseminated extensive narrowing of several arteries | ½ of coronary sinus | middle cardiac vein, anterior cardiac vein, posterior ventricular vein |

[a] variations in vessels distrubution

they can act as a safety valve if an artery is surgically connected to a vein or when clinical coronary vein retroperfusion tends to cause high intravascular pressures.

b) Medium v-v anastomoses were found in the epicardial system of cardiac veins (between the tributaries of these veins) or in the deep myocardial course of the cardiac veins forming intramuscular venous plexuses (ivp), (Fig. 13). Such plexuses are noted both in corrosion preparation of the myocardial veins, and on the micro-angioradiographs. The ivps situated in the myocardium of the left ventricle were multilayered, while ivp in the myocardium of the right ventricle were generally single layered. The ivp were connected directly with the veins situated subepicardially and with the Thebesian veins; the diameters of those vascular casts ranges from 500 to 1000 μm.

c) Small v-v anastomoses occured in the ventricular walls; the type of v-v anastomoses were visible only at a magnification of about 20-times; the diameters of those vascular casts ranged from 50 to 500 μm.

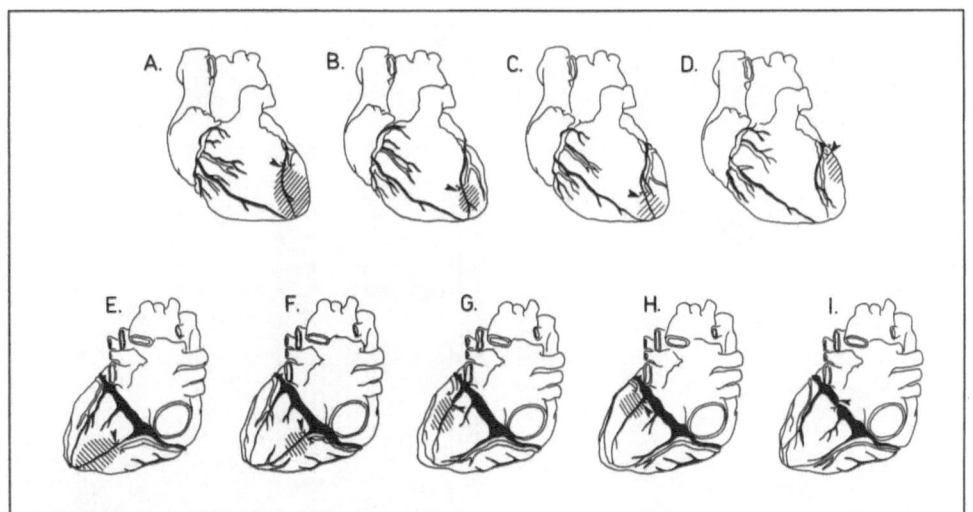

**Schema 3**

From the anatomical point of view, there is a great deal of controversy about the possible benefits of a retrograde perfusion of coronary vein or veins, e.g., with arterial blood. Obviously, if one perfused a vein that does not correspond to the ischemic area or if a wide anastomosis shunts the blood back into the atrium, no significant benefits will be derived. The appropriate vein to be used for retroperfusion should be determined in each case according to the localization of the coronary artery occlusion. For this reason, and taking into consideration the extent of venous vascularization, the regional veins and also the directions of venous outflow from the myocardium through v-v anastomoses, Pakalska and Kolff [51] proposed possible criteria for using an appropriate coronary vein for retrograde perfusion with arterial blood in cases of coronary arterial occlusion (Table 4: Schema 3).

### E. Arterial-venous anastomoses

In addition to the arterio-arterial, veno-venous and other types of anastomoses between myocardial blood vessels and/or cardiac chambers, the arterio-venous (a-v) anastomoses are yet another route of collateral circulation of fundamental importance in the pathology of the coronary vessels and in the treatment of heart disease [2, 8, 9, 37, 63, 67, 73].

Knowledge about a-v anastomoses in the human myocardium is insufficient, and their presence is still controversial. A-v anastomoses have been described by Prinzmetal et al. [60], Truex and Angulo [85], Archangielski (cited by [76]), James [37], Ludwig (cited by [3]), Provenza and Scherlis [58], Ratajczyk-Pakalska [64]. Other authors, including Zoll, Wessler and Schlesinger [64], and recently Baroldi (cited in [3]), do not confirm the existence of these anastomoses. However, in newer monographs [3] on the blood vessel of the heart, the view prevails that such anastomoses exist.

67

 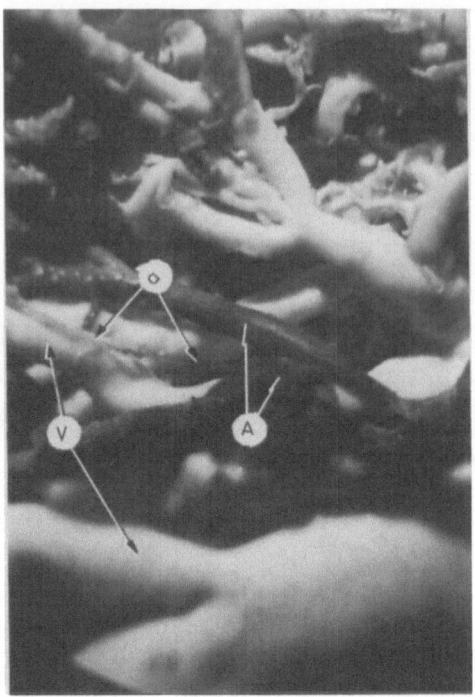

**Fig. 14.** Corrosion preparation of the coronary vessels of the left ventricular wall; a-v anastomoses.

**Fig. 15.** Corrosion preparation of the coronary vessels of the left anterior ventricular wall; a-v anastomoses.

In the study of the author of this chapter [64], a-v anastomoses were found to be present in the majority of the hearts, localized in the ventricular myocardium at various depths, usually near the endocardium (Figs. 14, 15). In agreement with the findings of Archangielski (cited in [76]), the current author [64] maintains that such anastomoses were more frequent in the walls of the left ventricle, indicating a better chance for developing a collateral circulation in arterial occlusive disease of the left coronary artery. A-v anastomoses were found [64] in all hearts with atherosclerotic lesions in the coronary arteries, the number of a-v vessels being significantly greater than were noted in normal hearts. This is in agreement with the view that the number of collateral vessels increases as a result of coronary atherosclerosis [3, 47, 60, 64, 72].

### F. Thebesian veins

Vieussens first described Thebesian veins in 1906 [88, 89], reporting on connections between coronary arteries and the heart cavities. Two years later, Thebesius [83] in his book entitled "Dissertatio Medica de Circulo Sanguinis in Corde" clearly characterized the existence of vascular structures between the myocardial veins and the heart cavities as the "venae cordis minimae" (subsequently named after Thebesius). Thebesius found these veins participated in providing blood outflow from the myocardium (in early systole) and noticed that the blood drained via these veins derived from the one-third internal portion of the heart walls.

The observations and discussions begun by Vieusses and Thebesius were continued by many authors, including Senac (cit. [34]), Verneyer (cit. [34]), Vinslow (cit. [34]), Lancisi (cit. [76]), Duvernie (cit. [30]), Luszka (cit. [30]), Lower (cit. [30]), Bohdalek [8], Langer [39], Cruvellier [14], Lannelongue (cit. [34]), Pratt [59].

Within the past dozen years, studies connected with ThV were reported by Esperanca Pina et al. [17–21, 77] in Portuguese, by Russian investigators Djavak-kishvili et al. [15, 16] and Tarasow [82], and by the author of this chapter in Poland [50, 52, 53, 63–70].

ThV will be discussed in terms of:
- formation of ThV
- course of ThV
- types of ThV
- connections and anastomoses of ThV orifices of ThV and their localization
- areas of outflow through the ThV
- number of ThV
- sphincters and valves in the walls of ThV.

Contrary to other veins of the heart, Thebesian veins ThV) collect blood from the veins of the myocardium and drain to the ventricular cavities at the endocardium,

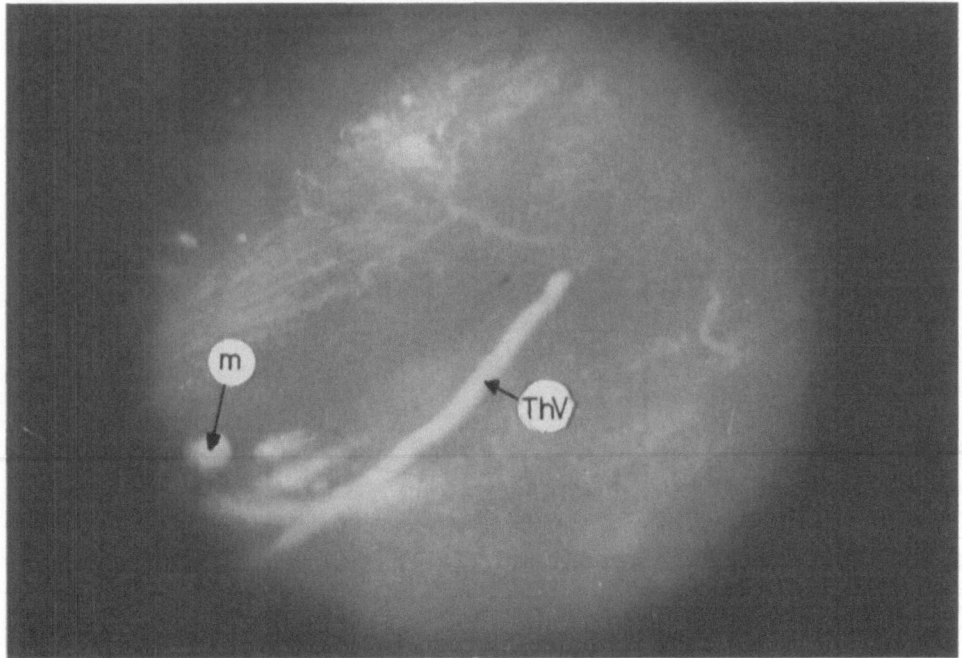

**Fig. 16.** Superficial ThV in the endocardium of the atrial septum from the side of the right atrium.

the blood escaping through orifices of ThV. It has been stated that ThVs were formed as continuation of the small myocardial veins, viz. in the ventricles as continuation of intramuscular venous plexuses, in atria as continuation of small veins of the atrial muscles, and in all areas of the cardiac walls in a more or less equal manner. Both the former and the latter veins form a continuation of venous capillary vessels.

According to some authors [28, 65, 82], ThV of the so-called deep coronary circulation in adults may be considered remnants of "venous lacunae" of the heart walls through which the heart bud of the fetus received its nourishment in the first 2–3 weeks [6, 26, 29, 30, 71, 72, 92].

One finds in the pertinent literature [82] a ThV classification, depending on their location and patency: "open" ThVs are those which occur in the atrium and in the ventricles and through which veins of the heart wall may be filled; "blind" or "closed" ThVs are vessels most often observed in the endocardium of the ventricular septum (and sometimes, in papillary muscles), hampering the filling of the veins of the heart walls.

Based on previous observations [65] concerning the venous layers of the ventricular walls, and taking into consideration the depth at which the ThV will form and their course in the atrial and ventricular walls, superficial and deep ThVs have been singled out [65]. Superficial ThV (Fig. 16) run parallel to the endocardial surface and enter the heart cavity at a right angle, whereas deep ThVs (Figs. 17, 18)

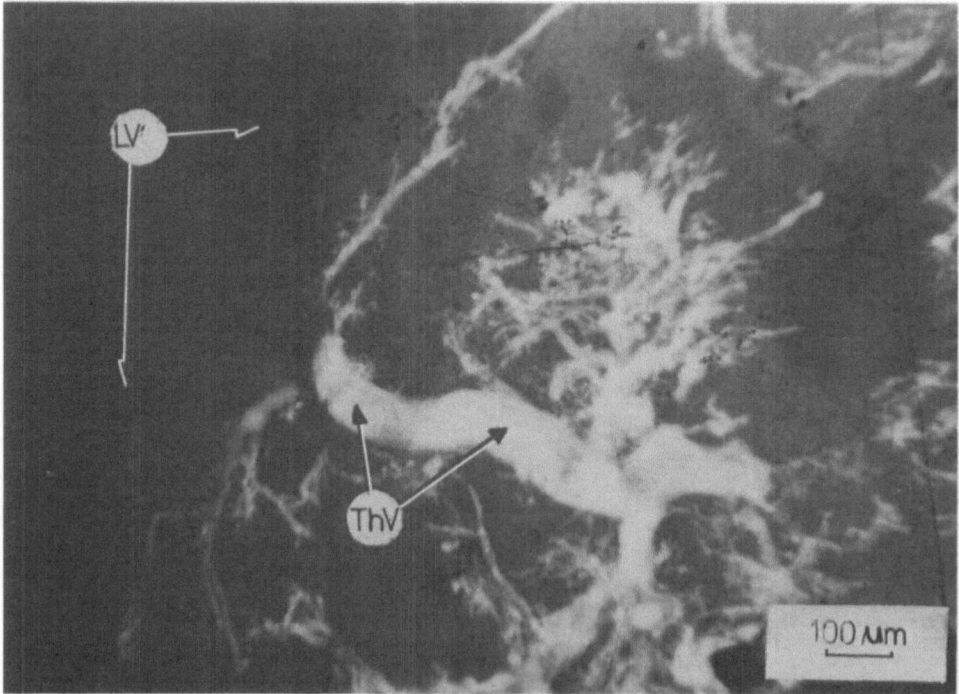

**Fig. 17.** Arboriform, deep-course-type of ThV. Microangiograph of the veins of anterior wall of the left ventricle (horizontal section).

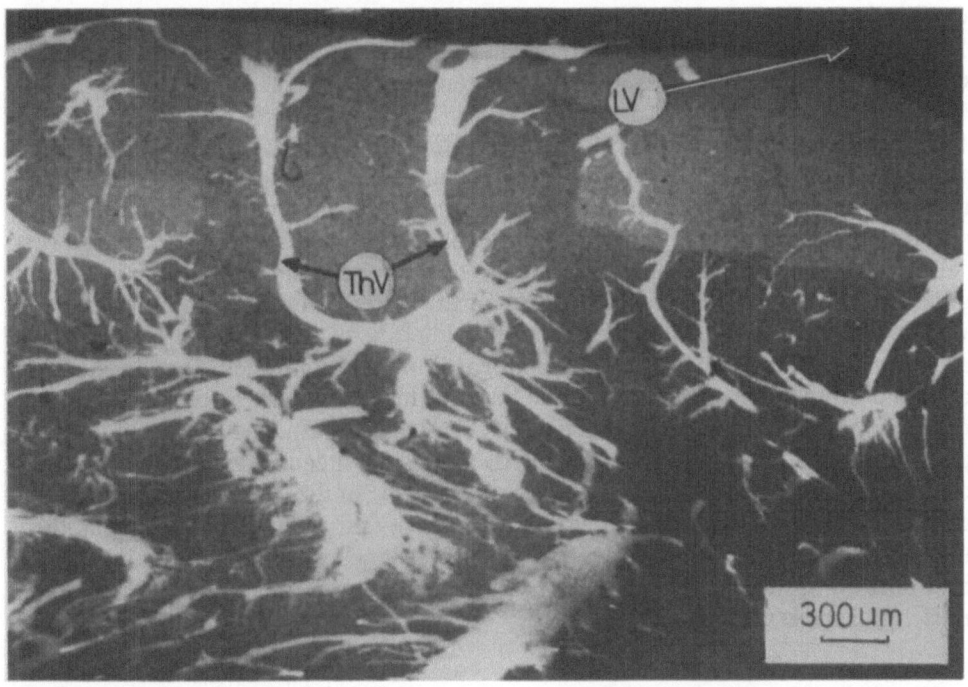

**Fig. 18.** Sinusoidal type of ThV (deep course). Microangiograph of the veins of ventricular septum (horizontal section).

run perpendicularly to the endocardial surface and discharge into the heart cavities at a right angle.

It has also been noted that the ThVs begin in various layers of the ventricular walls as:

a) deep ThV in the medial venous layer in the walls of the left ventricle, or more seldom as superficial ThV in the internal venous layer;

b) superficial and deep ThV in the internal venous layer in the walls of the right ventricle; and

c) superficial and deep ThV in a single atrial venous layer in the atrial walls.

From the anatomical point of view, the varied topography of the course of superficial and deep ThVs, in relation to the endocardial surface, suggests different conditions for ThV venous blood drainage from the myocardium.

Taking into consideration the course and tributaries of the ThV, Esperanca Pina et al. [19] described four types of ThVs, viz.: arboriform, sinusoidal, cannaliculate, and stellate, in the walls of the right atrium and right ventricle of the dog heart.

Observation of corrosion specimens and also microangiograms of the horizontal sections of the ventricular walls by Ratajczyk-Pakalska [65] revealed ThVs of arboriform and sinusoidal types in the walls of both ventricles, as well as in the walls of both atria, not only in man but also in other species studied. (Figs. 17, 18, 21). In addition, two other different types (viz.: bushlike, and threadlike) of ThV have been described [65]. The ThV of the thread-like type were rare and atypical, anas-

71

tomosed with the larger heart veins situated subepicardially (Fig. 23). Thread-like ThVs formed connections between the heart cavities and the larger tributary of the coronary sinus and may have special importance relative to coronary outflow (Figs. 17–22, 23).

The studies of Samoilova [76], Tarasow [82], Ilinskij [34], Esperanca Pina [21], and others [60, 64, 85, 91] confirm the presence of anastomoses between ThV and the coronary arteries. The author of this chapter [65] confirmed these connections and found, moreover, that the ThV anastomoses with: capillary venous blood vessels of the myocardium, superficial atrial veins, superficial ventricular veins and adjacent ThV (Fig. 21). The localization of the orifices of the ThV in the atria and ventricular myocardium indicated the following [65, 67, 32]. The orifices of ThV in the atria are grouped mainly on the endocardium of the auricles. Only in the right atrium do they form a grouping at the base of this atrium (along the projection of the right part of the coronary sulcus). In the atrial septum, the ThV orifices were distributed mainly around fossa ovalis from the side of each atrium. On both sides of the ventricular septum, mainly in the middle part of the septum there were ThV orifices: such orifices were also noted at the base and around the septal papillary muscles. In the walls of both ventricles, ThV orifices were located mainly at the mid-portion of these walls, and also in the vicinity and at the base of the papillary muscles of these ventricles.

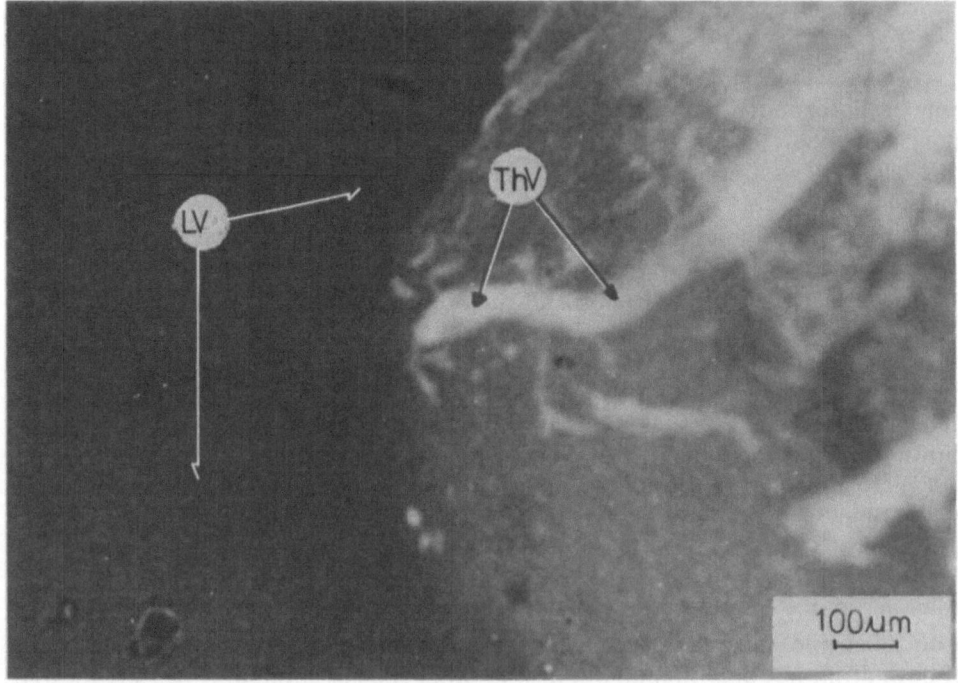

**Fig. 19.** Canaliculate type of ThV (deep course). Microangiograph of the veins of posterior wall of the right ventricle (horizontal section).

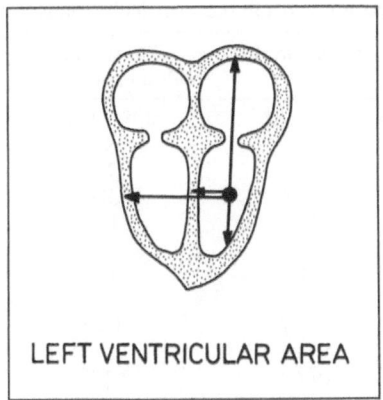

LEFT VENTRICULAR AREA

**Schema 4**

During observation of the displacement of volumes injected through the ThV orifices, the author of this chapter [65] revealed six areas of probable ThV outflow of the venous blood from the myocardium: Right atrial area, left atrial area, septal atrial area, right ventricular area, left ventricular area and ventricular septal area (schema 4).

Most interesting seemed to be the left ventricular zones of outflow. In this case, the filling mass injected to the orifices of the ThV in the left ventricular endocardium appeared not only in the adjacent ThV orifices, but also in the orifices of ThV, in the endocardium of ventricular septum, in the right ventricle and in endocardium of the left atrium. The number of ThV orifices in the endocardium of the ventricle cited by different authors varied from 20—80 in each of the atria, and from 5—30 orifices in each of the ventricles [32, 59, 82, 88, 36, 90, 97]. Clearly, there are numerous interconnections.

The number of ThV vessels was calculated for the first time by Ratajczyk-Pakalska [52, 65] on the microangioagraph, from serial cross-sections of the ventricular walls (statistical analysis was also employed). In ventricular walls, the following mean values of the number of ThV was noted: in human hearts − 152 veins, in the pig − 175 veins, in the sheep − 122 veins, in the dog − 146 veins (Table 5). Comparison of the number of ThV in both ventricles showed a high statistically significant difference in favor of the right ventricle, as follows: the number of ThV in the human right ventricular wall varied from 84 to 98 veins (mean value − 90 veins); in the human left ventricular walls it ranged from 48 to 88 veins (mean value − 62 veins). (Table 5).

Using an automated image analyzer (Quantimet 720) for measurement of the area of the ThV (on the microangiograms taken from the own sections of ventricular myocardium), it was also shown [68], that the share of the area of the ThV in regard to the combined area of all veins in the ventricular myocardium was $9.32 \pm 3.41\%$ in the human heart (Figs. 27, 28).

The significant ($9 \pm \%$) participation of the ThV in the entire venous vascularization of the left ventricular myocardium in the human (and also in animal hearts) corroborates the concept of the substantial importance of the ThV in blood outflow

**Table 5.** The parameters of number distrubution of the Thebesian veins in the ventricular walls

| Parameters | Homo sapiens L. | Canis familiaris L. | Ovis aries L. | Sus scrofa domesticus L. |
|---|---|---|---|---|
| Zone of variation | 132–177 | 108–175 | 97–162 | 128–216 |
| Range of feature | 45 | 67 | 65 | 88 |
| Mean value x̄ | 152 | 146 | 122 | 175 |
| Mean ± error | 3,64 | 5,40 | 6,38 | 3,88 |
| Standard deviation | 13,1 | 20,2 | 23,9 | 21,3 |
| Coefficient of skewness | +0,5 | −0,25 | −0,35 | −0,22 |
| Coefficient of excess | 0,58 | 0,58 | 0,72 | 0,61 |
| Coefficient of variability | 8,65% | 13,8% | 19,6% | 12,1% |

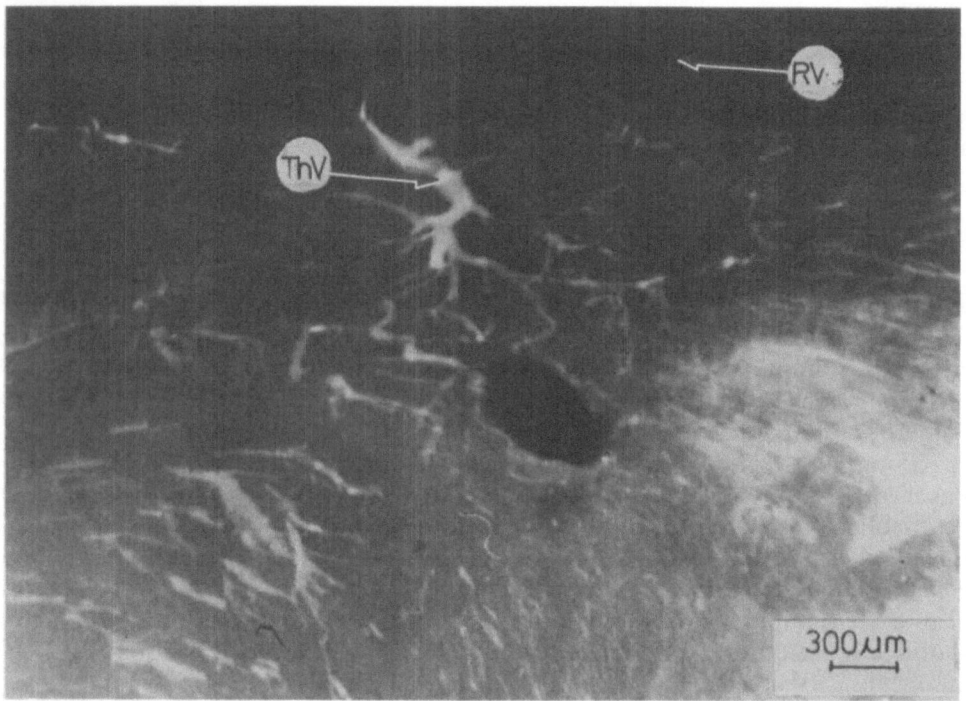

**Fig. 20.** Bush-like type of ThV. Microangiograph of the veins of ventricular septum (horizontal section).

**Fig. 21.** Sinusoidal type of ThV. Microangiograph of the veins of ventricular septum (horizontal section).

**Fig. 22.** Thread-like type of ThV. Microangiograph of the veins of ventricular septum and anterior wall of the left ventricle (horizontal section).

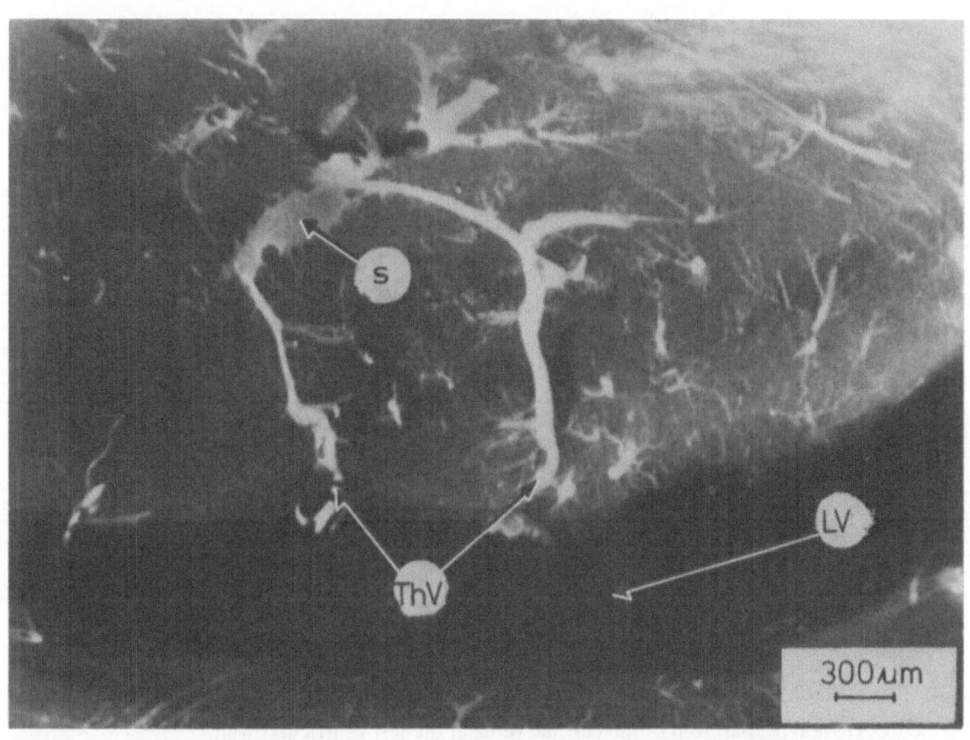

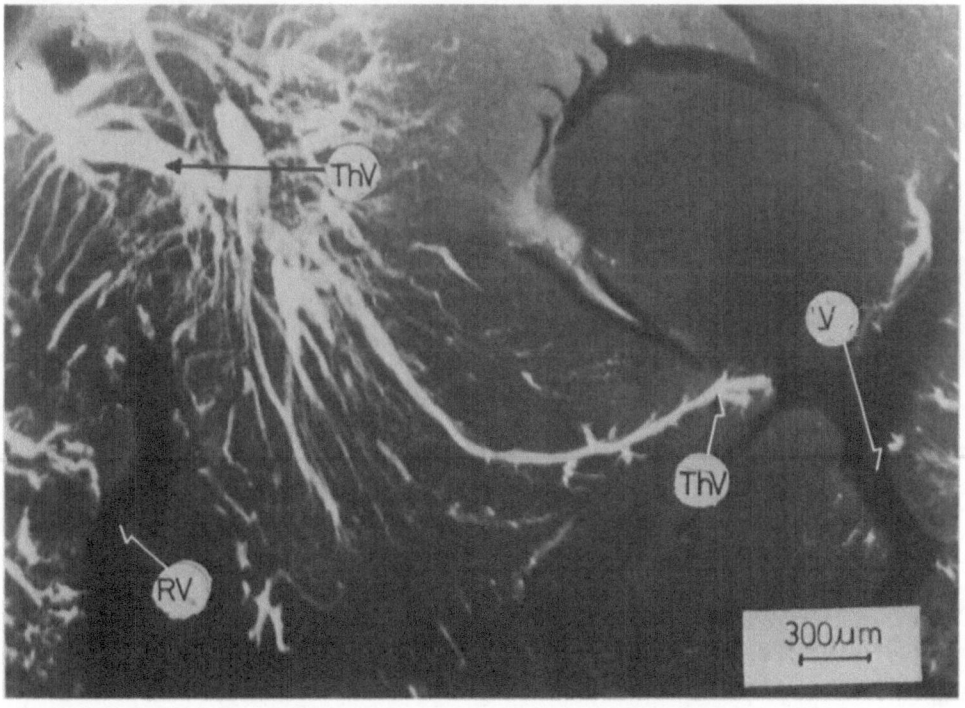

75

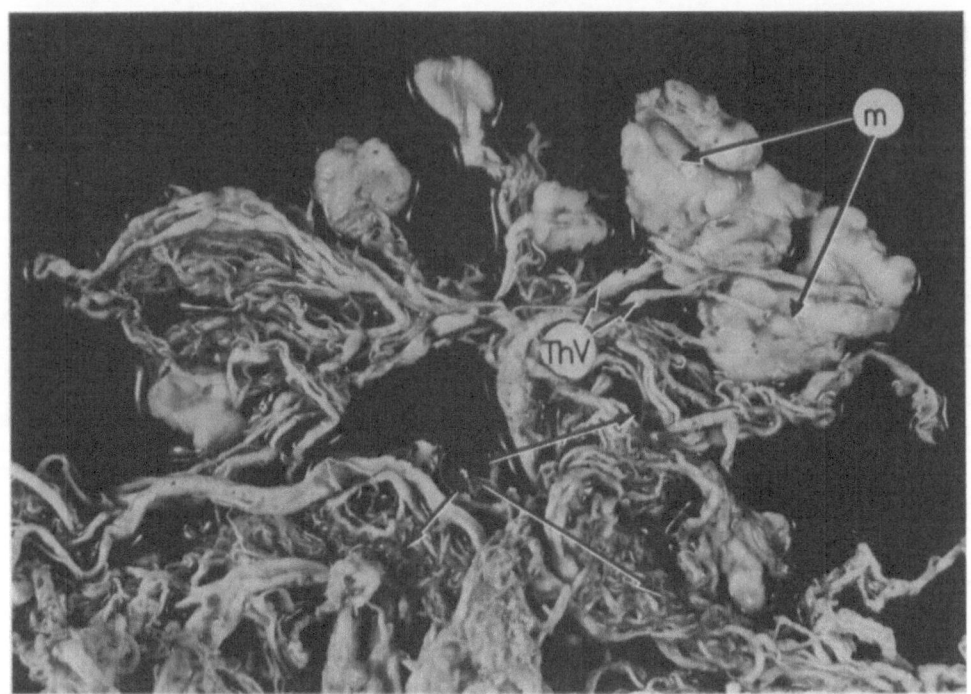

**Fig. 23.** Corrosion preparation of the cardiac veins of the left ventricular wall.

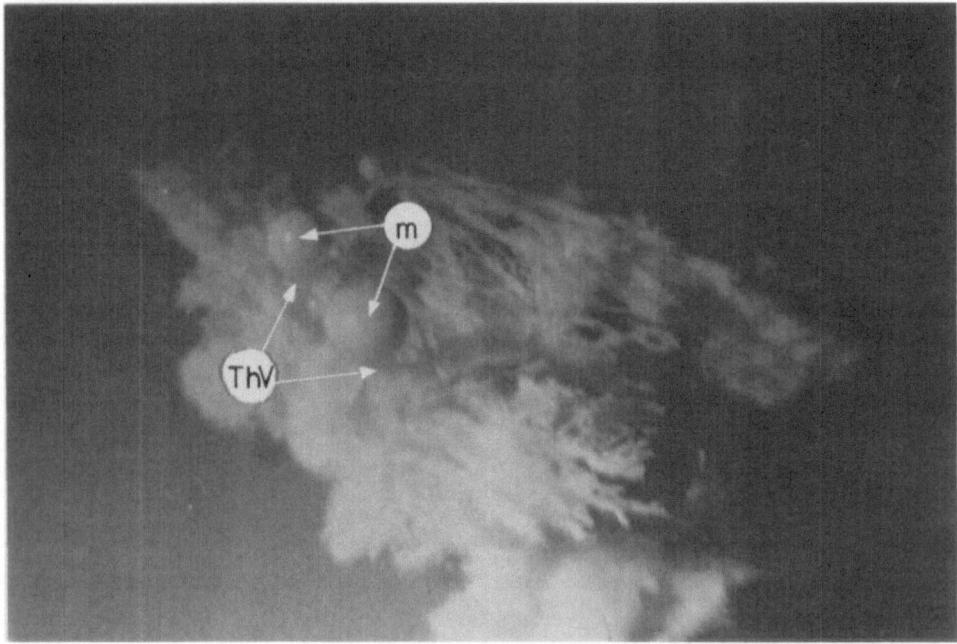

**Fig. 24.** Corrosion preparations of two connecting ThV in the right ventricular wall.

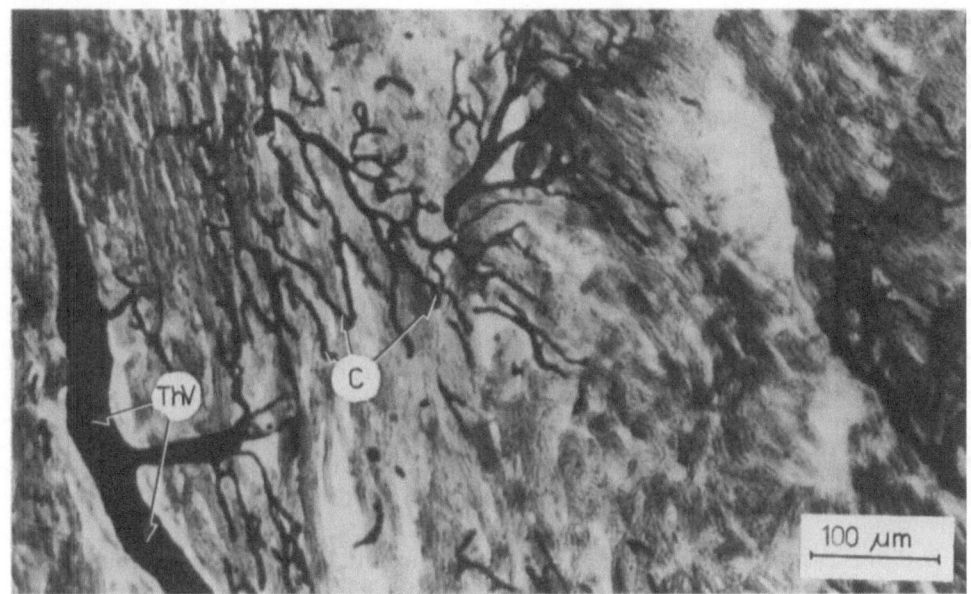

**Fig. 25.** Histologic preparation of cross-section of the left ventricular wall. Thebesian vein (canaliculate type) emptying into the left ventricular cavity (H–E).

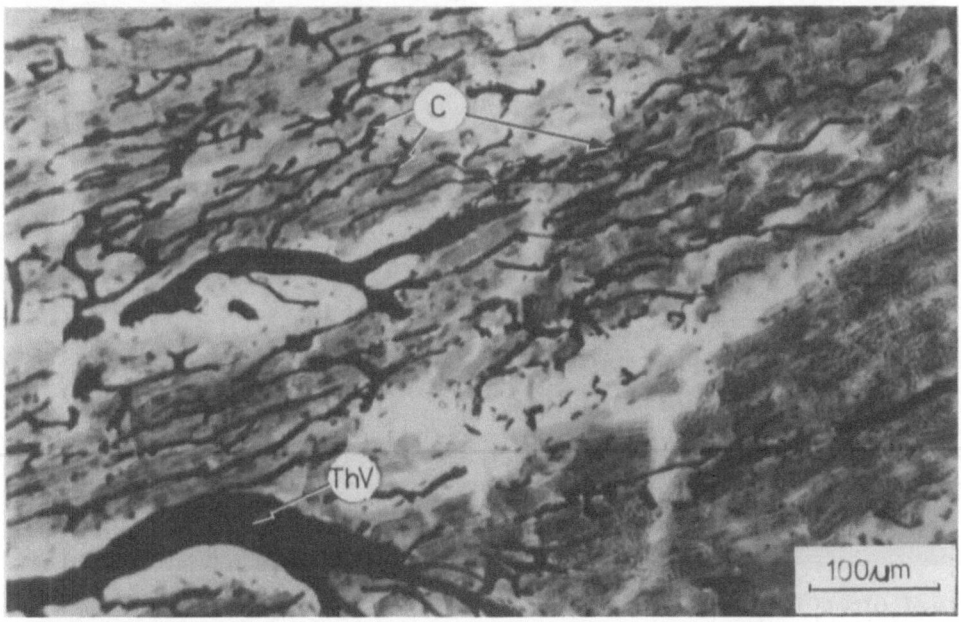

**Fig. 26.** Histologic preparation of cross-section of right ventricular wall. Thebesian vein (canaliculate type) emptying into the right ventricular cavity (H–E).

77

from the myocardium. While only five atherosclerotive hearts were analyzed in this study, and further unvestigation is indicated, it appears that there is a larger amount of ThV in cases of ischemic heart disease.

In microhistoangiographical studies of the wall of the ThV located in the left ventricular wall (at the site of vein outlet into the lumen of the ventricle), structures of thickened ThV wall were noted ([53, 65, 66]; Figs. 29–31). The smooth muscle cells located there were mainly circular (in regard to the lumen of the vessel); their nuclei had a more circular shape and were larger than similar nuclei of smooth muscle cells located in the distal part of the wall of the vessel. The appearance of those structures recall a type of sphincter in the walls of the ThV. In the walls of the ThV, other types of structures were described by the same authors [53, 65, 66]; Figs. 32–34). Microscopically, in serial sections, these structures looked like a duplication of the intima, a loose fibrous connective tissue with singular fibrocites, and fat cells as well as smooth muscle cells. Some of these "constructions" recall typical monocuspid valves of the other veins; they consist of intima only and are located either in the walls of the ThV (parietal valves) or in the outlet of the ThV (outlet valves).

The presence of valves in the ThV had been anticipated by researchers [72], although they did not support the existence of the valves by means of evidence from histological specimens. Ascertainment of the existence of monocuspid valves,

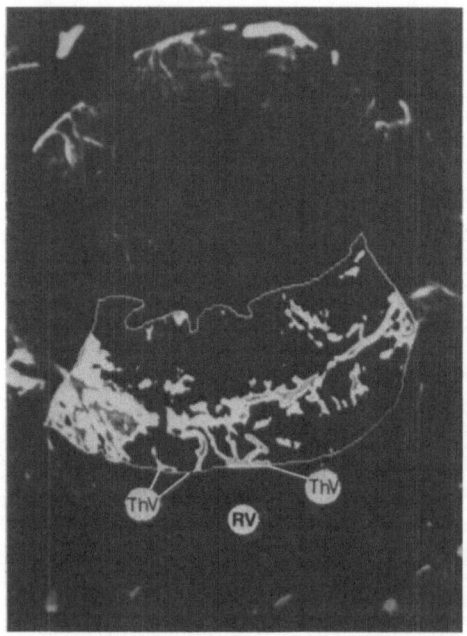

Fig. 27. Area of ThV of ventricular septum, emptying into the right ventricular cavity; radiograph of the cross-section of ventricular myocardium.

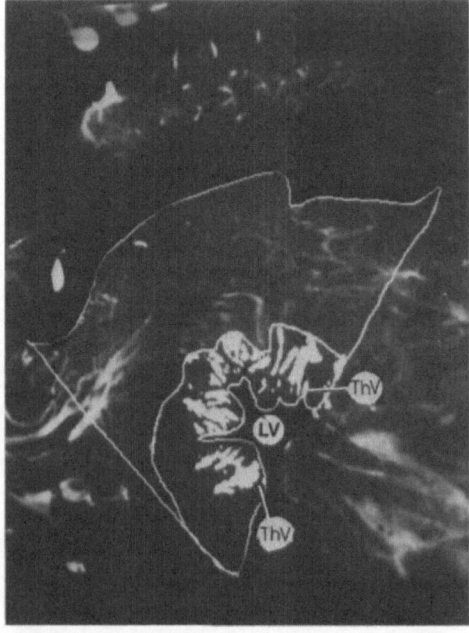

Fig. 28. Area of ThV veins of ventricular septum, emptying into the left ventricular cavity; radiograph of the cross-section of ventricular myocardium.

located in the outlets and in the walls of the ThV, as well as the presence of sphincters in the outlets of these veins, may explain some of the difficulties which appeared when attempting to fill the ThV with contrast from the chamber side of the ventricular endocardial openings. It may be that the so-called smaller blind ThV described by Tarasow [82], were the ThVs in which the sphincters and/or valves did not allow a retrograde filling. Schaper [78], discussing the problem of regulation of blood flow in the microcirculation, maintains that precapillary sphincters are also probably responsible for the different distribution of blood flow on the microcirculatory level.

From the anatomical point of view, it is believed that the valves and also the sphincters at the outlets of the ThVs play a particular function in the myocardial blood flow, as well as in the regulation of blood outflow from the myocardium into the ventricles (possibly also of the inflow of blood to the ThVs).

## G. Intramyocardial components of the venous system in the normal and diseased heart

Observing of cross sections of the right and left ventricular myocardium in the normal human heart (using the technique of microangiography), Ratajczyk-Pakalska et al. [50] described three zones of venous vascularization (Figs. 35–37):
1) the external-epicardial zone containing a thin, 0.1 mm external layer of the free walls of the ventricle,

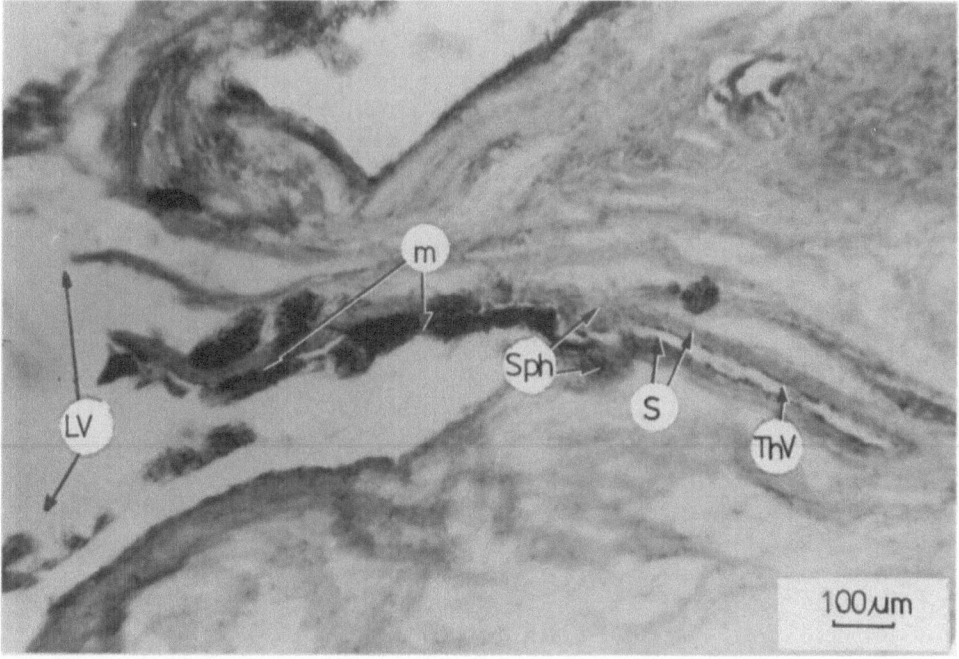

**Fig. 29.** ThV in the human ventricular septum. Visible sphincter in the vein outlet (H–E).

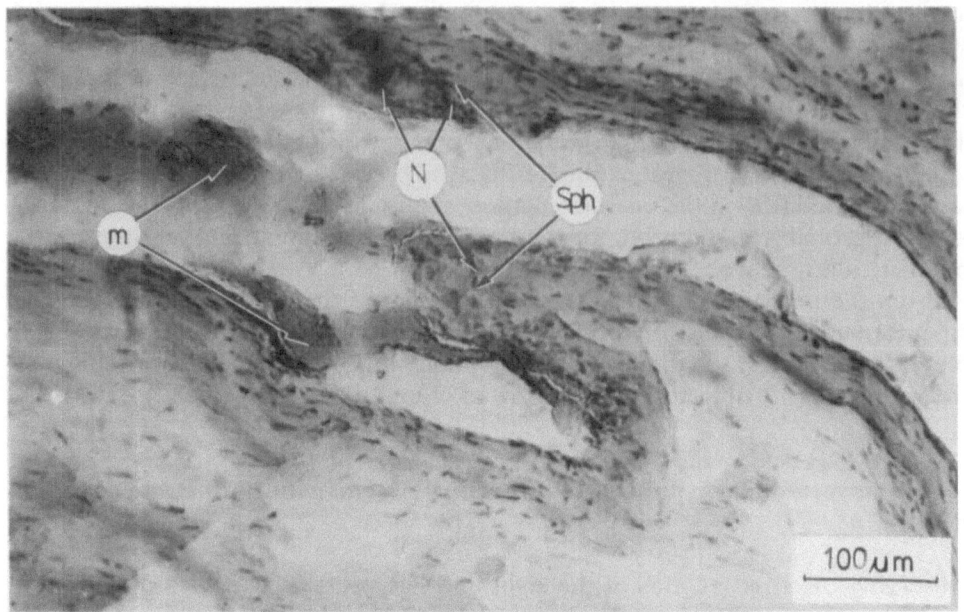

**Fig. 30.** ThV in the human anterior ventricular wall. Visible sphincter in the vein outlet (H−E).

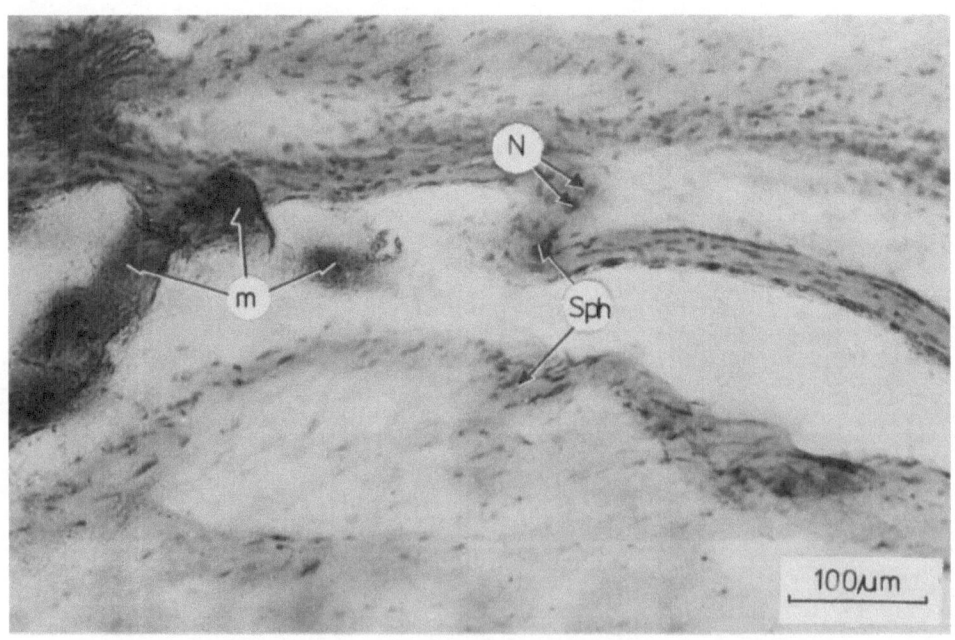

**Fig. 31.** ThV in the human ventricular myocardium. Visible sphincter in the vein outlet (H−E).

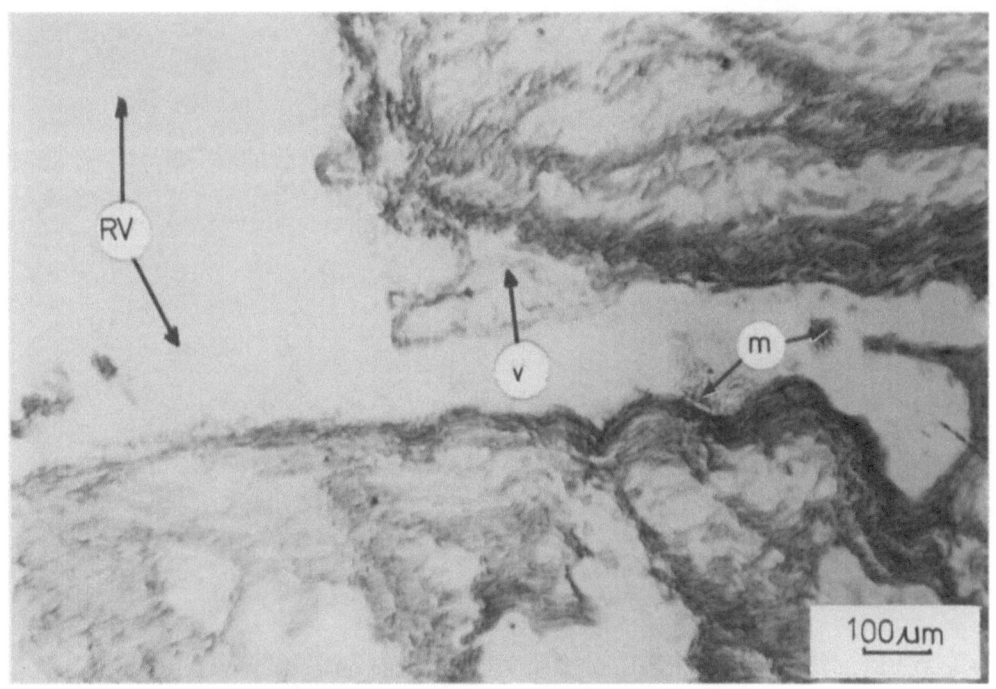

**Fig. 32.** ThV in the human ventricular septum. Visible outlet valve at the opening of the vein (H–E).

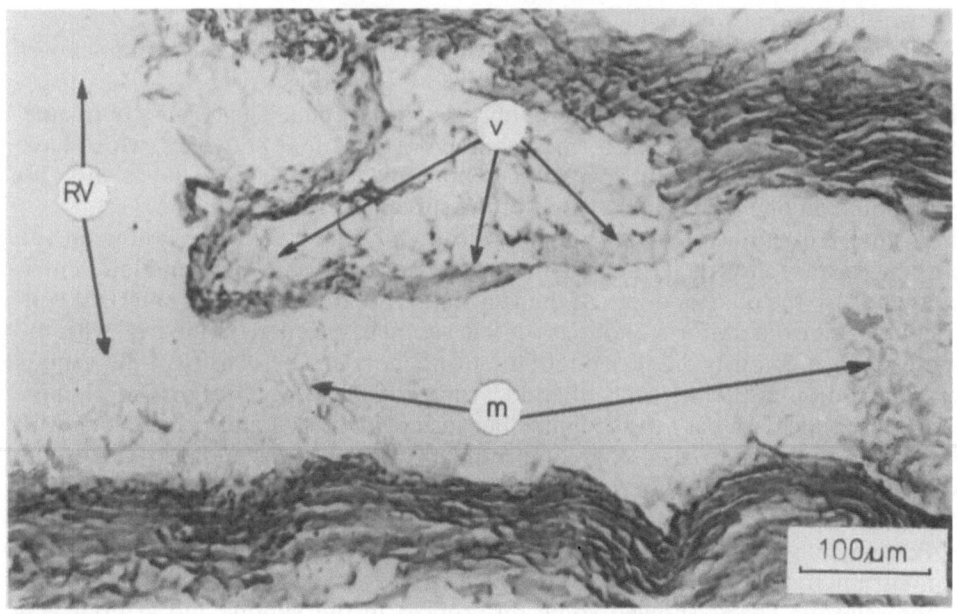

**Fig. 33.** ThV in the human ventricular septum. Visible outlet valve at the opening of the vein (H–E).

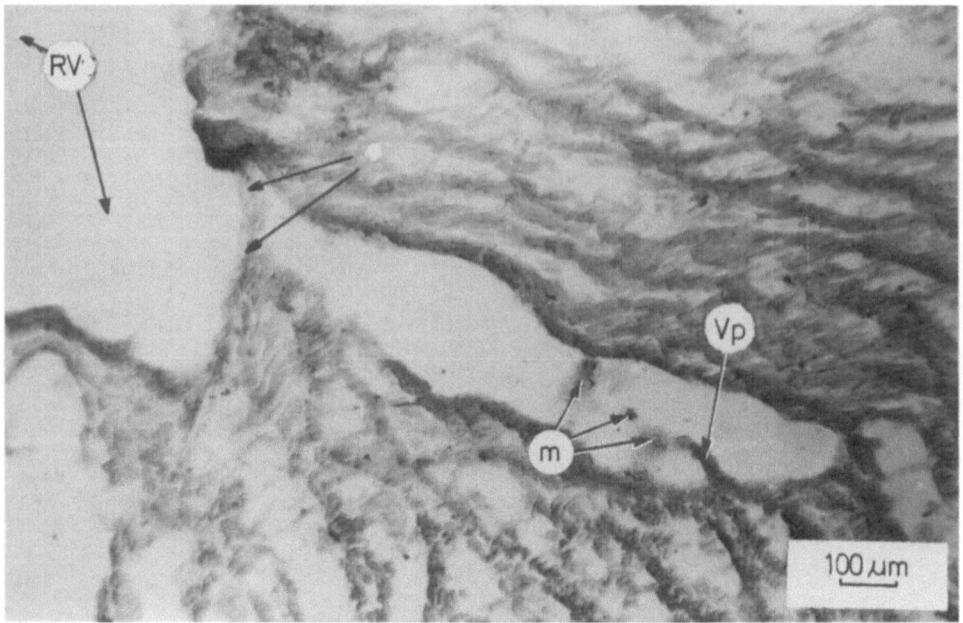

**Fig. 34.** ThV in the human ventricular septum. Visible parietal and outlet valve (H−E).

2) the medial thickest zone which, together with the external zone, make up about two-thirds of the ventricular or septal wall, and
3) the internal-subendocardial zone, which contains the rest of the wall thickness of the ventricles and ventricular septum.

The largest veins were situated in the subendocardial zone. The tertiary or quaternary tributaries of these veins formed characteristic venous networks, which have been identified with the intravascular venous plexus [50, 65]. Thebesian veins were noted among the minute vessels of the subendocardial zone.

Postmortem studies [69] on human hearts with atherosclerotic lesions in the coronary artery showed in the subendocardial zone (of both right and left ventricles, and also the quarter depth of the ventricular septum facing left and right ventricular cavities) a uniform and rich venous vascularization, as compared with the same zones previously described [50] in normal hearts. The density of the venous vascularization in the internal-subendocardial zone also depends on the Thebesian veins, which are easy to recognize because of their size and connections with the lumen of the ventricles. A greater number of small veins and also ThV in the ventricular walls of the atherosclerotic hearts than in normal hearts is in agreement with previous observations.

Postmortem investigations of Ratajczyk-Pakalska, Bloch and Kulig [70] on the venous microcirculations of the left ventricular myocardium in normal and atherosclerotic hearts showed some differences in the venous and arterial vascularization. The density of veins of the left ventricle is much greater than that of the arteries (Fig. 38 [16]).

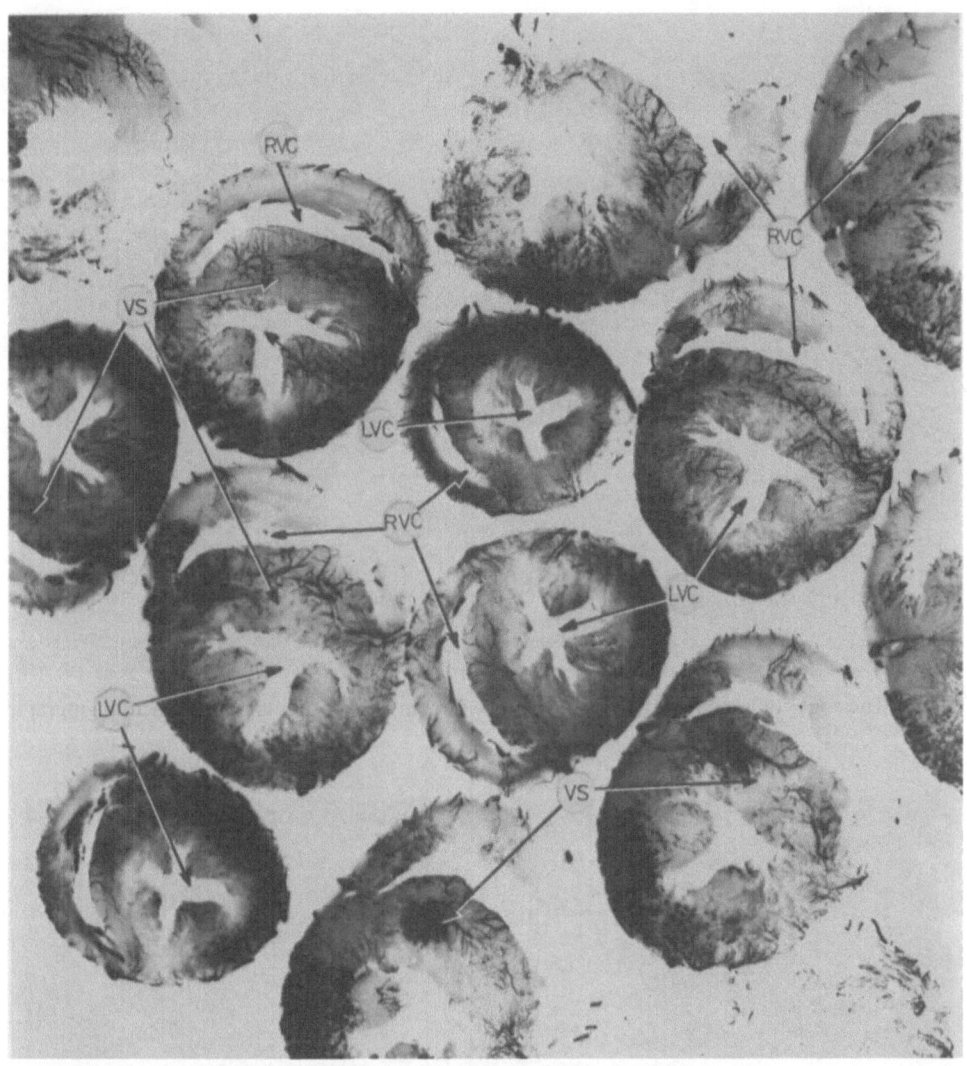

**Fig. 35.** Microangiographs of the veins of the ventricular walls (cross-section).

There are other differences in the venous and arterial vascularization. If there is a differentiation in the arteries (type A and B per Astes) of the ventricular walls of the heart, the venous vascularization of those two types in the ventricular myocardium is not noted (Fig. 39).

The ventricular septum of the normal heart is characterized by the location of the main veins, which are situated proximally to the right endocardium. Depending on the intraventricular pressure (lower in the right, higher in the left), there is a tendency to replace main veins in the direction of the right ventricle (Fig. 40).

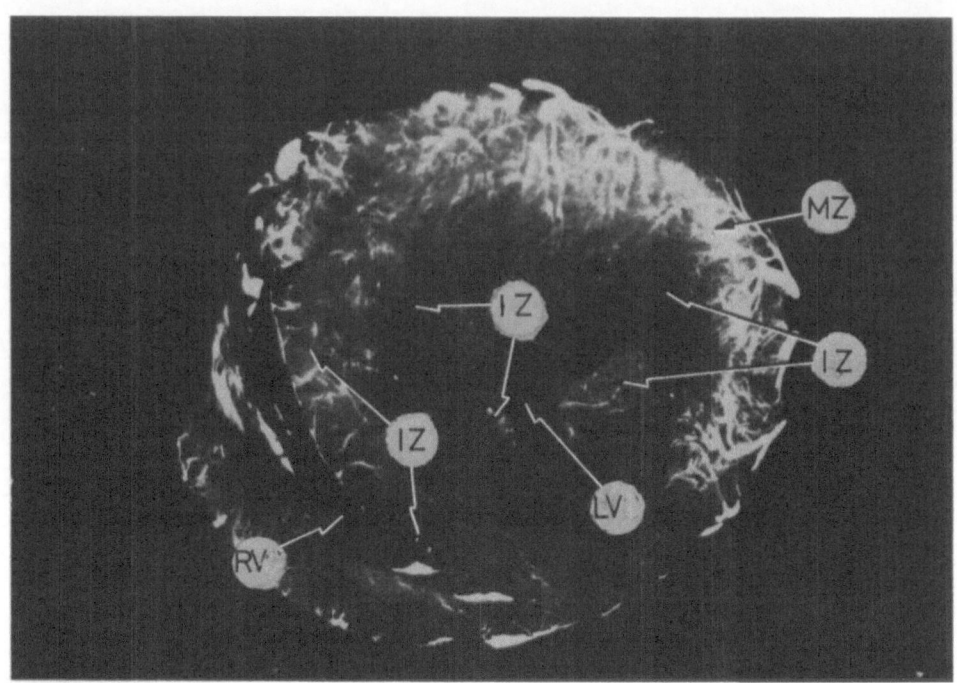

**Fig. 36.** The zones of venous vascularization in the human left ventricular walls. Microangiograph (cross-section).

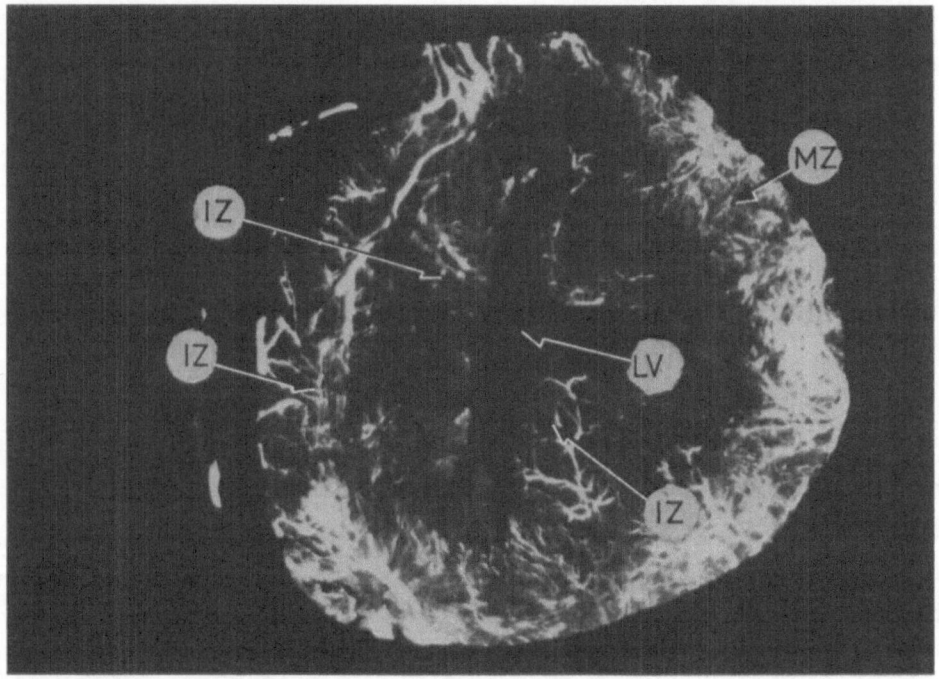

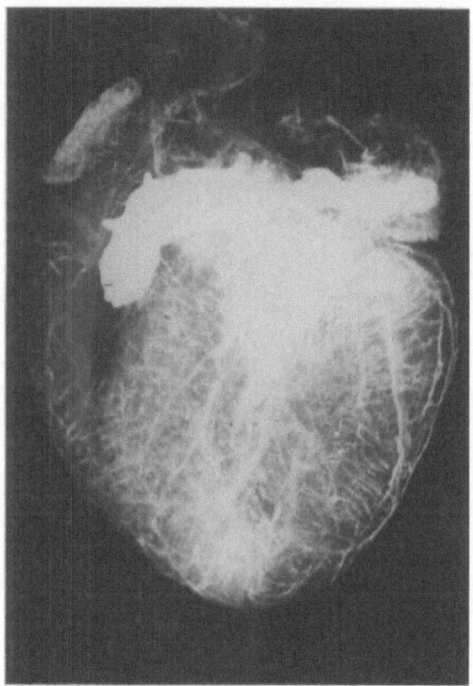

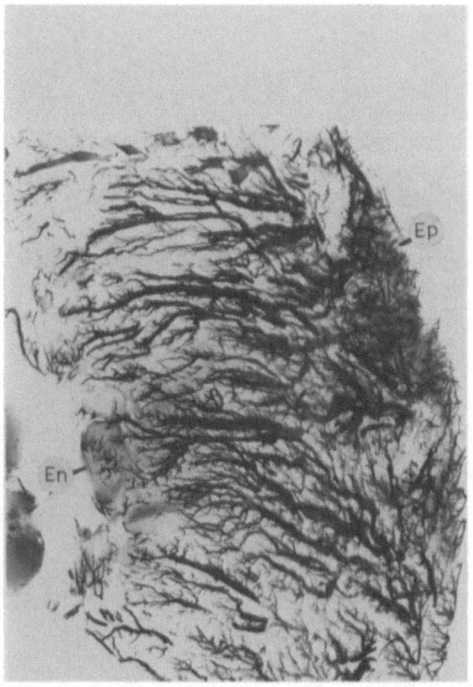

**Fig. 38.** Angiogram of the cardiac veins filled with radio-opaque microfil. Visible veno-venous anastomoses (large and medium).

**Fig. 39.** Microangiograph of cross-section of the left ventricular wall. Myocardial venous pattern of the left ventricle shows a rich density of veins as compare with the arteries (normal heart).

In the atherosclerotic hearts with small scars, the capillary pattern disappears – it loses its rhythm, and abnormal or chaotic outflows occur from this zone, most likely as a result of the repair process (Fig. 41).

In cases of severe insufficiency of the ventricle, the congestion appears mainly in larger veins. In contrast, in capillary vessels congestion in less, and in these cases, the outflow through ThV seems to be greater (subendocardial "empty spaces" are seen distinctly, Figs. 42, 43).

On the microangiograms obtained from cross-sections of the left ventricular myocardium, it could be noted that the larger cardiac veins and their tributaries in the subepicardial left ventricular myocardium are more abundantly filled with Microfil as compared with the subendocardial zone. During the filling of the cardiac veins through the coronary sinus, drainage of the Microfil (through the ThVs) into the left ventricular cavity has been observed. This appears to testify from an

◄─────────────────────────────────────────────

**Fig. 37.** The zones of venous vascularization of the human ventricular walls. Microangiograph (cross-section).

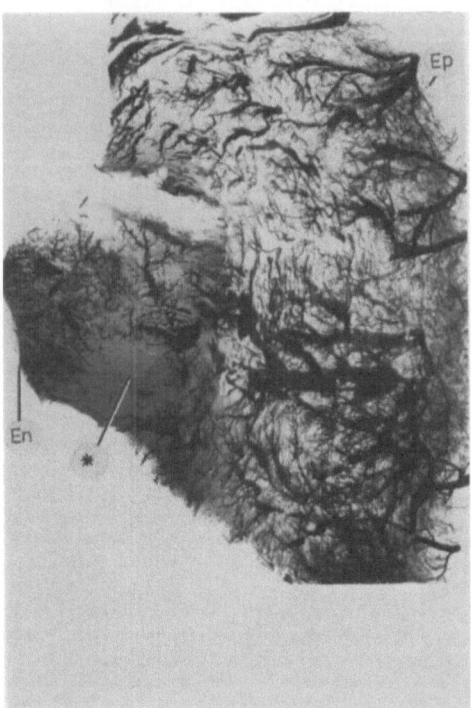

**Fig. 40.** Microangiograph of cross-section of the ventricular septum showing the localization of the main larger veins in the septum (toward the right ventricle).

**Fig. 41.** Microangiograph of cross-section of the left ventricular wall showing abnormal and chaotic outflow; there are no capillary vessels in small scars (artherosclerotic hearts).

anatomical point of view to an efficient outflow from the subendocardial zone, and it also would be the answer to absent hermorrhages in this zone after coronary venous retroperfusion.

The myocardial necrosis in the treated animal after retroperfusion is predominantly localized in the mid-myocardial and subepicardial regions of the left ventricle. This may be explained by distribution of blood which flows into the perfused vein and depends on the efficient veno-venous anastomoses, which direct the blood mainly to the larger veins and into the right atrium, and also through the ThV into the left ventricular cavity, instead of being delivered retrogradely to the capillaries.

On the other hand, a lack of v-v anastomoses of sufficient diameter can be responsible for hemorrhages in the subepicardial region.

As shown in experiments on venous retroperfusion during experimental ischemia in open chest pigs (Carlson et al.), no increase of ischemic zone blood flow was measured by the radioactive microsphere (14 µm) method. Since there was no evidence of an intracardiac shunt, the microspheres could only get to the systemic circulation through the left-sided ThV. The great number of microspheres appearing in the systemic circulation suggest that numerous venous-to-Thebesian shunts, larger than 14 µm in size exist, and that ThVs can open in response to coronary

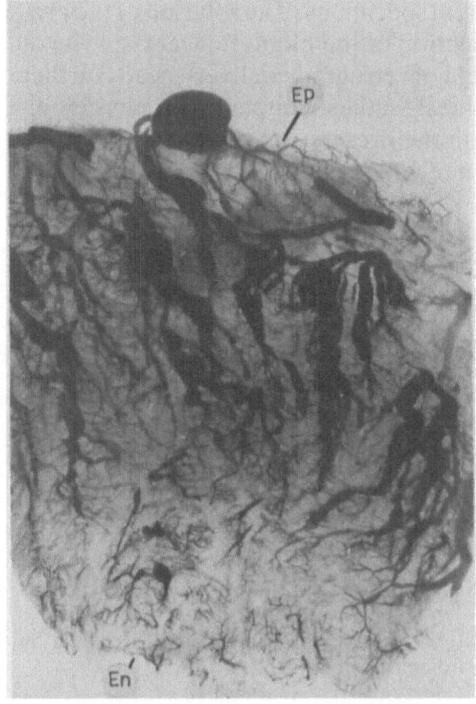

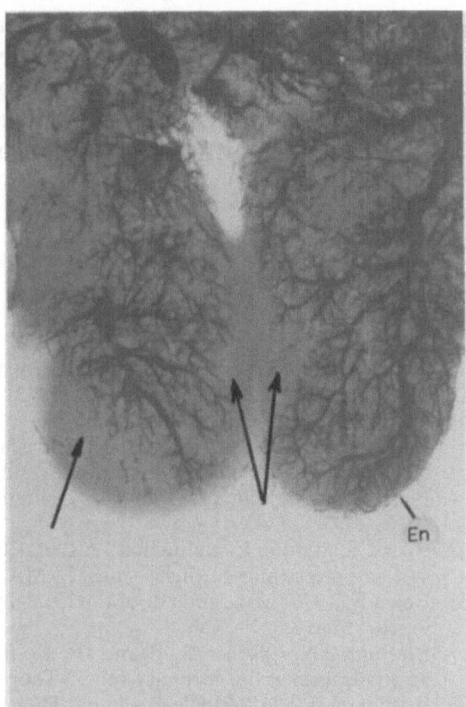

**Fig. 42.** Microangiograph of cross-section of the left ventricular wall in severe insufficiency of this ventricle. The congestion appears in mainly larger veins. The inflow through Thebesian veins seems to be greater (endocardial "empty spaces" are seen).

**Fig. 43.** Microangiograph of cross-section of papillary muscles in the left ventricle; the "empty spaces" are visible (the area of outflow through the ThV).

occlusion and coronary sinus or anterior interventricular vein occlusion with retroperfusion. This observation suggests that there is free access of the venous circulation to a mobilized left-sided Thebesian system that does not occur when venous pressure is normal and coronary arteries are patent.

Schaper [78], discussing the problem of regulation of blood in the coronary microcirculation, maintained that precapillary sphincters are probably responsible for the efficient distribution of blood flow on the microcirculatory level from the arterial side. The described [53, 65] sphincters as well as valves in the walls of the ThV probably play the particular role in the myocardial blood flow also in regulating the outflow from the myocardium into the ventricles, and may point to an important role of these structures after retroperfusion.

We do not yet know, but there also may be some sphincters or other "mechanism" on the level of the capillary vessels that are responsible for the regulation of the retrograde inflow of blood from the venous side toward the capillaries. These may not necessarily be anatomical sphincters, but a physiological response to specific physiologic and pharmacologic stimuli. Such effects would be very difficult and even impossible to ascertain in postmortem specimens. The solution of the problem should be expected in a dynamic evaluation of the blood flow at the level of capillary vessels. For this reason, the problem of retrograde delivery needs further histopathological and microhistoangiographical studies, coupled with physiological studies on the normal myocardium and in the presence of heart disease, when treated with coronary venous retroperfusion techniques.

## References

1. Adachi B (1933) Das Venensystem der Japaner. Kyoto, S 40–61
2. Arealis EG, Volder JGR, Kolff WJ (1973) Arterialization of the coronary vein coming from an ischemic area. Chest 63:462–463
3. Bajusz F, Jasmin G (1971) Methods and achievements in experimental pathology. Functional morphology of the heart, vol V. Karger, Basel
4. Beck CS, Leighninger D (1954) Operations for coronary artery diseases. JAMA 156: 1226–1233
5. Beck CS, Stanton E, Batiuchok W, Leiter E (1948) Revascularization of the heart by graft of systemic artery into coronary sinus. JAMA 137:436–441
6. Bellet S (1930) Nourishment of the heart by channels other than the coronary arteries. Cyclopedia Med 3:363
7. Bhayana YN, Olsen DB, Byrne JP, Kolff WJ (1974) Reversal of myocardial ischemia by arterialization of the coronary vein. J Thorac Cardiovasc Surg 67:125–132
8. Bochdalek VA (1868) Die Foramina Thebesii in Herzen. Arch Anat 302
9. Brown RE (1971) The pattern of the microvasculatory bed in the ventricular myocardium of domestic mammals. Am J Anat 116:355–374
10. Bucher C (1973) Cytologie Histologie und microskopische Anatomie des Menschen. Huber, Bern Stuttgart Wien
11. Bucher O (1947) Über den Bau der Blutgefäße des menschlichen Herzens. Acta Anat (Basel) 3:162–189
12. Carlson CJ, Ratajczyk-Pakalska E, Cogan JJ, Rapaport E (1985) Effect of venous retroperfusion on experimental myocardial ischemia in the open-chest pig. J Surg Res 38:105–112
13. Corday E, Tzu-Wang L, Meerbaum S, Gold H, Mirose S, Rubins S, Dalmastro M (1974) Closed chest model of intracoronary occlusion for study of regional cardiac function. Am J Cardiol 33:49–59
14. Cruveilhier J (1834) Anatomie descriptive. Paris
15. Djavakhishvili N, Komakhidze M (1970) Le courant sanguin retrograde dans le myocarde. CR Assoc Anat 149:733–736
16. Djavakhishvili N, Komakhidze M (1977) Potential of myocardial sinusoides. Bibl Anat 16/ 2:546–548
17. Esperanca Pina JA, Monteiro Trindade A (1977) Etude microangiographique des vaisseaux arterio-luminaux. Acta Anat (Basel) 98:334–339
18. Esperanca Pina JN (1973) Circulacao venosa cardiaca. Estudo anatomoexperimental; diss., Faculdade de Medicina, Lisboa
19. Esperanca Pina JA, Correira M, O'Neill G (1975) Morphological study on the thebesian veins of the right cavities of the heart in the dog. Acta Anat 92:310–320
20. Esperanca Pina JA (1973) Note sur les veines de thebesius d'apres reconstruction par la methode de Born. Arch Anat Pathol 21:331–335
21. Esperanca Pina JA, Monteiro Trindade A, Santos Fereira A (1973) Microangiographic study of arterio-luminal vessels of the heart. 7-th Europ. Conf. Microcirculation. Aberdeen, 1972, Part X. Bibl Anat 11:133–138
22. Evans CL, Starling EH (1913) The part played by the lung in the exudative process of the body. J Physiol 46:413
23. Farcot JC, Meerbaum S, Lang TW, Kaplan L, Corday E (1978) Synchronized retroperfusion

of coronary veins for circulatory support of jeoparadized ischemic myocardium. Am J Cardiol 41:1191–1201

24. Farrer-Brown G (1974) Patterns of the microvasculature in normal and diseased hearts. Acta Cardiol 19:119–127
25. Gorlin R (1968) Perforations and other cardiac complications. Circulation 37:36–38
26. Grant RT, Viko LE (1929) Observations on the anatomy of the Thebesian vessels of the heart. Heart 15:103
27. Gregg DE, Pritchard WH, Shipley RE (1947) Studies of the venous drainage of the heart. Am J Physiol 151:13
28. Gregg DB (1953) Coronary circulation in health and disease. Lea and Febiger, Philadelphia
29. Hammond GL, Davies AL, Austen WG jr (1967) Retrograde coronary sinus perfusion: a method of myocardial protection in the dog during left coronary artery occlusion. Ann Surg 39:166–170
30. Hammond GL, Moggio RA (1971) Function of microvascular pathways in coronary circulation. Am J Physiol 220:1463–1467
31. Hammond GL, Austen WG (1967) Drainage patterns of coronary arterial flow as determined from the isolated heart. Am J Physiol 212:1435–1440
32. Halonen P, Tossavainen P (1947) Beitrag zur Kenntnis der Venae Thebesii des Herzens. Acta Soc Med Fenn "Duodecim" 24:172–181
33. Hellerstein HK, Orbison JL (1951) Anatomic variations of the orifice of the human coronary sinus. Circulation 3:514–523
34. Ilinskij SP (1958) O sasudach Tebezija. Arkh Pathol 20:1–11
35. James T (1961) Anatomy of the coronary arteries. Hoeber, New York
36. James TN, Anderson K (1965) Anatomy of the coronary arteries in health and disease. Circulation 32:1020–1033, 32:808–810
37. James TN (1970) The delivery and distribution of coronary collateral circulation, Chest 58:183–203
38. Kolff J, Kolff WJ (1976) The history of coronary vein perfusion. Minn Med 59:648–650
39. Langer L (1980) Die Foramina Thebesii in Herzen des Menschen. SB Akad Wess Wien 82:25
40. v. Lüdinghausen M, Ratajczyk-Pakalska E, Tschabitscher M, Maurer G, Glogar D, Mohl W (1984) Nomenclature: Venae cardiacae – cardiac veins, In: Mohl W et al. (eds) The Coronary Sinus, Steinkopff, Darmstadt, pp 1–4
41. v. Lüdinghausen M (1986) Nomenclature and distribution pattern of cardiac veins in man. In: Mohl W et al. (eds) Clinics of CSI, Steinkopff, Darmstadt, pp 13–32
42. Lunkenheimer P, Merker J (1974) Morphologic und Funktion eines intramyokardialen, "sinussoidalen" Strömungsnetzes. Thoraxchirurgie 22:26–35
43. Meerbaum S, Tzu-Wang L, Osher JV, Hashimoto K, Lewis G, Feldstein A, Corday E (1976) Diastolic retroperfusion of acutaly ischemic myocardium. Am J Cardiol 37:588–598
44. Mierzwa J, Kozielec T (1975) Variation of the anterior cardiac veins and their orifices in the right atrium in man. Fol Morph (Warsz) 34:125–132
45. Mohl et al. (1984) The coronary sinus. Steinkopff, Darmstadt
46. Mohl et al. (1986) Clinics of CSI. Steinkopff, Darmstadt
47. Moll JW, Dziatkowiak A, Rybiński U, Edelman M, Ratajczyk-Pakalska E (1973) Arterialisierung des sinus coronarius. Indikationen, Technik, Ergebnisse. Thoraxchirurgie 21:295–301
48. Moll JW, Dziatkowiak A, Edelman M, Iljin W, Ratajczyk-Pakalska E, Stengert K (1975) Arterialization of the coronary veins in diffuse coronary arteriosclerosis. J Cardiovasc Surg 16:520–525
49. Moll JW (1976) Ratajczyk-Pakalska E, Dziatkowiak A, Zienkiewicz J (1976) Experimentelle und klinische Erfahrungen mit der Arterialisierung der Herzvenen. Zentralbl Chir 101:112–117
50. Pakalska E, Golab B (1978) Coronary circulation of the venous side in the human hearts with the special references to the venae cordis minimae. Adaptability of Vascular Wall. Avicenum and Springer, Prague and Heidelberg, pp 679–681
51. Pakalska E, Kolff WJ (1980) Anatomical basis for retrograde coronary vein perfusion (Venous anatomy and veno-venous anastomoses in the hearts of human and some animals). Minn Med 63:795–801

52. Pakalska E, Kolff WJ, Golab B (1980) The number of the smallest cardiac veins in ventricular myocardium in human and some animals. XI Congreso International de Anatomia, Ciudad de Mexico, 1980, Abstract
53. Pakalska E, Fortak W, Golab B (1981) An occurrence of sphincter and regulator — like systems in the smallest cardiac veins of the human heart. Folia Morphol (Praga) 29:213—215
54. Parsonnet V (1953) The anatomy of the veins of the human heart with special reference to normal anastomotic channels. J Med Soc (New Jersey) 50:446—452
55. Plotz M (1957) Coronary heart discase. Hoerber, New York
56. Polacek P (1960) Mikroskopocke sledovani svalcych mustku a poutek lidskich vencitych tepnach. Ceskoslovenska Morfologie 4:345—360
57. Poulhes J, Trovetto L, Lacomme J (1958) Systematisation des veines de Thebesius (venae cordis minimae). CR Assoc Anat 100:621—628
58. Provenza V, Scherlis S (1959) Demonstration of muscle sphincters as a capillary component in the human heart. Circulation 20:35—41
59. Pratt HF (1891) The nutrition of the heart through the vessels of Thebesius and the coronary veins. Am J Physiol 1:86—103
60. Prinzmetal M, Simkin B, Bergman HC, Kruger HE (1947) Studies on the coronary circulation. II. The collateral circulation on the normal heart by coronary perfusion with radioactive erythrocytes and glass spheres. Am Heart J 33:420—442
61. Ratajczyk-Pakalska E (1970) Heart veins. Folia Med (Lodz) 10:46—72
62. Ratajczyk-Pakalska E (1973) Veins of the interventricular septum in the human heart. Folia Morphol (Warszawa) 32:331—338
63. Ratajczyk-Pakalska E, Moll JW, Edelman M, Chetkowska E (1974) Badanie unaczynienia żylnego mięśniówki komór serca po podwiązaniu tętnic wieńcowych. Kardiol Pol 17:133—141
64. Ratajczyk-Pakalska E (1975) Arterio venous anastomoses in the human myocardium. Folia Morphol (Warszawa) 34:285—292
65. Ratajczyk-Pakalska E (1978) Variation of the smallest cardiac veins. Folia Morphol (Warszawa) 37:415—418
66. Ratajczyk-Pakalska E, Fortak W, Golab B (1984) Sphincters and valves in the walls of the smallest cardiac veins. Folia Morphol (Warszawa) 2:121—127
67. Ratajczyk-Pakalska E, Kolff WJ (1984) Anatomical basis for the coronary venous outflow, In: Mohl et al. (eds) The coronary sinus. Steinkopff, Darmstadt, pp 40—46
68. Ratajczyk-Pakalska E, Wagiel J, Kamiński M, Kolff WJ (1984) Automated image analysis for measurement of area of the smallest cardiac veins in the ventricular myocardium, In: Mohl et al. (eds) The coronary sinus. Steinkopff, Darmstadt, pp 33—39
69. Ratajczyk-Pakalska E (1986) Thebesian veins in the human hearts with atherosclerotic lesions in the coronary arteries, In: Mohl et al. (eds) Clinics of CSI. Steinkopff, Darmstadt, pp 141—145
70. Ratajczyk-Pakalska E, Bloch P, Kulig A: Venous microcirulation of the left ventricular myocardium in the normal and atherosclerotic hearts. (Manuscript in preparation). Department of Pathology MMA, Clinic of Cardiology Medical Academy in Lodz
71. Ravin MB, Epstein RM, Malm JR (1965) Contribution of thebesian veins to the physiologic shunt in anesthetized man. J Appl Physiol 20:1148—1152
72. Roberts J (1959) Encyclopedia of the cardiovascular system. Cardiology. McGraw Hill, New York
73. Roberts JT (1943) Experimental studies on the nourishment of the left ventricle by the lumenal (Thebesial) vessels. Fed Proc 2:90—91
74. Roberts JT, Spencer RF, Browne RS (1943) Drainage of myocardium by cardial lumen (Thebesian) vessels of the left ventricle. Fed Proc 2:90—91
75. Robertson HF (1941) The physiology, pathology and clinical significance of experimental coronary sinus obstruction. Surgery 9:11—15
76. Samoilova SV (1970) Anatomy of the cardiac blood vessels. Medicina, Leningrad
77. Santos Ferreira A, Esperanca Pina JA (1971) Systeme veineux du coeur: Veins de Thebesius. Arch Anat Anthrop 35:95—125
78. Schaper W, Schaper J (1977) The coronary circulation. Am J Cardiol 40:1008—1012
79. Sherf L, Ben-Shaul Y, Lieberman Y, Neufold HN (1977) The human coronary microcirulation. An electron microscopic study. 39:599—607

80. Smith GT (1962) The anatomy of the coronary circulation. Am J Cardiol 9:327–342
81. Tandler J (1913) Anatomie des Herzens, Bardelebens Hdb der Anatomie des Menschen, 3. Bd. Abt. 1
82. Tarasow LA (1961) Materialy k dopolnitielnoj sasudistnoj sistemie serdca (veny Tebezija-Vessena). Grudd Khir 2:44–50
83. Thebesius AC (1708) Dissertatio medica de circulo sanguinis in corde. Lagduni Batavorum, Amsterdam Elsevier
84. Truex RC, Schwartz MJ (1951) Venous system of the myocardium with special reference to the conduction system. Circulation 4:881–889
85. Truex RC, Angulo AW (1952) Comparative study of the arterial and venous systems of the ventricular myocardium with special reference to the coronary sinus. Anat Rec 113:467–492
86. Tschabitscher M (1984) Anatomy of coronary veins. In: Mohl W et al. (eds) The coronary sinus, Steinkopff, Darmstadt, 8–25
87. Unger K (1916) Beitrag zur Kenntnis der venae cordis minimae (Thebesii) des menschlichen Herzens. Z Anat Entwicklungesch 108:365–375
88. Vieussens R (1706) Nouvelle decouvertes sur le comur. Paris
89. Vieussens R (1715) Traite nouveau de la structure er causes du nouvement natural du coeur. Toulouse
90. Wearn J (1928) The role of the Thebesian vessels in the circulation of the heart. J Exp Med 47:293–304
91. Wearn JT, Mettier SR, Klumpp TE, Zschiesche LJ (1933) The nature of the vascular communications between the coronary arteries and the chambers of the heart. Am Heart J 9:143
92. Young DAB, Fell BF (1962) Vascular connections between the coronary circulation and the ventricles of the rat heart. Anat Rec 144:149–154

**List of abbreviations used in the figures**

| | | | | |
|---|---|---|---|---|
| CS | – Coronary sinus | a | – Arterio-venous anastomosis |
| AIV | – Anterior interventricular vein | $a_v$ | – Veno-venous anastomosis |
| LCV | – Left coronary vein | vP | – Intramuscular venous plexus |
| PIV | – Posterior interventricular vein | RV | – Right ventricular cavity |
| PLV | – Posterior left ventricular vein | LV | – Left ventricular cavity |
| RMV | – Right marginal vein | c | – Capillary vessels |
| ARV | – Anterior right ventricular veins | m | – injection mass (Vinidur, or Latex, |
| LAO | – Left atrial oblique vein | | or Microfil) |
| RAV | – Right atrial veins | Sph | – Sphincter in the wall of the ThV |
| LAV | – Left atrial vein | w | – wall of the ThV |
| SCV | – Small cardiac vein | IZ | – Subendocardial zone |
| ThV | – Thebesian vein | MZ | – Subepicardial zone |
| LCA | – Left coronary artery | $V_0$ | – Outlet valve |
| RCA | – Right coronary artery | $V_p$ | – Parietal valve |
| OV | – Left ventricular oblique vein | En | – Endocardium |
| ASV | – Anterior septal vein | Ep | – Epicardium |
| PSV | – Posterior septal vein | | |

# Microanatomy of the human coronary sinus and its major tributaries

M. v. Lüdinghausen, C. Schott

## Introduction

There are four separate intercommunicating systems of veins in the human heart: a) the system of the tributaries of the coronary sinus (c.s.); b) the system of the anterior cardiac veins; c) the system of atrial veins, and d) the system of Thebesian veins. The largest system is that of the tributaries of the c.s., which collects all the important cardiac veins – great, left marginal and/or posterior ventricular vein, middle and (in one-third of cases) small cardiac vein – and issues into the c.s. which empties into the right atrium (Figs. 1, 2). These veins drain, almost exclusively, the left ventricular myocardium [7]. This pattern of tributaries of the coronary sinus is called a pentade of large cardiac veins [13, 16], which is described and illustrated in textbooks and most anatomic atlases. Thus the c.s. is inserted between the systematic cardiac veins (except anterior cardiac veins) and the right atrium. Due to the concepts and experiments of Beck [2, 6] the c.s. seems anatomically to be the ideal location for the placement of a catheter for purposes of retrograde perfusion or other cardiological procedures [4, 6]. Retroperfusion techniques of the c.s. or regional cardiac veins are based on the generally assumed pattern in which 95% of the ventricular myocardium is drained by the c.s. tributaries. Such a relatively simple general organization does not however fully conform with reexamination in a large number of heart specimens. In fact, individual coronary venous systems feature a variety of anatomical configurations, demanding selective considerations and clinical adjustment of retrograde coronary venous interventions. In practical cardiology in 10% – 20% of cases there is a failure of catheterization of the c.s. and cardiac veins [1], such as local subendocardial and mural hemorrhages, disturbances of the conduction system or even perforation of the atrial wall. On the strength of these experiences it is assumed that quite often the procedure of cannulation of the c.s. and cardiac veins is limited by anatomical variations, irregularities, anomalies, and malformations. A few of these, constituting a wide spectrum of variations will be demonstrated.

## Material and Methods

This study encompasses 150 human hearts (66 male, 84 female) obtained from the student gross anatomy dissection course, and 24 corrosion casts of hearts (11 male, 13 female) from a pathology dissection course.

The specimens were from individuals between 64 and 88 years of age.

Supplemental information was gathered from carefully performed microanatomical preparations using a suitable equipped stereomicroscope (C. Zeiss, Oberkochem, FRG).

Photographs were taken by a Nikon-camera AS and Medical-Nikkor C Auto 1:5,6, f = 200 mm.

## Results and Discussion

Venous drainage of the myocardium of the left ventricle is primarily performed through the c.s. and its major tributaries.

We recently classified and described in detail the various types of venous myocardial drainage into the c.s. [9, 10]. Summarizing these findings, and additional results of one of the authors (M.v.L.), the following incidences should be noted (Fig. 2): in 11% of the hearts the c.s. is found to collect all the cardiac veins, equally from the walls of both the left and right ventricles; in 25% (the common textbook situation) the anterior cardiac veins do not issue into the c.s., but rather they drain directly into the right atrium; in 61% the anterior cardiac veins, as well as the right marginal vein, do not drain into the c.s. but open into the right atrium; in 2% of the hearts, neither the anterior cardiac veins, the right marginal vein, or even the middle cardiac vein reached the c.s., and drained instead directly into the right atrium; finally in one case the great cardiac vein was seen to leave the anterior interventricular sulcus, pass the arterial conus and join with the anterior cardiac veins in the right atrium (Fig. 17). In this case the c.s. collects only a few small left marginal veins.

The c.s. itself continues the course of the great cardiac vein in the left coronary or atrioventricular sulcus. The identification at the juncture of these two structures beneath the epicardium happens at the entrance of the oblique vein of Marshall, which descends along the posterior wall of the atrium; and which in many cases is translucent and visible through the epicardium. Opposite the entrance of this vein is the valve of Vieussens, which also marks the beginning of the c.s. The myocardial cover, because of its irregular extension, is impractical for determination of the division of the great cardiac vein and the c.s. On the other hand it is impossible to fix the exact location of the atrial ostium of the c.s. on the diaphragmatic epicardiumcovered surface from outside. There is no conspicious landmark at the posterior coronary sulcus to determine its entrance into the right atrium, which is guarded by a more or less complete and sufficient fold, the Thebesian valve.

The tables in the following subchapters contain our findings, compared with results of other authors. (The numbers in parentheses refer to the author's listing among the references.)

I. The dimensions and parameters of the c.s. determined or measured are: length, shape, diameter, cross-sectional area, circumference, volume, curvature, myocardial cover, ostial (Thebesian) valve.

94

*Length and shape of the c.s. (Figs. 3—5)*

The length was measured between the entrance of the oblique vein of Marshall and the base of the Thebesian valve. Both length and shape showed numerous variations.

**Table 1.** Length and shape of c.s.

| | |
|---|---|
| Length — averages: 15—70 (37,0) mm, one extreme case 82 mm | |
| —20—65 mm, in 75% 30—50 mm | [16] |
| —20—40 mm | [11] |
| —17—65 (37,8) mm | [15] |

Length and shape:
short c.s. (<20 mm) with compact, bellied form 17%
medium-long c.s. (20—40 mm) with narrow, cylindrical form 74%
long c.s. (>40 mm) with tubular, stretched form 18%
similar results [12]

*Diameter, cross-sectional area, circumference, and volume of c.s. (Figs. 3—5)*

These measurements were very carefully performed by a group of morphologists [5, 14, 15, 16]. Some authors [14, 15] distinguished hearts with normal weight (i.e., 351 g) and with increased weight (i.e., 458 g). The results of the latter are marked with an asterisk. In patients with chronic congestive heart failure the c.s. is noted to be enlarged and the results of measurements are significantly increased.

**Table 2.** Diameter, cross-sectional area, circumference, volume of c.s.

| | | | |
|---|---|---|---|
| Origin of c.s. Cross-sect. area | | 13—38 (20.6) mm$^2$ | [15] |
| Origin of c.s. Cross-sect. area | — * | 13—79 (28) mm$^2$ | |
| Midcor. sinus Diameter | | 6—16 mm | [16] |
| Midcor. sinus Cross-sect. area | | 26—58 (38.8) mm$^2$ | [15] |
| Midcor. sinus Cross-sect. area | — * | 29—109 (52.4) mm$^2$ | |
| Atrial ostium Diameter | | 5—14 (8.6) mm | [15] |
| Atrial ostium Diameter | — * | 7—16 (9.7) mm | |
| Atrial ostium Diameter | — | 4.5—7.3 (6.2) mm | [14] |
| Atrial ostium Diameter | — | 7—19 (9.9) mm | [5] |
| Atrial ostium Cross-sect. area | | 20—154 (61.5) mm$^2$ | [15] |
| Atrial ostium Cross-sect. area | — * | 38—201 (77.1) mm$^2$ | |
| Atrial ostium Cross-sect. area | — | 16—42 (31) mm$^2$ | [14] |
| Atrial ostium Circumference | | 14—23 (20) mm | [14] |
| Atrial ostium Circumference | — * | 17—53 (35) mm | |
| Volume | | 0.67—2.09 (1.26) mm$^3$ | [15] |
| | — * | 0.99—3.52 (1.76) m$^3$ | |

*Curvature of c.s. (Figs. 4, 5)*

The c.s. and the great cardiac vein form a gentle curve in the left coronary sulcus, embracing the posterior margin of the left atrium nearly parallel to the attachment

95

of the mural (posterior) leaflet of the mitral valve. The narrowness of that curve partly depends on the degree of the ostial angle (angulus ostii sinus coronarii, Fig. 24) and upon the absolute length of the c.s.

**Table 3.** Curvature of c.s.

| | |
|---|---|
| One-quarter of circle | 33% (Fig. 4) |
| One-eighth of circle | 29% |
| A gentle curvature | 29% (Fig. 5) |
| Nearly straight course | 9% |

## Position of c.s. (Fig. 6)

The well known position of the c.s. is the left half of the posterior coronary sulcus. In most specimens there was a transferred position of c.s. to the posterior wall of the left atrium. An extreme elevation up to 15 mm above the coronary sulcus and an arched course of the c.s. were observed in nearly one-third of the cases. The more remarkable the elevation of the c.s. is found to be, the more a sigmoid course of the terminal portion of the great cardiac vein is developed (Fig. 15).

**Table 4.** Position of the c.s. and its incidence

| | |
|---|---|
| In the left posterior coronary sulcus | 12% |
| Elevation (1−3 mm) | 16% |
| Moderate elevation (4−7 mm) | 50% |
| Extreme elevation (8−15 mm) | 22% (Fig. 6) |

## Myocardial cover of the c.s. and the terminal portions of its tributaries (Fig. 7)

The c.s. in its total length is covered by a surrounding myocardial coat. The myocardial fibers on the surface − part of the left atrial myocardium − mostly follow a longitudinal course. Therefore the distal margin of the myocardial coat is not sharp and exact but irregular in a zone of 1−2 mm width. In a few cases this myocardial coat not only covers the c.s. but the terminal portions of its tributaries as well. Interestingly the myocardial coat of the terminal portion of the middle vein fastens the vein to the dorsal right atrial wall. In principle two or three myocardial belts or fiberbundles of 2- or 3-mm-width may possibly lie on the surface of all the terminal portions of the tributaries of the c.s.

**Table 5.** C.s. myocardial covering and its widening on the terminal portions of the cardiac veins

| | |
|---|---|
| Myocardial covering of the c.s. | 100% |
| Great cardiac vein (<3 mm) | 11% |
| Great cardiac vein (>3 mm) | 2% (Figs. 8, 9) |
| Middle cardiac vein (bulbous) | 2% (Fig. 10) |
| Isolated myocardial belts | |
| Great cardiac vein | 5% |
| Left posterior ventricular vein | 2% |

*Right ostial (Thebesian) valve (Fig. 11)*

The Thebesian valve secures the right atrial ostium more or less completely [15, 18]. In 19% the ostium is valveless. The incidence of others is shown in brackets. In very rare cases a congenital complete ostial occlusion (atresia) of the c.s. is reported [10].

**Table 6.** Shape and incidence of the Thebesian valve

| | | |
|---|---|---|
| Complete circular valve | 31% | (30.7% [5], 20% [15]) |
| Incomplete cribriform valve (Fig. 12) | 7% | (5.3% [5]) |
| Incomplete semilunar valve (Fig. 13) | 7% | (6% [5]) |
| Incomplete cresecentic valve | 34% | (38% [5]) |
| Incomplete threadlike valve (Fig. 14) | 2% | (5.3% [5]) |
| Valveless orifice | 19% | (14.7% [5], 16% [15]) |

## II. Major tributaries of the coronary sinus: the cardiac veins

The primary tributaries of the c.s. are as follows: great cardiac, left marginal, left posterior, and middle cardiac vein.

*Topography*

The great cardiac vein runs in very close relationship with the left anterior descending branch of the left coronary artery in the anterior interventricular sulcus, and the middle cardiac vein in the posterior interventricular sulcus is accompanied by the right posterior descending branch of the right coronary artery (in 10% of cases a branch of the circumflex branch of the left coronary artery). On the lateral and posterior wall of the left ventricle there is a great variability of the subepicardial vascular distribution pattern.

For instance, in most cases the veins of this part of the left ventricle, the left marginal and the posterior ventricular vein, in their courses and fields of drainage do not correspond to the supplying arteries. The special relations and interactions of arteries and veins in this area as well in the interventricular septum will be discussed in a latter report.

As a rule, the openings of the cardiac veins entering the coronary sinus are secured by single or double valves. In most of our specimens they seem to be insufficient. In about 50% of cases there are ostial valves, but in the other half the valves are situated in a distance of 1–15 mm distal to the openings. This means that in these cases the venous terminal portion is included in the space of the c.s.

### Great cardiac vein (V. cardiaca magna)

The great cardiac vein consists of two main parts: the interventricular and the basal part. The first arises from the superficial venous network of the cardiac apex, and ascends into the anterior interventricular groove until it reaches the atrio-ventricular sulcus, following which, it turns to the left. The second continues the first part and passes the atrio-ventricular (coronary) sulcus as it empties into the c.s. Along its course the great cardiac vein directly receives ventricular veins, which drain the anterior walls of both the right and left ventricle. In addition, 4–6 septal veins, which drain the anterior one or two thirds of the interventricular septum, also join the interventricular part of the great vein.

*S-shaped (sigmoid) course (Fig. 15)*

The terminal portion of the great vein runs in one-third of the cases, in an s-shaped course out of the coronary sulcus to the posterolateral wall of the left atrium, where it empties into the c.s. The valve of Vieussens marks the juncture of great vein and c.s.

*Intramural course*

An intramural or intramyocardial course below crossing myocardial bridges in the anterior interventricular part of the great vein was found in a length of 2–10 mm and in 2% of the cases.

*Aberrant course*

There was one case of an aberrant course of the great vein, originating in the distal anterior interventricular sulcus, leaving that groove to the right, passing the atrial conus and root of the pulmonary trunk, and finally emptying into the right atrium.

**Table 7.** Great cardiac vein and its varieties

| | |
|---|---|
| Entering the c.s. | 99%' |
| Terminal s-shaped course | 32% (Fig. 15) |
| Myocardial belt(s) of terminal portion | 6% (Fig. 7–9) |
| Intramural course | 2% (Fig. 16) |
| Aberrant course | 1% (Fig. 17) |
| Cross-sectional area of terminal part | 15 (7–37) mm$^2$ [7] |
| Crossing over the stem or all the branches of the left main coronary artery | 49% |
| Running underneath the diagonal branch | 9% |
| Running underneath circumflex branch | 21% |
| Running underneath both branches | 20% |

*Clinical comment:* In cases of aberrant, intramural or terminal s-shaped course and extensive compression of the great cardiac vein by crossing of a sclerosed and calcified stem or branch of the left coronary artery, in particular, catheterization of that vein may not be possible.

*Valve of Vieussens (Fig. 16)*

The valve of Vieussens secures the origin of the c.s. or covers the end of the great cardiac vein on an average of 87%. In 13% of cases it does not exist.

**Table 8.** Shape and Incidence of the valve of Vieussens

| | |
|---|---|
| Unicuspid valve | 62% (Fig. 17) |
| Bicuspid valve | 25% (Figs. 18, 19) |
| Valveless orifice | 13% (12% [15], 22% [12]) |

## The oblique vein of the left atrium (Marshall) (Fig. 3, 4, 7−9)
## (V. obliqua atrii sinistri, oblique vein of Marshall)

This is a small vessel entering the c.s. directly distal to the valve of Vieussens. The vein of Marshall has been used as the landmark designating the beginning of the c.s. It descends from the lateral and posterior wall of the left atrium. Embryologically the oblique vein represents the residue of the left superior cardinal vein and left superior vena cava. Its entrance into the c.s. in most cases is guarded by a small endothelial fold or in a few cases by a competent or incompetent pair of ostial semilunar valves.

On the contrary, other authors have never observed any ostial valves [16]. In 84% of cases the Marshall vein was found to be a small vessel filled with blood, whereas in 12% of cases we could not find a true vein, but noted a chord of connective tissue. The angle between the long axis of the Marshall vein and the c.s. varied between 25° and 50°.

In rare cases the left superior vena cava persists. These patients are symptomless when there is an open communication between left subclavian vein and the c.s. [10].

**Table 9.** Oblique vein of Marshall

| | Incidence |
|---|---|
| Developed with a diameter of approximately 0.5 mm and a length of 2−3 cm | 84% |
| Completely regressed to a fibrous cord | 12% |
| Not exactly verifiable | 3−4% |
| Persistent left superior vena cava | <0,5% |

## Middle cardiac vein (V. cardiaca media) (Figs. 20−23)

The middle cardiac vein originates from the superficial venous network at the cardiac apex and runs upward in the posterior interventricular sulcus to the crux cordis (i.e., confluence of posterior interventricular and coronary sulcus). The middle cardiac vein drains the posterior walls of the both the right and left ventricle, as well as the apical area and with its posterior septal veins the posterior one- or two-thirds of the interventricular septum.

The terminal portion of the middle vein of 1−3 cm length turns to the left in a more or less oblique course and empties into the c.s. The ostium is secured by different ostial valves. In a few cases of a terminal portion with myocardial coverage, and only ectopic (but missing ostial) valvulae, this part of the middle vein is included in the space of the c.s. ("accessory coronary sinus").

**Table 10.** Middle cardiac vein: its frequency, terminal portion, mode of opening and ostial valvulae

| | | |
|---|---|---|
| Frequency | – single | 75% |
| | – double (confluent terminal portion) | 24% |
| | – treble | 1% |
| Terminal portion | – straight course | 18% |
| | – moderate angulation | 46% (Fig. 20) |
| | – severe angulation | 35% (Fig. 10) |
| | – bulbous dilatation | 30% (Figs. 20, 22) |
| | – myocardial cover | 2% (Fig. 10) |
| Opening | – into c.s. | 90% |
| | – into right atrium | 2% |
| | – in between the leaflets of the Thebesian valve | 1% |
| | – not determined in cases without Thebesian valve | 8% |
| Mean cross-sectional area at ostial opening | | 6–25 (13) mm$^2$ [7] |
| Ostial valve | – single velum | 50% (Fig. 27) |
| | – couple of vela | 11% |
| | – ectopic velum | 19% (Figs. 28, 29) |
| | – missing valve | 39% |

*Clinical comment:* Because of its usual termination in the c.s. very close to the atrial ostium and its ostial valve, the important *middle cardiac vein* and small cardiac vein (if developed) generally will not be retroperfused by any of the clinical c.s. intervention techniques. This comment also applies to an atypical middle cardiac vein (2% of our cases) which empties directly into the right atrium, or (1% of the cases) features a special drainage opening in between the two layers of a Thebesian valve at the c.s. ostium.

## Small cardiac vein (V. cardiaca parva)

The small cardiac vein is developed in only one-third of our specimens as a vessel of approximately 0.5 mm diameter and more. In these cases it drains the posterior and lateral wall of the right ventricle, runs parallel (above or below) to the right coronary artery in the right coronary sulcus and empties into the middle cardiac vein or into the c.s. In a few cases the small cardiac vein might be the continuation of the right marginal vein [13]. In rare cases it even collects some other anterior cardiac veins [9]. In one-third of cases the vein is absent; sometimes it is reduced to a fibrous cord [8].

**Table 11.** Small cardiac vein: Frequency, mode of opening and size

| | | |
|---|---|---|
| Opening | – into middle cardiac vein | 29% |
| | – into c.s. | 4% |
| | – mean cross-sectional area | 1.0 mm$^2$ [7] |
| Terminal anastomozing loop | | 4% |
| Less than 0.5 mm diameter and/or fibrous cord | | 31% |
| Absent | | 36% |

## Left posterior ventricular vein(s) (V. posterior ventriculi sinistri) (Fig. 24)

The left posterior ventricular vein (V. posterior ventriculi sinistri) might consist of one large vessel or of two or three smaller ones. It (they) drain(s) the lateral and posterior wall of the left ventricle. As a rule, the left posterior ventricular vein(s) open(s) into the proximal second and third centimeter segment of the c.s. or it lead(s) − in a few cases − into the terminal portion of the great cardiac vein. Also in a few cases there is no left posterior ventricular vein existing but a strong left marginal vein draining the posterior wall of the left ventricle.

**Table 12.** Frequency and mode of opening of the left posterior ventricular vein

| | |
|---|---|
| Frequency | |
| Single developed strong vessel (3 mm ∅) | 64% |
| Double developed medium sized vessel (3 mm ∅) | 23% |
| Multiple (3−5) small vessels (1 mm ∅) | 9% |
| No posterior ventricular but strong left marginal vein | 4% |
| Mode of opening | |
| Into the c.s. with ostial valvulae | 87% (Fig. 24) |
| Into the c.s. with ostial valvulae | 77.7% [12] |
| 5−15 mm distal ostial valve | 57% |
| 15−25 mm distal ostial valve | 17% |
| 25−35 mm distal ostial valve | 8% |
| 35−45 mm distal ostial valve | 5% |
| Into the great cardiac vein without ostial valvulae | 13% |
| Terminal bulb | 5% |
| Myocardial belts of the terminal portion | 2% |

The *left posterior ventricular vein* features, in most cases, a sufficient valve at its opening (5−25 mm distal to the Thebesian valve) which renders retrograde perfusion of vessels of the posterior and lateral walls more difficult or impossible.

## Conclusion

The coronary sinus (c.s.) and its major tributaries were explored by microanatomical procedures and with corrosion casts in 174 human heart specimens. The observations and measurements encompassed length, shape, diameter, cross-sectional area, volume, position, curvature, myocardial covering, and ostial valves of the c.s. Relative to the c.s. tributaries, the following items were identified: terminal course and mode of opening into the c.s., incidence, location and form of the ostia and, in few cases, ectopic valves. To a large extent, much of this detailed information has not be previously reported in the literature.

Detailed knowledge of the anatomy and variability of the c.s. and its major tributaries may reduce irregularities and failures of catheter-placement and thus enhance the effectiveness of coronary venous interventions.

# References

1. Baim DS (1988) Percutaneous placement of coronary sinus catheters. 3rd international symposium (June 1988, Cambridge, Massachusetts, USA) on myocardial protection via the coronary sinus
2. Beck CS (1948) Revascularization of the heart. Ann Surg 128:854–864
3. Beck CS, Stanton E, Batiuchok W, Leiter E (1948) Revascularization of heart by a graft of systemic artery into the coronary sinus. JAMA 137:436–442
4. Grossmann W (1985) Cardiac catheterization and angiography. 3rd ed Lea & Febiger, Philadelphia
5. Hellerstein HK, Orbison JL (1951) Anatomic variations of the orifice of the human coronary sinus. Circulation 3:514–523
6. Hochberg MS, Austen WG (1980) Selective retrograde coronary venous perfusion. Ann Thorac Surg 29:578–588
7. Hood WB (1968) Regional venous drainage of the human heart. Br Heart J 30:105–109
8. James TH, Sherf L, Schlant RC, Silverman ME (1982) Anatomy of the heart. In: Hurst JW (ed) The heart, 5th ed, McGraw-Hill, New York
9. v. Lüdinghausen M (1987) Clinical anatomy of cardiac veins, Vv. cardiacae. Surg Radiol Anat 9:159–168
10. v. Lüdinghausen M, Lechleuthner A (1988) Atresia of the right atrial ostium of the coronary sinus. Acta Anat 131:81–83
11. Malhotra VK, Tewari SP, Tewari PS, Agarwa SK (1980) Coronary sinus and its tributaries. Anat Anz (Jena) 148:331–332
12. Maros TN, Racz L, Plugor S, Maros TG (1983) Contribution to the morphology of the human coronary sinus. Anat Anz (Jena) 154:133–144
13. Mechanik N (1934) Das Venensystem der Herzwände. Z Anat Entw Gesch 103:813–843
14. Potkin BN, Roberts WC (1988) Size of coronary sinus at necropsy in subjects without cardiac disease and in patients with various cardiac conditions. Am J Cardiol 60:1418–1421
15. Silver MA, Rowley NE (1988) The functional anatomy of the human coronary sinus. Am Heart J 115:1080–1084
16. Tandler J (1926) Lehrbuch der systematischen Anatomie, 3 Bd. Das Gefäßsystem. Vogel, Leipzig, pp 26–95
17. Tschabitscher M (1984) Anatomy of coronary sinus. In: Mohl W, Wolner E, Glogar D (eds) The coronary sinus. Springer Berlin Heidelberg New York, pp 8–25
18. Tschabitscher M (1986) The so-called "silent zone" of the coronary sinus. In: Mohl W, Faxon D, Wolner E (eds) CSI – A new approach to interventional cardiology. Springer Berlin Heidelberg New York, pp 11–14
19. Yater WM (1929) Variations and anomalies of the venous valves of the right atrium of the human heart. Arch Pathol 7:418–441

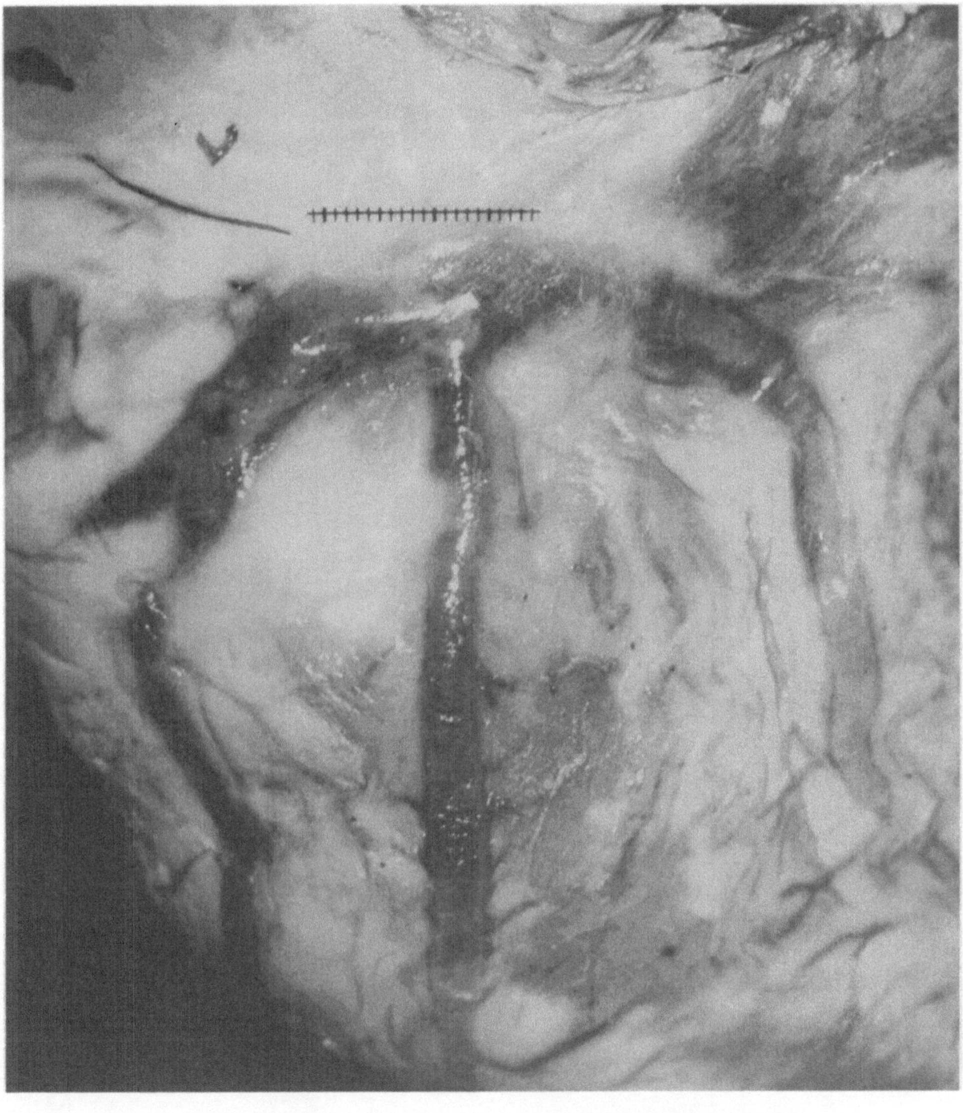

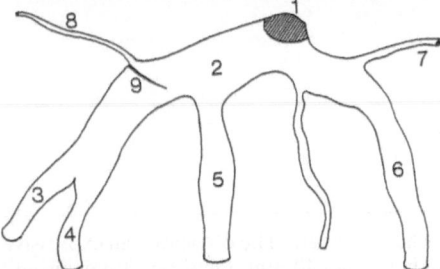

Numbers correspond to:
1: atrial ostium of the coronary sinus
   (in projection)
2: coronary sinus
3: great cardiac vein
4: left marginal vein
5: posterior left ventricular vein
6: middle cardiac vein
7: small cardiac vein
8: left atrial oblique vein of Marshall
9: valve of Vieussens

**Fig. 1.** Diaphragmatic surface of the human heart. The coronary sinus (c.s.) and its tributaries great and middle cardiac veins, left posterior ventricular vein and the oblique vein of Marshall, are visible through the translucent epicardium.

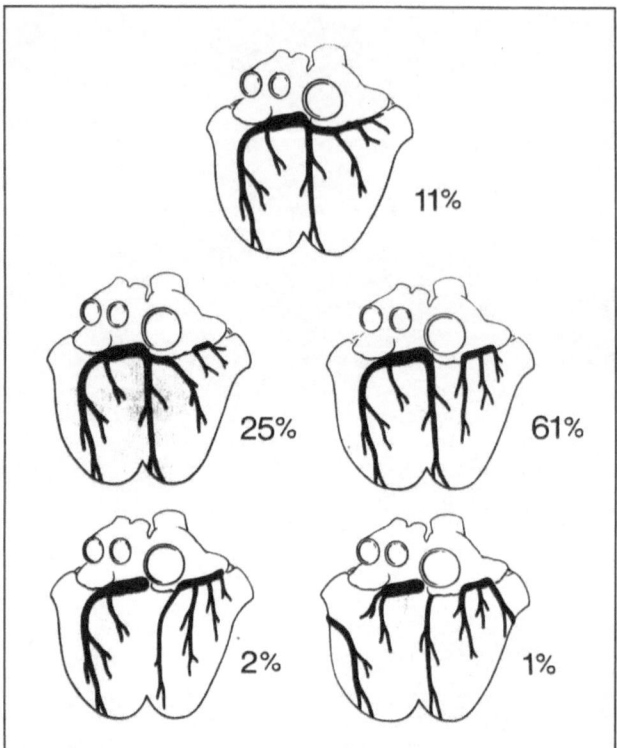

**Fig. 2.** Scheme of the distribution pattern of the cardiac veins and their relationship to the c.s.

**Fig. 4.** Cranial aspect of a corrosion cast of the c.s. of a human heart. The c.s. shows an expressive curvature, which conforms to a quarter-circle. Its length is about 40 mm, the shape is narrow and cylindrical, the diameter at origin is about 5.5 mm, the midcoronary diameter is about 10 mm and the ostial diameter about 11 mm. In this case there is no ostial angle of the c.s. (Numbers correspond to Fig. 1.)

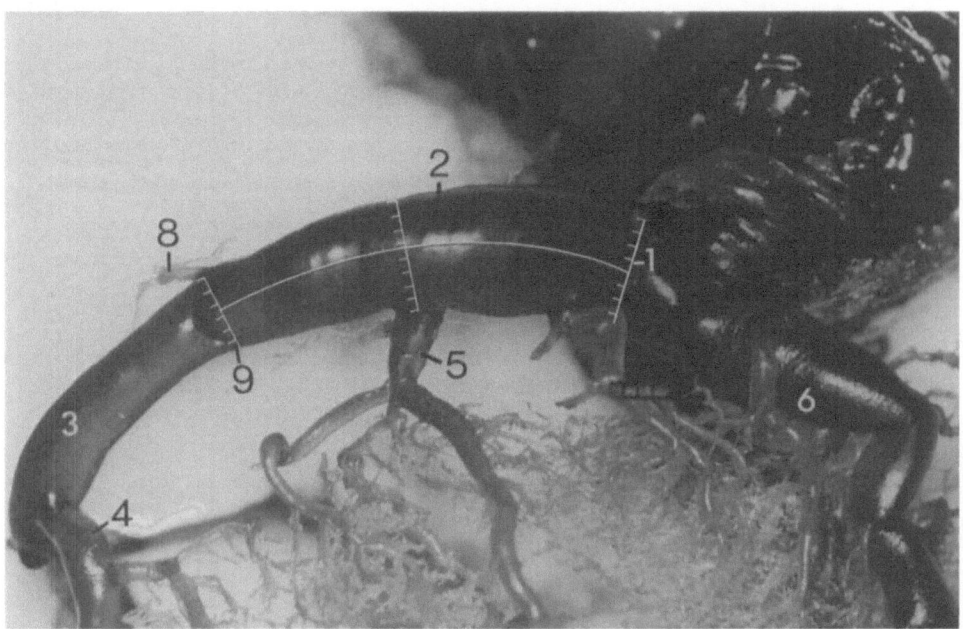

**Fig. 3.** Posterior surface of a corrosion cast of the c.s. of a human heart. The c.s. has a length of about 40 mm and a cylindrical shape. Its diameter at origin is about 6 mm, the midcoronary diameter is about 8 mm and the ostial diameter about 8 mm as well. (Numbers correspond to Fig. 1.)

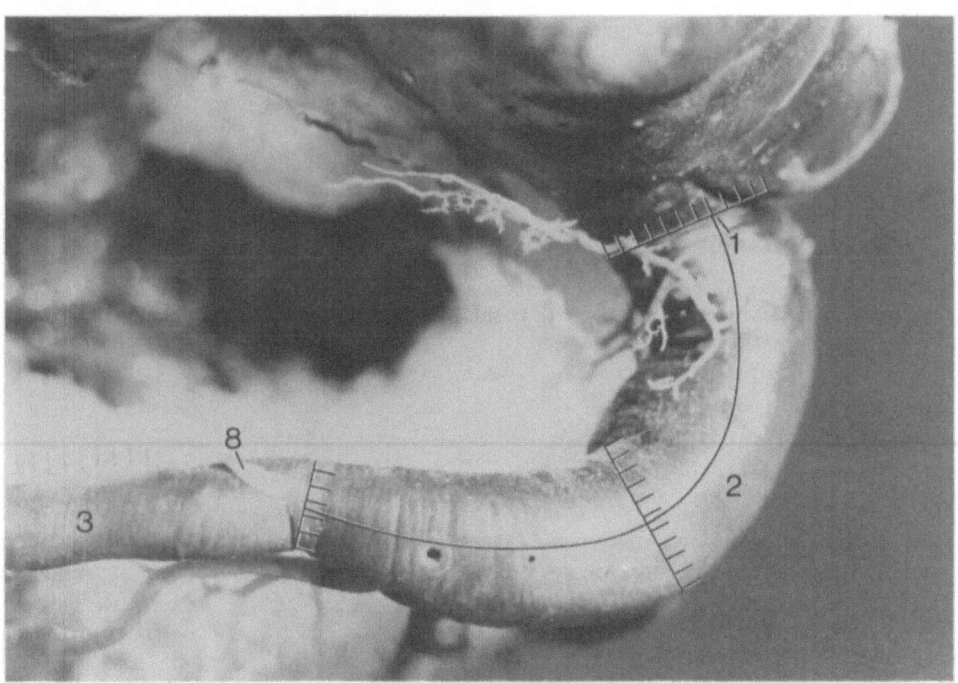

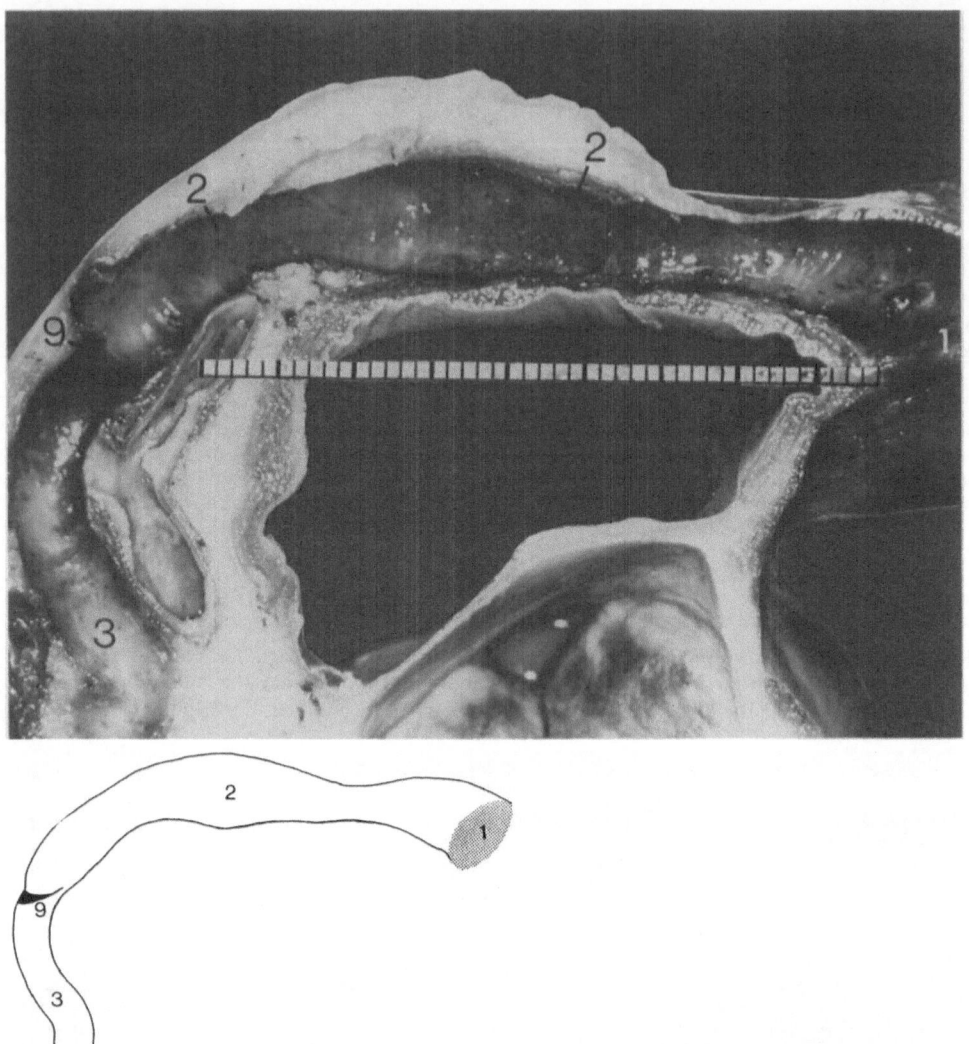

**Fig. 5.** Inferior aspect of an opened c.s., which has a length of about 55 mm and shows a gentle curvature in its course. At the atrial ostium (right side) there is a remarkable ostial angle of the c.s. (arrow). (Numbers correspond to Fig. 1.)

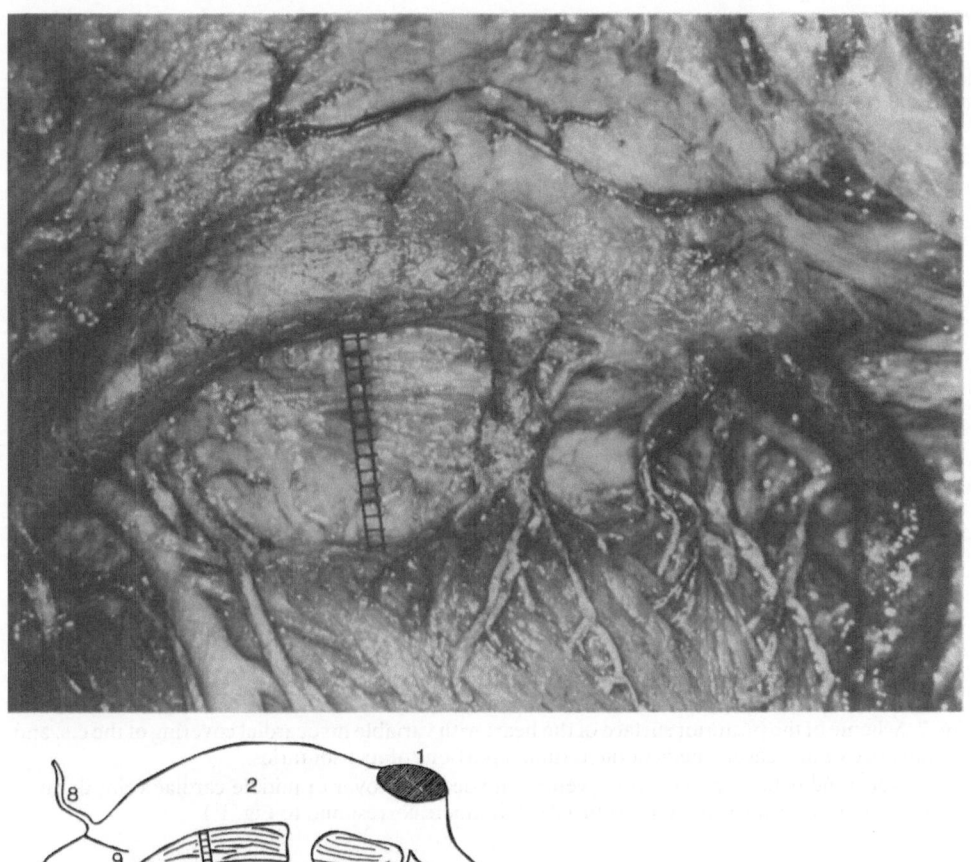

**Fig. 6.** Posterior surface of a human heart specimen with severe elevation of the c.s. to a high position of about 14 mm above the posterior coronary sulcus. (Numbers correspond to Fig. 1.)

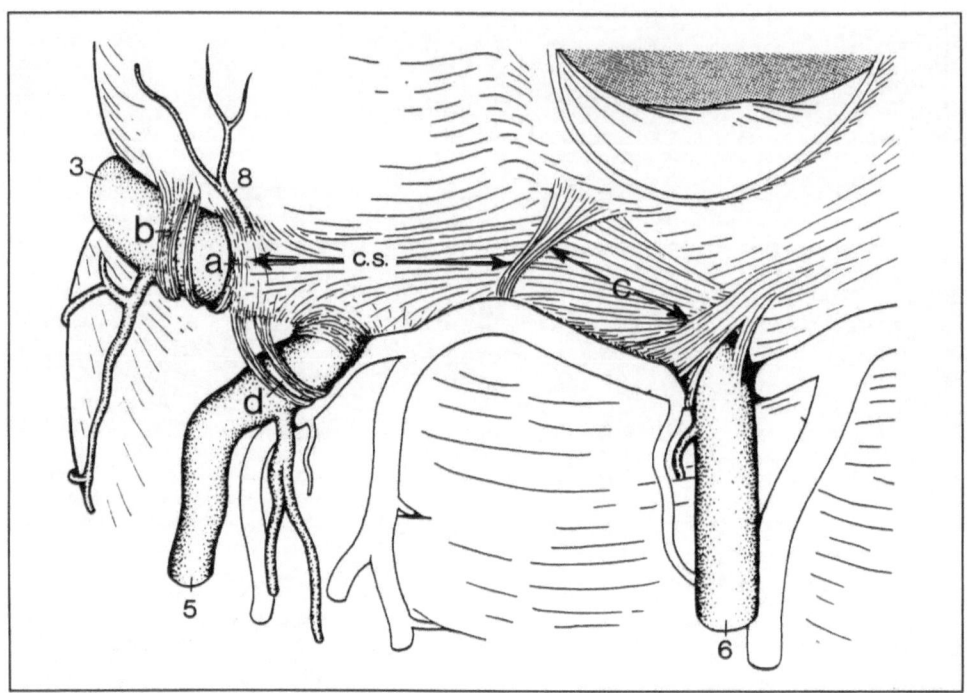

**Fig. 7.** Scheme of the posterior surface of the heart with variable myocardial covering of the c.s. and variable size of myocardial belts of the terminal portions of its tributaries.

a, b: myocardial belts of great cardiac vein; c: myocardial cover of middle cardiac vein; d: myocardial belt of posterior vein of left ventricle. (Numbers correspond to Fig. 1.)

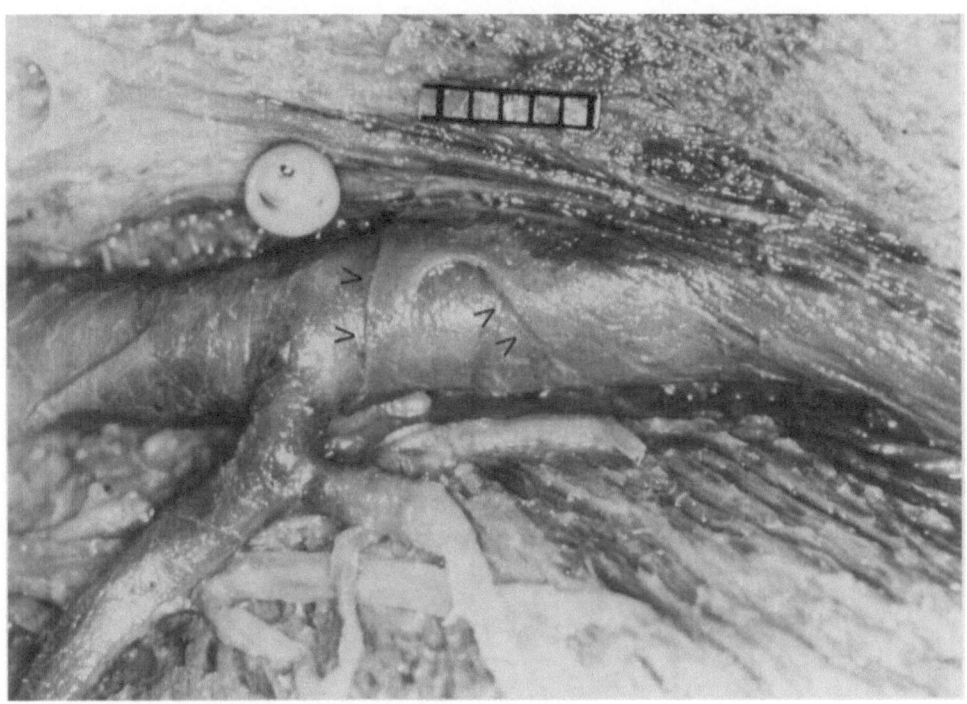

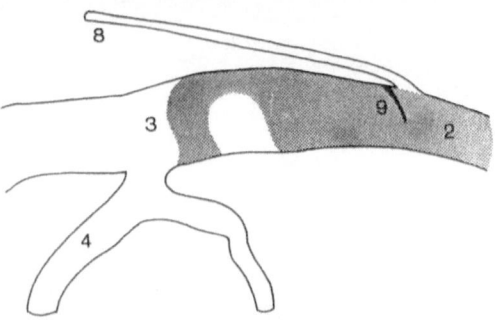

**Fig. 8.** Myocardial belts of the terminal great cardiac vein, distal of 2 mm, proximal of 11 mm breadth (arrows), with partial or total connection with the myocardial coverage of the c.s. (Numbers correspond to Fig. 1.)

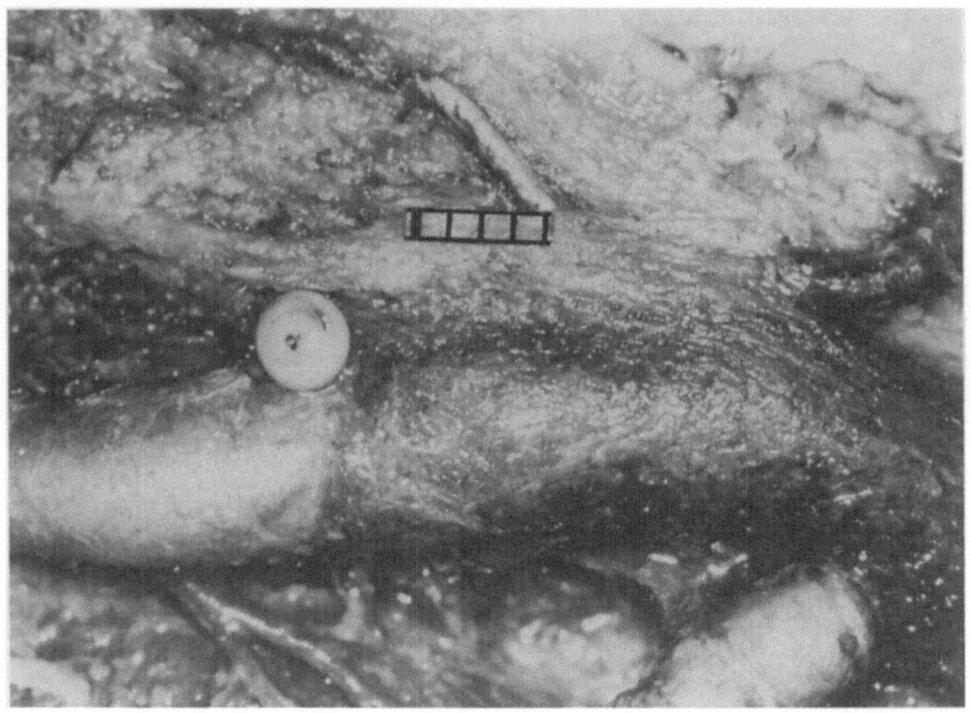

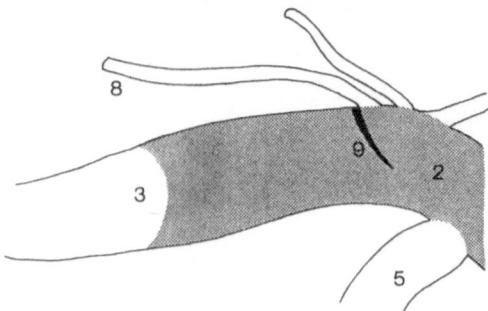

**Fig. 9.** Broad (12 mm) myocardial belt of the terminal great cardiac vein. (Numbers correspond to Fig. 1.)

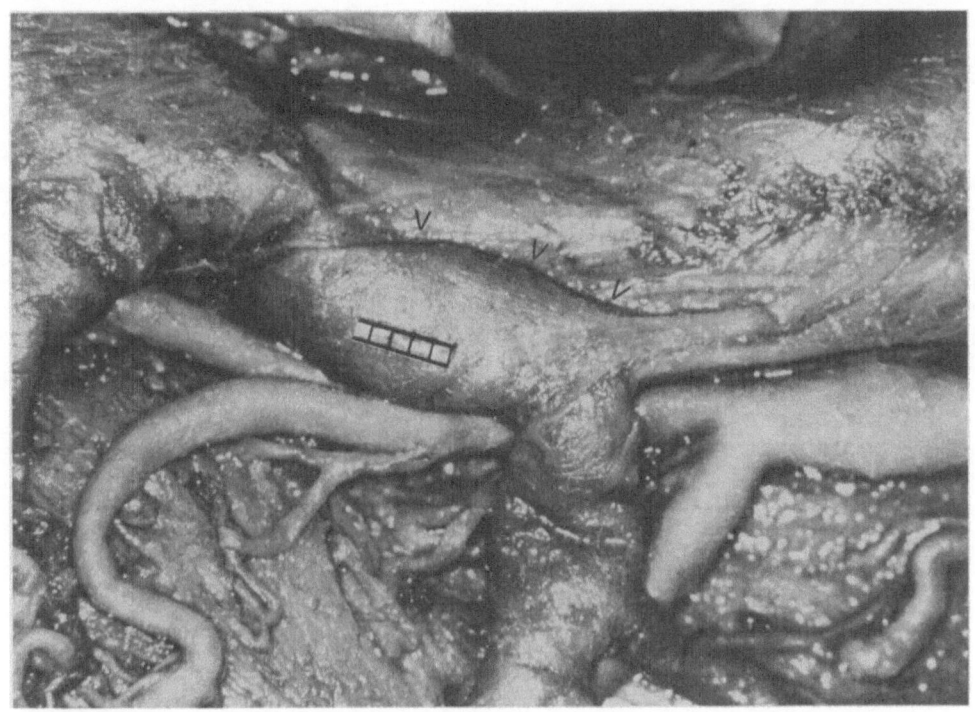

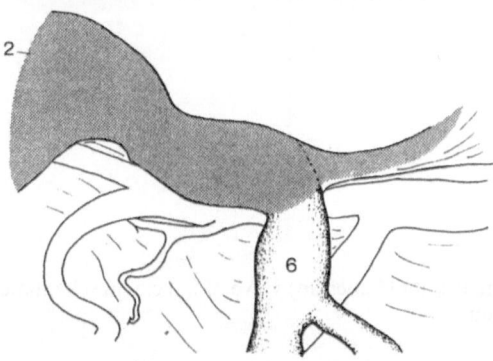

**Fig. 10.** Myocardial coverage of the terminal segment of the middle cardiac vein; fastening of the vein to the posterior right atrial wall. (Numbers correspond to Fig. 1.)

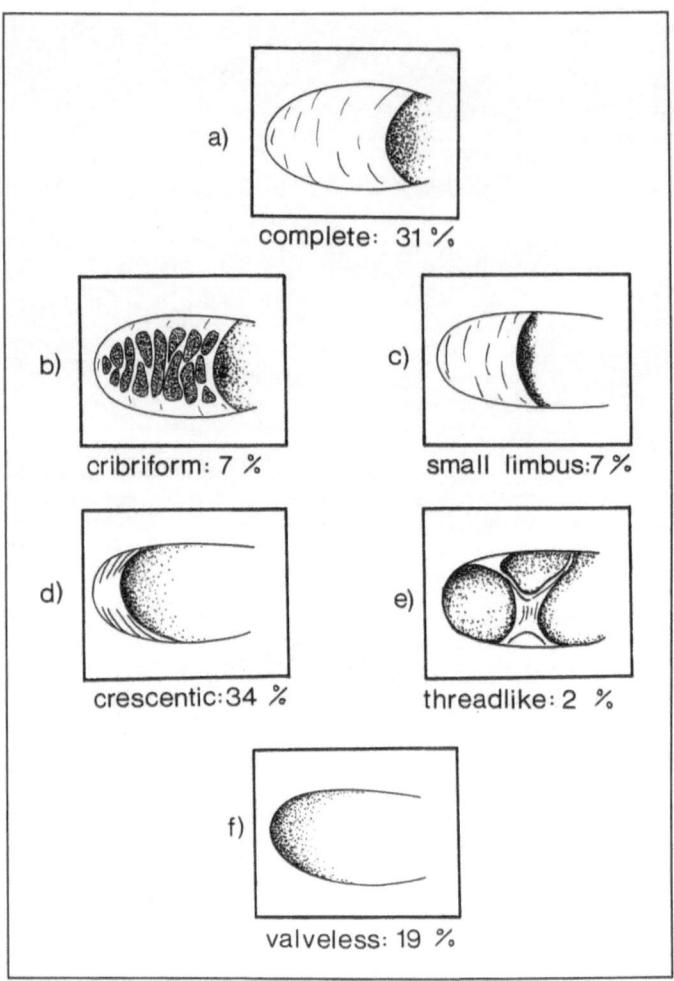

**Fig. 11.** Various formations and frequency of the ostial (Thebesian) valve of the c.s. in schematic presentation. All aspects are from the right atrium.

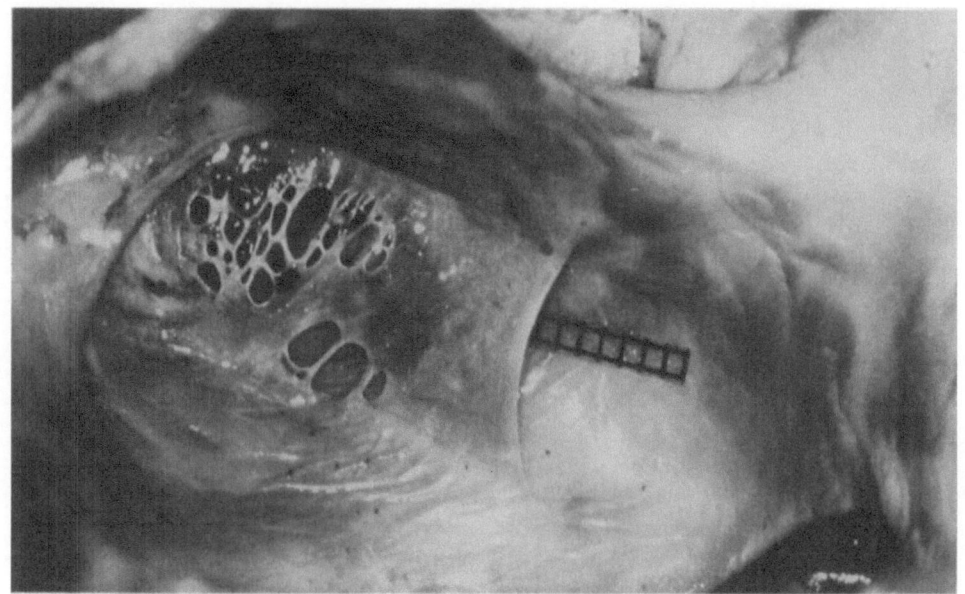

**Fig. 12.** Cribrate valve at the atrial ostium of c.s. (The photograph corresponds to scheme b in Fig. 11).

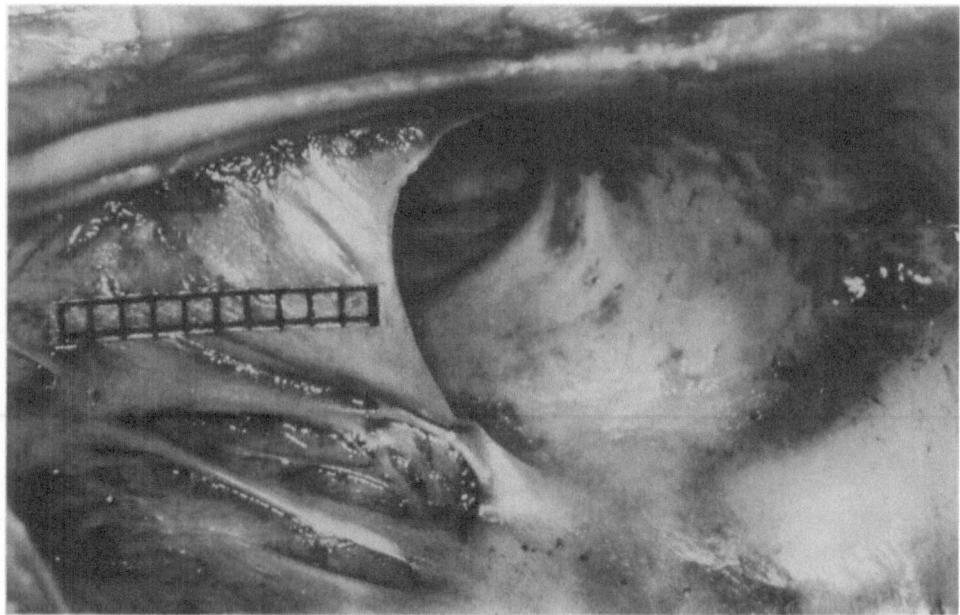

**Fig. 13.** Incomplete semilunar valve (The photograph corresponds to scheme c in Fig. 11).

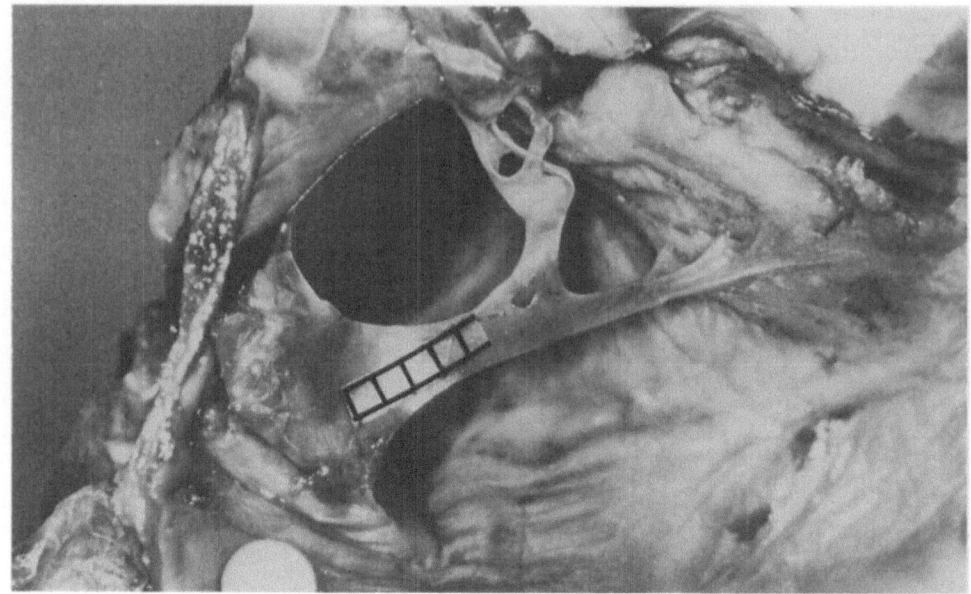

**Fig. 14.** Threadlike or rhomboid shaped valve (The photograph corresponds to scheme e in Fig. 11)
Fig. 11).

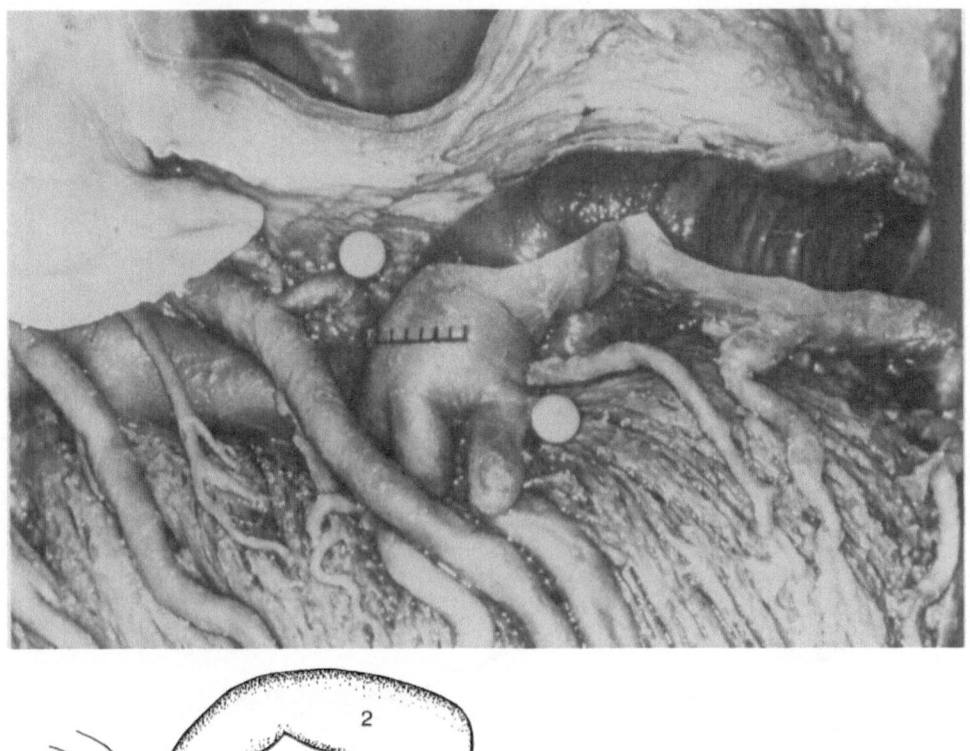

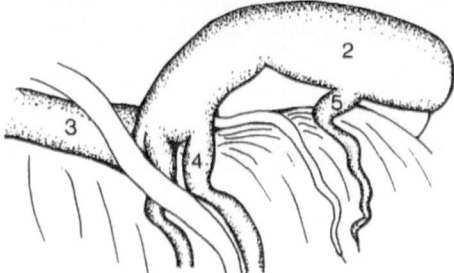

**Fig. 15.** S-shaped course of the terminal portion of the great cardiac vein.

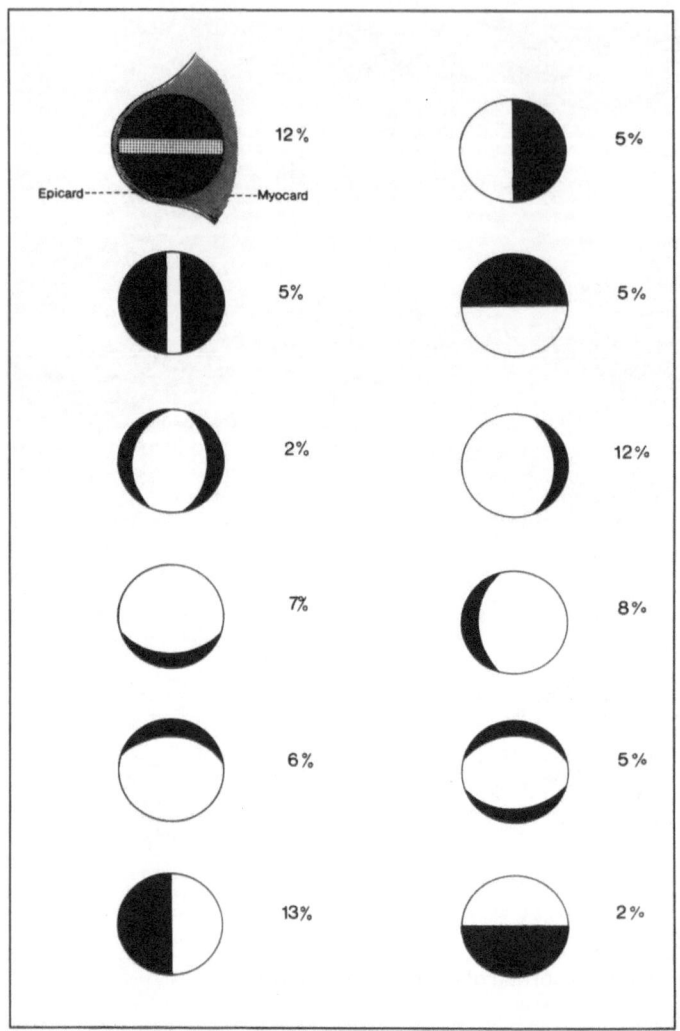

**Fig. 16.** Frequency of various shapes of the Vieussenian valve, which may be formed either by a single velum or two vela, closing the orifice more or less sufficiently.

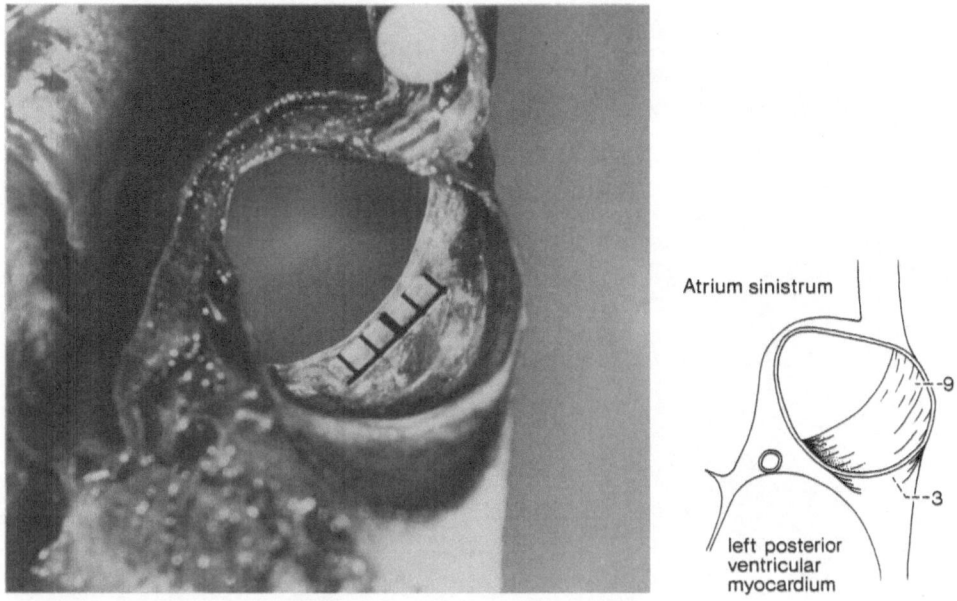

Atrium sinistrum

left posterior
ventricular
myocardium

**Fig. 17.** Univalvular Vieussenian valve on a cross-section of the great cardiac vein. (Numbers correspond to Fig. 1.)

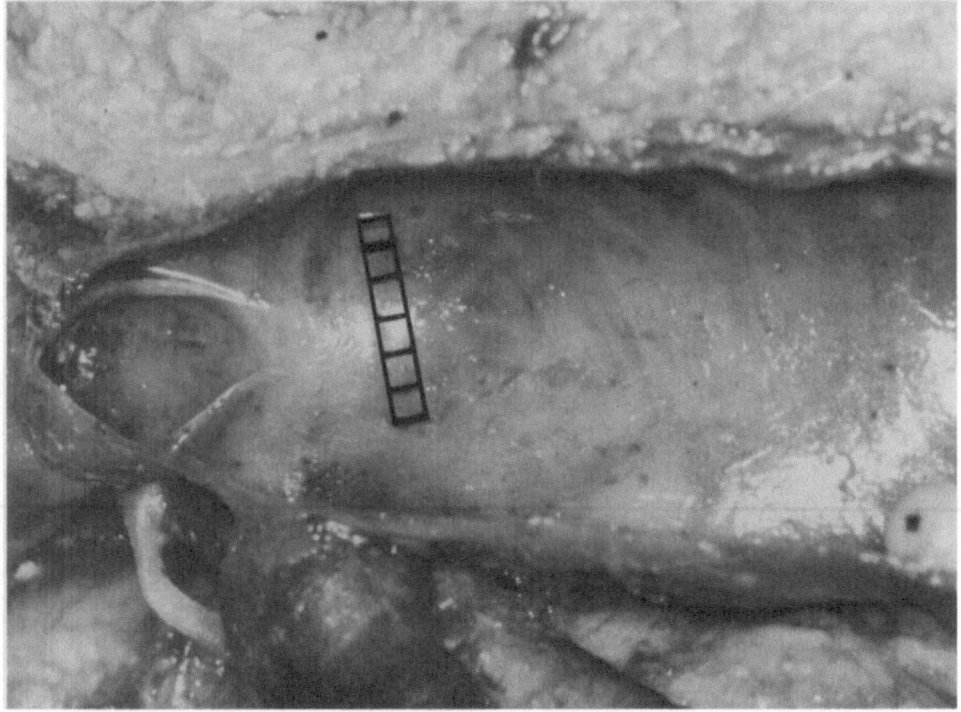

**Fig. 18.** Bivalvular formation of the Vieussenian valve (arrows).

**Fig. 19.** Corrosion cast of the c.s.: at its origin there are impressions of two endothelial folds, i.e., vela of the Vieussenian valve (arrows).

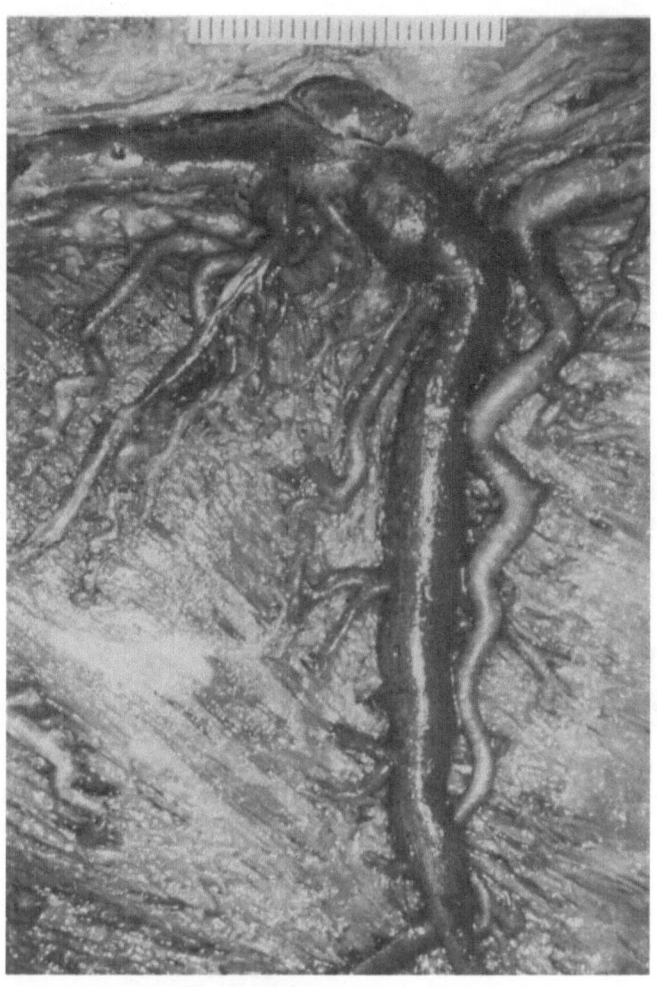

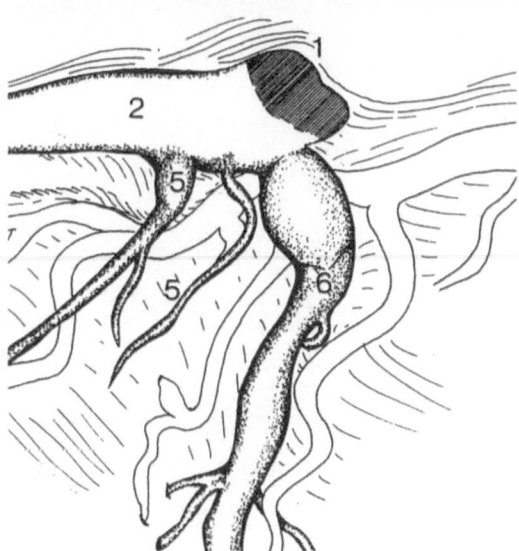

**Fig. 20.** Middle cardiac vein with moderate angulation and a remarkable bulb at its terminal portion before opening into the c.s.

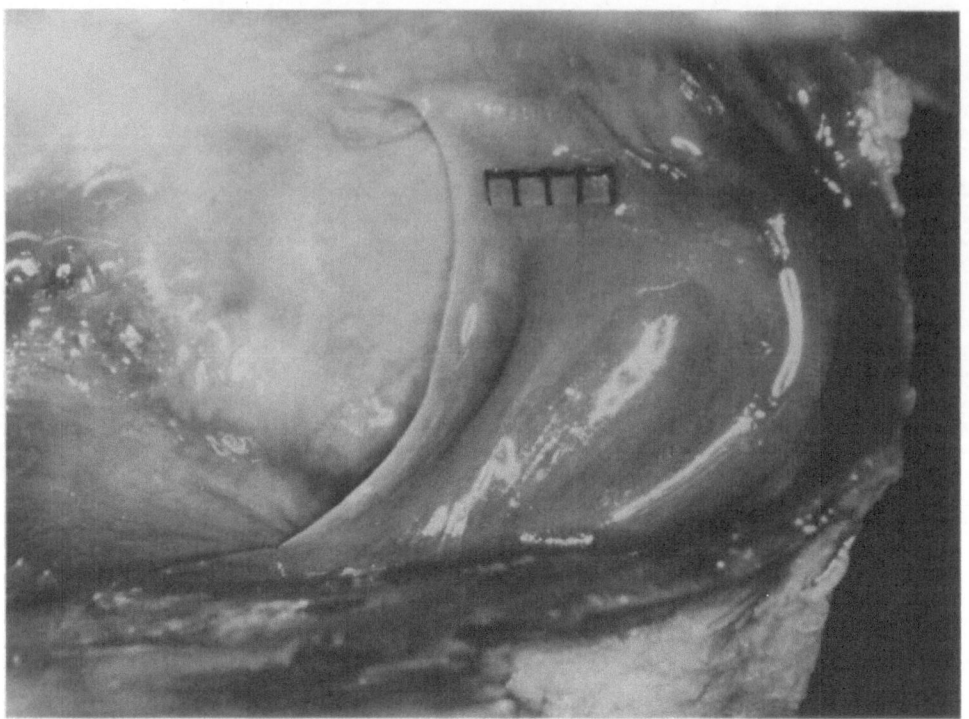

**Fig. 21.** Cranial aspect of an opened c.s. and at orifice of the middle cardiac vein. The ostium of the vessel is almost completely closed by a sufficient valve.

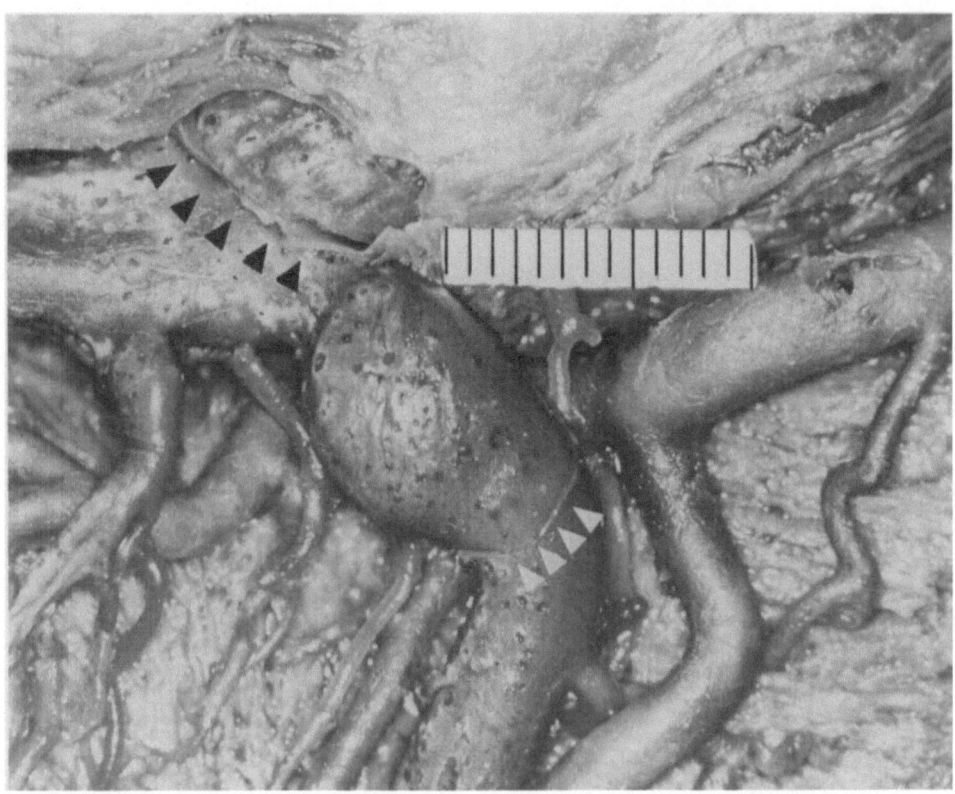

**Fig. 22.** Terminal portion of the middle cardiac vein with bulbous enlargement and a peripheral "ostial" valve (white arrows). The atrial ostium of c.s. is secured by an almost complete Thebesian valve (black arrows). Magnification of Fig. 26.

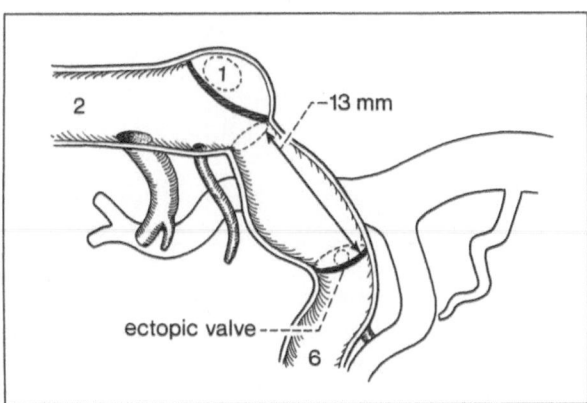

**Fig. 23.** Scheme of a middle cardiac vein with an ostial valve at about 13 mm distance to the opening into the c.s.

121

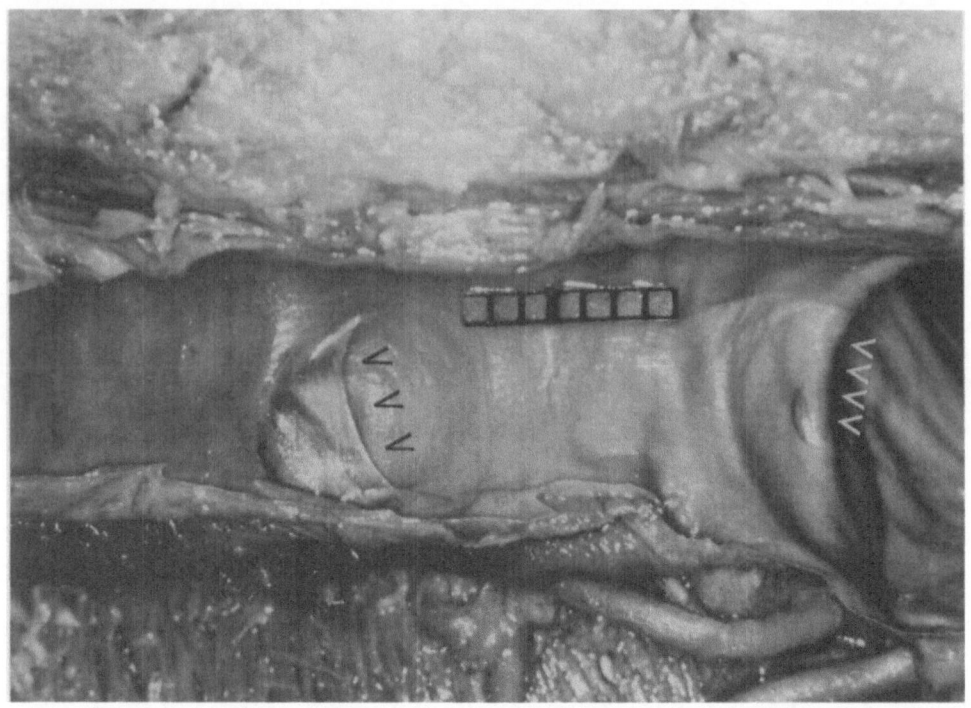

**Fig. 24.** Opened c.s. after replacement of its roof: a complete unicuspid valve secures the ostium of left posterior ventricular vein (black arrows). Marked ostial angle of the c.s. (white arrows).

# Principles of coronary venous interventions

S. Meerbaum

After updating information on the coronary venous system, this chapter will begin consideration of coronary venous interventions by discussing some of the general principles and issues. This will be followed by more specific reports and evaluation of effects of coronary sinus occlusion, and then the evolving retrograde treatment system will be reviewed. A number of published reviews provide additional perspectives and offer comprehensive lists of references [1–8].

## Underlying rationales

In contrast with coronary arteries, the profuse and complex coronary venous system is not involved in the atherosclerotic process. This, in itself, would motivate investigators to examine the possibility of a retrograde approach to jeopardized myocardium. Coronary venous interventions involve obstruction of the coronary sinus or subsidiary coronary veins, without or with supplemental retroinfusion. Considerable efforts have been aimed at devising systems and protocols which would assure safety by facilitating adequate coronary venous drainage. Once the latter is satisfied, e.g., through a clinically oriented intermittent phased retroperfusion, one can take advantage of the often relatively small functional impedance to a retrograde perfusion via the coronary veins, and the infrequent hindrance of catheterization by any residual valves. Thus, when the normal antegrade supply is severely restricted, existing alternate pathways allow retrograde delivery of blood (and/or pharmacologic agents) to an underperfused zone of the heart, hopefully into its myocardial capillaries where the essential nutritional exchanges take place.

As already pointed out, the coronary venous channels are very complex, and both epicardial as well as intramyocardial coronary veins feature highly variable and proliferating anastomoses. Given, for example, a systemic artery-to-coronary vein bypass (i.e., instituted surgically or clinically with interposed pump and catheter) for delivery into coronory veins subserving a jeopardized myocardial region, the effectiveness of a retroperfusion will depend upon the prevailing anatomy and a number of physiologic factors. The latter generally include the retrograde coronary veno-arterial pressure gradient and flow, the resistance and capacitance of coronary veno-venous shunts, the particular site of retroinfusion in relation to ischemic zone, and the magnitude of residual antegrade coronary flow and pressure beyond the coronary artery obstruction. Retrograde delivery may also be compounded by any arterial-venous communications, intramyocardial sinusoids, and limited capacity of Thebesian drainage into the cardiac chambers. Significantly

123

underperfused regional myocardium can often be reached via the coronary veins, provided an adequate retrograde input pressure is applied and coronary veno-venous shunts do not compromise the retrodelivery. Safety and reliability require avoidance of vascular trauma or myocardial damage, which might ensue as a result of excessive coronary vein blood pressure and flow, particularly when the retrograde intervention competes with substantial residual antegrade perfusion.

**Effects of coronary venous anatomy**

The importance of the prevailing anatomic configuration in relation to retroperfusion is obvious. As pointed out in the preceding chapter, venous drainage of left ventricular myocardium occurs largely via the coronary sinus through its ostium into the right atrium. Roughly speaking, the great cardiac vein and the anterior interventricular veins subserve the anterior region of the left ventricle, perfused by the anterior descending coronary artery. The middle cardiac vein, issuing into the coronary sinus at its ostium (or less frequently, separately into the right atrium), ahnd the posterior interventricular veins entering the coronary sinus near its junction with the great cardiac vein, drain the posterior aspects of the left ventricle, subserved by the circumflex coronary artery. The anterior cardiac veins, covering the anterior wall of the right ventricle, collect venous blood from regions fed by the right coronary artery and drainage is generally into the right atrium.

Because of their important effects, the epicardial coronary veno-venous anastomoses described by Dr. Pakalska in chapter 4 as vessels as large as 1−2 mm in diameter, deserve repeated emphasis. These variable shunts are responsible for the frequent difficulty to predict or control the coronary vein pressure and flows for effective retrograde intervention. Pakalska also presents some evidence for existence of arterial-venous pathways, which could play a role in either antegrade or retrograde perfusion. Arterial-luminal vessels as well as intramyocardial sinusoids have also been described, however, of particular significance are the intramural Thebesian veins, which do provide direct drainage from the coronary venous system into the cardiac chambers (mostly the right ventricle). Obstruction of the usual efflux via the coronary sinus tends to magnify the functional Thebesian outflow. The possibility of any nutritional contribution by the various non-capillary microvessels remains in questions.

A pertinent practical point relative to retrograde interventions via epicardial coronary veins concerns the actually achievable placement of catheters and occlusive balloons. Rather small differences in the site of the coronary sinus catheter tip can greatly influence the shunting-off of the retrogradely propelled blood from its intended path, thus potentially compromising retroperfusion performance. Conversely, though, the same coronary veno-venous shunts are believed to importantly delimit excessive retroperfusion-induced coronary vein blood pressures.

**Interspecies differences**

It is important to keep in mind differences between the human vasculature and that of the dog and pig, the primary animal species used in experimental research inves-

124

tigations. In addition to the previously described differences, e.g., relative to coronary artery collaterals, species-related coronary venous anatomy must also be considered. Thus, in some species, the hemiazygous vein empties into the proximal coronary sinus; it has been employed for experimental retroinfusion studies in the domestic pig. Comparing primate, pig, dog, and sheep hearts, chapter 4 suggests that only sheep exhibit a minor amount of coronary veno-venous shunting. In human hearts, the most important and large coronary veno-venous anastomoses were those between the great cardiac vein system and the anterior cardiac veins, and also between the great cardiac vein and the middle cardiac vein, which often empties into the coronary sinus at its ostium. Such coronary veno-venous connections are said to increase with age, and as a result of ischemic history. Finally, compared to dogs or other species, more of the human coronary veno-venous anastomoses were noted to connect to the anterior cardiac veins. Although uncertain, one might thus infer that myocardial edema and hemorrhages, encountered during some healthy animal coronary occlusion or retroperfusion studies, should be less common in human applications.

## Coronary venous drainage: Implications in normal and ischemic states

As early as 1938, Katz [9] found in human hearts that the outflow through the coronary sinus and the Thebesian veins was highly variable. Hammond [10] concluded that 75% of the antegrade coronary flow in dogs is eventually delivered to the right atrium after passing through the myocardial capillary system, the remainder draining in part via shunts into the ventricular chambers through Thebesian veins and/or myocardial sinusoids. Heiss [11] found that in anesthetized closed chest dogs the fraction of left ventricular myocardial blood inflow, which entered the coronary sinus, amounted to 636%. Scharf reported similar results, the mean percentage of coronary sinus outflow being 68% [12]. Hood [13] concluded that the preponderance (96%) of left ventricular and septal myocardial blood flow was drained via the coronary sinus.

Information was also sought relative to coronary venous drainage during myocardial ischemia. In 1970, Eliska [14a] examined the arterial and venous circulation in regionally ischemic dog myocardium, and measured − in particular − anterior interventricular vein blood flow under normal conditions as well as after ligation of the left anterior descending coronary arteries. Following LAD ligation, anywhere from 16.6−70.5% of the normal preocclusion coronary artery blood flow rate was measured in the anterior interventricular vein. The origin of this blood appeared twofold: 1) residual arterial blood flow from the ischemic area, due to arterio-arterial coronary collaterals; and (apparently) also 2) venous blood from nonischemic myocardium redistributed through coronary veno-venous anastomoses into the low pressure drainage vessels of the ischemic area. Collecting (through a low resistance cannula) blood from the "peripheral stump" of the ligated LAD, it was found that the anterior interventricular vein blood flow exceeded the above collected arterial flow significantly, by an average of 147%. Other studies by Eliska [14b] in animals with minimal coronary artery collaterals, revealed that during an LAD coronary artery occlusion, most of the blood flow in the anterior interventricular vein could be attributed to redistribution from the nonischemic zone.

More recent studies (Chen, personal communication) sought to understand changes in drainage caused by coronary venous retroinfusion of arterial blood into an occluded coronary sinus (e.g., ostial ligation). Obviously, normal coronary sinus drainage cannot be maintained in the presence of a coronary vein obstruction, and/or with retrograde infusion. With the coronary arteries vented in an in vitro heart preparation, radionuclide microsphere measurements indicated that retro-infusion into a coronary sinus obstructed at its ostium resulted in as much as half of the retroinfusate passing through the regional left ventricular microcirculation (eventually collected as desaturated blood from the coronary arteries). Given the known coronary veno-venous shunts, it was not at all surprising that retroinfusion performed from an occlusion site located within the coronary sinus vessel and some distance from the ostium (or else within an obstructed great cardiac vein), yielded a substantially lesser retroinfusate delivery to regional myocardium, again measured by in vitro collection from the respective vented coronary arteries.

The situation is certainly different in the clinically germane setting, when the coronary artery exhibits not only a definite back-pressure and impedance to retro-grade perfusion, but may actually feature levels of residual distal pressures and flows, which would significantly interfere with and counteract the retroperfusion. In the experimental animal model, only limited retroinfusate penetration (often as low as a few percent) into the regional microcirculation has been demonstrated. Depending (among other factors) on the prevailing veno-arterial pressure gradients, much of the retroinfusate from the coronary sinus may simply not penetrate beyond relatively large intramyocardial coronary veins, and is then drained via epicardial shunts and Thebesian channels. In a retrograde treatment of acute regional ischemia, the ischemic myocardium distal to a LAD coronary artery occlusion is most effectively reached by the retroinfusate when the retrodelivery catheter tip is positioned in the vicinity of the jeopardized region, e.g., in the great cardiac vein or even in the less accessible anterior interventricular coronary veins.

Presumably, capacity of the highly compliant epicardial coronary veins also plays a role in this regard. With persisting obstruction of an epicardial coronary vein, and higher pressures prevailing in nonischemic as compared to ischemic zones, drainage from the ischemic myocardium appears to be accomplished to a large extent through Thebesian vessels (perhaps in a ratio of 3:1 as compared to drainage via coronary veno-venous shunts). In the intermittent or synchronized retrograde intervention systems (6 and 8), a higher myocardial delivery has been reported than was the case when the retroperfusion featured a persisting coronary vein occlusion. Systolic drainage from the ischemic myocardium is, of course, deliberately facilitated in the phased diastolic retroperfusion system, and washout of metabolites is also promoted through intermittent decompression of coronary veins.

Given the inevitable constraints on coronary venous drainage in the presence of coronary vein occlusion and retroinfusion, we are left with the general question: how effective could retroperfusion be in the "best" of circumstances, i.e., without any compromising epicardial coronary veno-venous shunting? Based on current limited experience, a realistic estimate is that about 25% of the retroperfusate might be expected to pass to the capillaries, and that most of the myocardial venous drainage would then take place through sinusoids and Thebesian veins directly into the right ventricle. During most coronary venous interventions, in a variety of physiologic settings, the Thebesian passages assume a very important function of providing for alternate drainage. Their anatomy and physiologic controls are yet to be fully defined, and while thin-walled and apparently coursing very close to myocardial muscle fibers, Thebesians may or many not be sufficient to facilitate a nutritionally beneficial retrograde circulation.

## Steady vs phased coronary venous interventions

The older surgical aorta-coronary sinus bypass (retroperfusion) procedure developed by Beck and his contemporaries [15-21] was applied in both animals with coronary artery obstructions and in patients with coronary artery disease. The procedure consisted of a continuous arterial blood retroinfusion through a shunt from the aorta to the coronary sinus, whose ostium was then obstructed so as to generate an increased coronary venous blood pressure potential for retrograde delivery. Currently, the most promising application of continuous retroperfusion during heart surgery is carried out during cardiac arrest by coronary sinus administration of cardioplegic agents, and is aimed at intraoperative protection of the jeopardized myocardium [7−8]. Another steady retroperfusion mode is applied for brief periods in the beating regional ischemic heart. It involves temporary obstruction of a coronary vein into which arterial blood or drugs are continuously and selectively administered for treatment of ischemic myocardium [22].

In contrast, most of the clinically oriented retrograde interventions, developed during the past decade, feature a discontinuous mode of coronary vein occlusion and retroperfusion. In the case of fixed timing or pressure-controlled intermittent coronary sinus occlusion (ICSO or PICSO), an optimized schedule of balloon occlusions of the coronary sinues permits alternating pressure buildup over several cardiac cycles, followed by a brief balloon deflation or "washout" period in which antegrade flow conditions are reestablished. These two phases of PICSO are to promote flow redistribution from nonischemic toward ischemic zones, while promoting washout of accummulated metabolites (see chapter 8). In synchronized retroperfusion (SRP), retrograde arterial blood infusion into a regional coronary vein is limited to diastole, and carefully controlled balloon deflation provides near normal systolic drainage (see chapter 10). When carried out in each cardiac cycle, SRP should fully avoid any of the previously encountered problems of injury due to excessive damming up of blood and elevated pressures in the coronary veins. Both PICSO, by virtue of its very brief coronary sinus occlusions, and SRP because of its facilitated systolic drainage, represent systems which emphasize safety of the coronary venous interventions. In terms of effectiveness and potential applications, the advantages of those systems remain to be clinically demonstrated. While the principal methods and developments will be described in subsequent sections, any further discussion of SRP and PICSO interventions requires first a definition of these two modalities.

Figures 1 [3] and 2 [23] illustrate graphically the principles of these methods, the SRP system being shown in its expanded role with potential supplemental

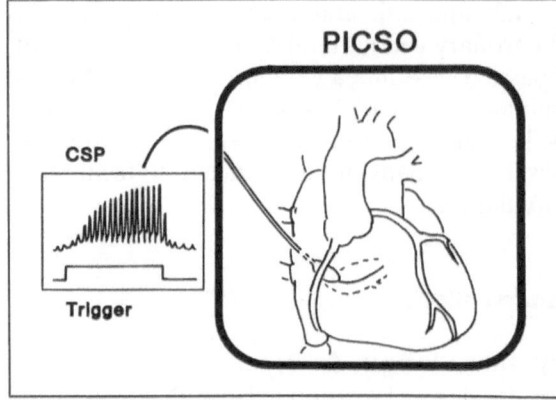

**Fig. 1.** Schematic of PICSO. A pneumatic pump activates a balloon, positioned into the coronary sinus. Obstruction of venous drainage is timed to the dynamics of the coronary sinus pressure during occlusion. For details see text. (By permission, from: Mohl W (1987) Coronary sinus interventions from concepts to clinics. J Card Surg 2:467–493.)

hypothermia and/or pharmacologic retroinfusion. Note that in the PICSO system, the hoped-for balloon occlusion site of the coronary sinus vessel is placed close to its right atrial ostium, while the tip of the SRP balloon catheter may be placed in the great cardiac vein (for effective treatment of LAD coronary artery occlusions). In both systems, it is desirable to monitor the resulting coronary venous pressures, to

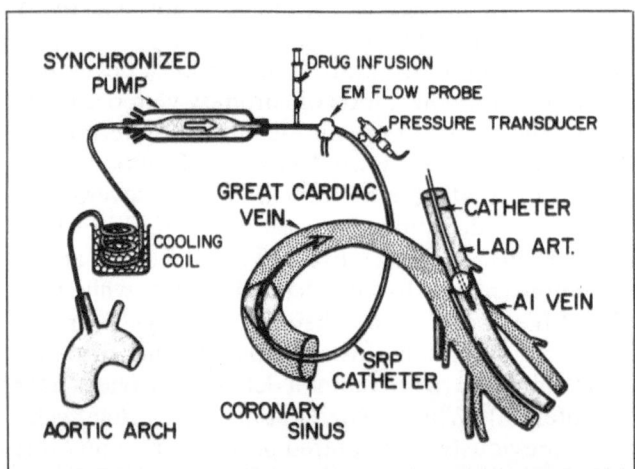

**Fig. 2.** Hypothermic synchronized retroperfusion (HSRP). Arterial blood is shunted from the aortic arch and then cooled and delivered into the synchronized bladder pump. In diastole, blood is propelled into the balloon-tipped synchronized retroperfusion (SRP) catheter and through it into the regional coronary veins. Forward flow stops in systole, causing the balloon of the SRP catheter to promptly collapse and allowing coronary venous drainage around the catheter into the coronary sinus. The system also provides measurement of infusate pressure and flow. SRP treatment is applied through the great cardiac vein and the anterior interventricular (AI) vein to perfuse myocardium made acutely ischemic by intracoronary balloon occlusion of the left anterior descending (LAD) coronary artery. EM = electromagetic. (By permission, from: Meerbaum S et al. (1982) Hypothermic coronary venous phased retroperfusion: a closed chest treatment of acute regional myocardial ischemia. Circulation 65(7):1435–1445.)

128

assure safety and as an aid in controlling PICSO timing or SRP retroinfusion flow rates. SRP features an ECG synchronized pump and an arterial blood shunt to the regional coronary vein for effective retrograde perfusion and treatment of ischemic myocardium. PICSO is simpler since it uses only coronary sinus occlusion without retroinfusion, but higher coronary vein pressures are developed and careful adjustments in coronary vein occlusion-release timing are required to achieve a safe yet effective circulation within and drainage from the ischemic zone. Both methods have been demonstrated in animal experiments, and have been undergoing trials which indicate feasibility in both clinical and surgical settings. A combination of features of both of these techniques is currently under consideration.

## Retrograde intervention in normal physiologic states

The retroperfusion effects in normally perfused myocardium must be appreciated, even though it is not normally the objective of the retrograde techniques to treat the heart in the absence of significant coronary artery obstruction. A reduction in forward myocardial perfusion would be expected to follow any unscheduled occlusion of the coronary veins, for example, as a result of coronary sinus thrombosis. Deliberate protocols or accidental coronary vein manipulation, e.g., during mobilization of one of the retrograde systems, could extend the effects of coronary vein occlusion and retroinfusion to myocardial territories with normal antegrade coronary supply. An "erroneous" situation might be encountered following incorrect placement of the retroinfusion catheter tip and balloon within coronary veins subserving an unoccluded coronary artery, or accidental dislodgement of the catheter during the retroperfusion procedure. Retroinfusion of arterial blood (for circulatory support of ischemia), may perhaps be programmed to commence prior to and extend temporally beyond brief ischemic events, such as those produced during a PTCA balloon inflation in the coronary artery. In general, extended regional retrograde coronary venous treatment would seem to be lacking in rationale whenever the corresponding coronary artery provides sufficient forward perfusion. A substantial antegrade blood flow represents a significant competitive impedance to retrograde coronary venous delivery, retroperfusion is unlikely, and – if persisted – retrograde intervention may lead to serious myocardial injury as well as dysfunction.

More specifically, in a controlled state featuring relatively uniform baseline pressures and flows in the coronary arteries and veins, one would not expect significant pressure gradients or mobilization of available coronary collateral supply, and other inter-regional blood shunting would also be minimal. Obstructing the coronary sinus elevates the impedance to normal outflow, and generally causes a significant increase in coronary vein pressures. The actual degree of this impedance to coronary venous outflow depends on the number and size of the previously described coronary venous shunt vessels, particularly those at the epicardial level. For example, shunts between the left and right sides of the heart could be so pronounced that even a total coronary sinus occlusion may be associated with imperceptible effects. On the other hand, coronary venous occlusion often results in a significant reduction of the coronary arterial inflow, as well as in a compromised contractile and metabolic function.

With PICSO, both systolic and diastolic coronary venous pressures are increased over a number of cardiac cycles during the coronary sinus occlusion phase. In particular, systolic "squeezing" of the microvessels during coronary vein occlusion leads to substantial elevation of coronary venous blood pressure. It is said that the occlusion-release phasing can be optimized so that the systolic pressure rise (during the repetitive brief coronary sinus occlusions) is limited to filling the coronary venous system capacities, i.e., without causing derangements or injury to normal myocardium. This has yet to be fully demonstrated, particularly when the PICSO application is protracted. It is also believed that preexisting coronary veno-venous shunts and Thebesian vessels would redistribute blood during the coronary sinus occlusion phase, in a manner to lessen any negative impact of PICSO on the normally supplied myocardium. Nonetheless, even relatively brief multiple coronary sinus occlusion and excessive coronary vein pressure elevation can be detrimental in territories with uninterrupted antegrade flow and pressure. The potentially deleterious consequences are: traumatic engorgement of the coronary venous vasculature, myocardial edema and hemorrhages, interference with the normal myocardial perfusion, and functional derangements in nonischemic myocardium.

The basic difference when using SRP is that this technique limits retroinfusion and coronary vein occlusion to diastole, and aims to allow normal systolic coronary venous drainage, promoted by contractile muscle squeezing of the myocardial vessels. Under ideal SRP conditions, retrograde myocardial perfusion through the coronary veins should thus not be much hindered and hazards of myocardial edema should be minimized. Once again, though, in the presence of uniform substantial forward supply to all regions of the heart, extended application of SRP would appear to be unjustified.

### Retrograde intervention during myocardial ischemia

Coronary obstruction and regional myocardial ischemia are the very setting in which retrograde coronary venous interventions may be applied to preserve myocardial viability and maintain function, at least during support periods. When a coronary artery becomes severely obstructed, the corresponding regional myocardial blood flow and pressures within the associated coronary vessels are depressed. With severe coronary artery stenosis or its occlusion, distal pressure within the coronary artery can be profoundly reduced. Thus, during LAD or left circumflex coronary artery occlusions (e.g., experimentally, or during PTCA balloon dilatation), measured distal pressures are quite low, with diastolic levels (3–12 mmHg measured in dogs) largely reflecting the degree of coronary arterial collateral supply, while systolic pressures may be relatively high as a result of the heart's contraction and myocardial wall stress.

Depending on availability of a sufficient collateral circulation and — of course — on the severity of the native coronary artery obstruction, an acute ischemic myocardial injury can be profound and its progression rapid. As discussed previously, coronary occlusions of a mere several minute duration are already associated with protracted dysfunction, while occlusions beyond 20–30 min often lead to some irreversible myocardial damage, which then propagates rapidly from endocardium toward the epicardial layer. It is well established that the mechanical, metabolic and electrophysiologic responses after acute coronary occlusion commence within seconds. Thus, with severe coronary stenosis or occlusion, time is of essence if the retrograde coronary venous intervention − or for that matter, any treatment − is to be of value. In may instances, such retroperfusion treatment would be directed at residual reversibly injured zones, while some portion of the myocardium may already be irreversibly damaged.

In the setting of injurious and progressing myocardial ischemia, the general objective of a coronary vein intervention is to rapidly retrogradely deliver sufficient circulation and nutrient supply to the jeopardized regional myocardium, so as to at least extend tissue viability until a more definitive correction of the underlying cause is possible. Alternately, with very short periods of acute ischemia, the aim may be to use the retrograde support to minimize dysfunction and related complications during a treatment such as PTCA.

Consider now a relatively short period (several minutes) of coronary sinus obstruction, and also assume the coronary venous shunting to be such as to minimize the problems of vascular trauma and edema alluded to above. If the coronary vein obstruction relates specifically to the myocardial region subserved by a significantly lowered coronary arterial supply, some of the hazards may be reduced by virtue of significantly reduced regional flows and pressures. It has been reported that, in the arrested heart during antegrade cardioplegia, coronary sinus obstruction alone benefits the preservation of myocardial regions situated distal to a major coronary arterial stenosis or occlusion. In the beating heart, use of persisting coronary sinus occlusion is definitely controversial. With myocardial squeezing occuring principally in systole, it is likely that emptying of the venous blood via coronary veins would be interfered with during continuous coronary vein occlusion.

With PICSO, the ischemic regions respond to the intermittent coronary sinus occlusions by a corresponding rise in distal coronary artery pressures. This pressure rise reflects the increased impedance to outflow, and perhaps also a supplemental inflow of venous blood or shunting into the lower pressure ischemic region from the nonischemic zone. Coronary arterial collateral supply to the ischemic region, e.g., from the left circumflex to the occluded LAD, might actually be decreased during the ICSO-induced elevation of the systolic and diastolic distal (occluded) coronary artery pressure. In the coronary sinus release phase of ICSO all coronary pressures drop sharply, and this appears to facilitate catabolite washout from the ischemic region. It has been suggested that such washout can be very effective in alleviating the metabolic derangements of acute ischemia (and reperfusion). The combination of PICSO occlusion phase "redistribution" and release phase "washout" may be responsible for the experimentally demonstrated benefits, such as infarct salvage and moderately improved ischemic zone function. Unless optimal ICSO occlusion-release phasing can limit effects of the occlusion to coronary vein filling, ICSO may induce some dysfunction in the nonischemic zone of the heart.

SRP provides both diastolically phased obstruction and diastolic retroinfusion of arterial blood via the regional coronary vein. Systolic release of the coronary vein obstruction along with interrupted retroinfusion facilitate normal drainage. This more complex system is phased in a similar manner as various intraaortic balloon pump systems, and should − in principle − avoid some of the potential problems connected with intermittent or persisting systolic coronary vein occlusion. Figure 3 [24] shows the selective diastolic pressure augmentation in the regional coronary vein, without significant systolic elevation. It would appear from experimental observations that SRP applied during acute ischemia (e.g., LAD coronary artery occlusion) results in only a small rise in distal coronary artery pressure. The diastolically propelled retroinfusate seems to cummulatively mix and concentrate arterial blood in the smaller and intramyocardial coronary veins, and there is evidence of selective retrograde perfusion of the ischemic zone, with particular benefits to its endocardial layer. It has also been shown that SRP causes significantly increased washout from the ischemic zone [25]. Both of these effects are thought to be responsible for the demonstrated SRP reduction in infarct size and significant improvements in regional cardiac function. Edema and vascular damage appear to be minimized in this technique.

## Limitation of experimental data

Before proceeding to a description of the encouraging specific experimental evidence regarding retrograde coronary venous interventions, it is necessary to point to limitations and urge caution against simply transferring conclusions from animal studies to the clinical domain. Significant anatomic differences between species alone would suggest extremely careful interpretation of data from the laboratory,

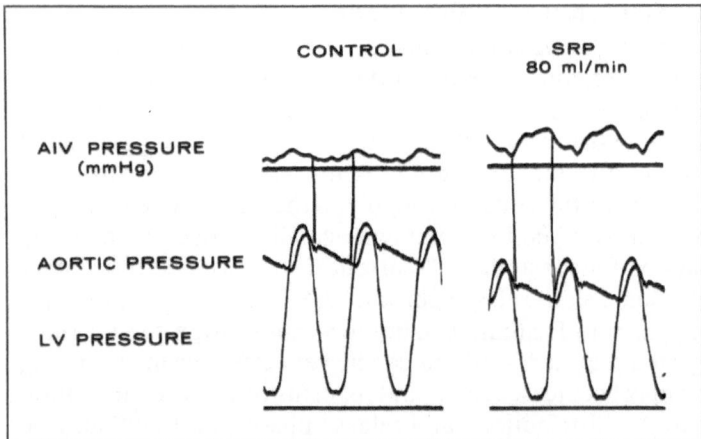

**Fig. 3.** Regional coronary venous pressure dynamics during synchronized retroperfusion. Illustration of the effect of synchronized retroperfusion (SRP) on the pressure measured within the anterior intraventricular vein (AIV) that adjoins the occluded left anterior descending coronary artery. Application of synchronized retroperfusion causes a rise in the diastolic level of the anterior intraventricular vein pressure, enhancing retrograde delivery of arterial blood into the ischemic myocardium. In contrast, synchronized retroperfusion does not significantly alter the systolic phase of the venous pressure, indicating that near normal systolic drainage is produced with the new technique. LV = left ventricular. (By permission, from: Farcot JC (1978) Synchronized retroperfusion of coronary veins for circulatory support of jeopardized ischemic myocardium. Am . J Card 41:1191−1201.)

before applying the experience to the complex and extremely varied settings of coronary artery disease. As previously pointed out, there are instances (e.g., pharmacologic infarct salvage) where clinical studies simply did not bear out the projections made from animal research. Nonetheless, most of the significant developments in cardiology and cardiovascular surgery are in some way related to the extremely useful role played by animal experiments. Such experiments clarified physiologic mechanisms and factors contributing to treatment effectiveness as well as safety, and also helped develop operational criteria for upgraded equipment, and selection of most appropriate measurement techniques.

The unfortunately frequent divergence of experimental results is not merely due to species differences. Among many other reasons for varying reports covering apparently similar methods of treatment, attention must be given to the experimental model used, details of the particular experimental protocol, and the validity of measurements underlying conclusions. Thus, major differences have been demonstrated between conscious and anesthetized preparations, or closed vs open-chest dog investigations. An intervention applied in the regionally ischemic animal prior to any permanent injury cannot be equated with treatment of myocardium when a portion of the left ventricle is already irreversibly damaged. Ischemia should certainly be differentiated from anoxia (with maintained circulation). Acute coronary artery occlusion should be recognized as much more severe than moderate or even substantial degrees of stenosis, and effects of a significant arterial coronary collateral supply need to be appreciated. In view of the rapidly changing scene of noninvasive as well as invasive measurement techniques, it is wise to critically assess the validity and reproducibility of all data, even before analyzing results for statistical significance. In a word, a critical attitude should prevail in comparing experimental investigations, and again, in extrapolating results of animal studies to the clinical scene.

## References

1. Meerbaum S (1984) The Beck era: a springboard for renewed research of coronary venous retroperfusion aimed at treatment of myocardial ischemia. In: Mohl W et al. (eds) The coronary sinus. Steinkopff, Darmstadt, pp 320–327
2. Meerbaum S (1986) The promise and limitations of coronary venous retroperfusion: lessons from the past and new directions. In: Mohl W, Faxon D, Wolner E (eds) CSI – A new approach to interventional cardiology. Steinkopff, Darmstadt, pp 40–60
3. Mohl W (1987) Coronary sinus interventions from concepts to clinics. J Card Surg 2:467–493
4. Faxon DP, Jacobs AK (1988) Coronary sinus retroperfusion and intermittent occlusion. In: Acute copronary intervention. Liss, Inc., pp 250–267
5. Mohl W (1988) The momentum of coronary sinus interventions clinically. Circulation 77(1):6–12
6. Mohl W, Roberts AJ (1985) Coronary sinus retroperfusion and pressure-controlled intermittent coronary sinus occlusion (PICSO) for myocardial protection. Surg Clin North Am 65(3):477–495
7. Mohl W (1987) Retrograde cardioplegia via the coronary sinus. Ann Chir Gynaecol 76:61–67
8. Lazar H (1988) Coronary sinus interventions during cardiac surgery. Ann Thorac Surg 46:475–482
9. Katz LN, Jochim K, Weinstein W (1938) The distribution of the coronary blood flow. Am J Physiol 122:236

10. Hammond GL, Davies AL, Austen WG Jr (1967) Retrograde coronary sinus perfusion, a method of myocardial protection in the dog during left coronary artery occlusion. Ann Surg 39:166
11. Heiss HW, Hensel I, Kettler D, Tauchert M, Bretschneider JH (1973) Über den Anteil des Koronar-Sinus-Ausflusses an der Myokard Durchblutung des Linken Ventrikels. Z Kardiol 62(7):593–606
12. Scharf SM, Bromberger-Barnea B, Permutt S (1971) Distribution of coronary venous blood. J Appl Physiol 30(5):657–662
13. Hood WB (1968) Regional venous drainage of the human heart. Br Heart J 30:105–109
14a. Eliska O, Eliskova M (1970) Arterial and venous circulation in the ischemic myocardium in young animals. Angiologica 7:77–83
14b. Eliskova M, Eliska O (1966) Subepicardial veins of the dog's heart and their anastomoses. Acta Univ Carol [Med] (Praha) 12:21–30
15. Beck CS (1948) Revascularization of the heart. Ann Surg 128:854
16. Beck CS, Stanton E, Batilechuk W et al. (1941) Venous stasis in the coronary circulation. Am Heart J 21:767
17. Beck CS, Leininger DC (1954) Scientific basis for the surgical treatment of coronary artery disease. JAMA 159:1264
18. Bailey CP, Truex RC, Angulo AW et al. (1953) The anatomic (histologic) basis and efficient surgical technique for the restoration of the coronary circulation. J Thorac Surg 25:143
19. Bakst AA, Adam A, Goldberg H et al. (1955) Arterialization of the coronary sinus in occlusive coronary artery disease: III. coronary flow in dogs with aortico-coronary sinus anastomosis of 6 months' duration. J Thorac Surg 29:188
20. Eckstein RW, Smith G, Aleff M et al. (1952) The effect of arterialization of the coronary sinus in dogs on mortality following acute coronary occlusion. Circulation 6:16
21. Eckstein RW, Leininger DS (1954) Chronic effects of aorto-coronary sinus anastomosis of beck in dogs. Circ Res 2:60
22. Corday E, Meerbaum S, Drury K (1986) The coronary sinus: an alterate channel for administration of arterial blood and pharmacologic agents for protection and treatment of acute cardiac ischemic. JACC 7(3):711–714
23. Meerbaum S, Haendchen RV, Corday E, Povzhitkov M, Fishbein MC, Y-Ritt J, Lang TW, Uchiyama T, Aosaki N, Broffman J (1982) Hypothermic coronary venous phased retroperfusion: a closed chest treatment of acute regional myocardial ischemia. Circulation 65(7):1435–1445
24. Farcot JC, Meerbaum S, Lang TW, Kaplan L, Corday E (1978) Synchronized retroperfusion of coronary veins for circulatory support of jeopardized ischemic myocardium. Am J Card 41:1191–1201
25. Chang BL, Drury JK, Meerbaum S et al. (1987) Enhanced myocardial washout and retrograde blood delivery with synchronized retroperfusion during acute myocardial ischemia. J Am Col Cardiol 9:1091

# Effects of coronary vein occlusion alone

S. Meerbaum

We now proceed to consider one of the basic elements of most of the proposed coronary venous interventions, namely the relatively brief and continuous, or else more persistent obstruction (or occlusion) of the coronary veins, most often the coronary sinus or the great cardiac vein. As already indicated, the ostium of the coronary sinus drains a major portion of the left ventricular myocardial blood. This drainage occurs primarily in systole, largely by virtue of cardiac contraction, essentially propelling into the coronary veins the myocardial blood volume accumulated during a prior diastolic arterial input. By virtue of very significant shunting, coronary vein occlusion actually entails lower occluded pressures than would otherwise be expected. Among other anatomic specifics to be recalled are the frequent communications between the coronary sinus vessel and the anterior cardiac veins which drain the right coronary artery perfused territory, and also the Thebesian channels which, in circumstances of major coronary vein occlusion, can be mobilized to provide substantial venous drainage directly into the cardiac chambers. Species differences need to be considered, and in a recent study in sheep (extensive Thebesian system) elevated coronary sinus pressure had little effect [1], as compared to significant changes found in dogs.

Clearly, obstruction of coronary veins, or alternate significant elevation of coronary vein pressure, constitutes an abnormal manipulation which may be associated with deleterious effects. Conversely, such pressure augmentation might be helpful in the setting of severely depressed regional myocardial blood flow following a coronary artery obstruction, in which case antegrade perfusion would be minimal. If indeed certain potential benefits can be derived during myocardial ischemia from coronary venous occlusion alone (e.g., through redistribution of the circulation toward, and washout from the ischemic myocardium), it is obviously important to ascertain that the procedure does not result in vascular trauma, edema and myocardial damage. It has been reported that mean coronary sinus pressures must not exceed 60 or 40 mmHg, or even lower threshold levels [2, 3]. But these practical limits represent generalizations, and safe systolic or diastolic pressures probably depend on particular anatomy and physiologic circumstances.

Coronary vein obstruction was initially proposed on the basis of very limited experimental studies suggesting benefits. As will be seen this was followed by additional studies, many of which countered these early claims. Although several current intermittent coronary vein occlusion techniques tend to avoid the above mentioned hazards, the effects of non-phased coronary vein obstructions also remain of some interest (e.g., during pharmacological retroinfusion).

## Studies of persistent coronary sinus occlusion

In one of the earliest coronary investigations Gross et al. [4] first studied normal canine hearts, partially or fully occluding the coronary sinus at its ostium. Mortality within 24 h was 10% and 42%, respectively. This compared with a 53% mortality following left proximal anterior descending (LAD) coronary artery occlusion alone. The LAD occlusion alone was associated with significant infarcts. Although mortality was not significantly reduced when the LAD occlusion was preceded by several weeks of coronary sinus occlusion, infarction in the survivors was found to be greatly reduced. Occlusion of both the coronary veins and arteries seemed to reduce the apparent perfusion defect, compared to occlusion of the coronary artery alone. Based on their observations, these authors felt that coronary sinus stenosis or occlusion somehow modified the ischemic coronary circulation, leading to reduction in infarction. Yet, their own evidence and that by others [5], was not helpful in rationalizing the reported benefits and, in fact, they generally agreed that coronary sinus obstruction, when superimposed on coronary artery stenosis or occlusion, resulted in an undesirable further reduction of coronary artery blood supply to the ischemic zone.

Pertinent coronary investigations were carried out by Gregg and DeWald [6], during relatively brief dog-experiments. They found an occluded coronary sinus pressure increase, which was reflected in higher pressures distal to the coronary artery occlusion, from a mean 10 mmHg to an aortic systolic level, and to 20−40 mmHg in diastole. The coronary artery inflow was found to be reduced. They strongly felt that the source of this high peripheral intracoronary blood pressure was to be found in the nonoccluded coronary arteries. The flow from distal to the LAD occlusion (collected to the outside) was greatly increased, and was also highly unsaturated (3−4 volume % of oxygen vs 8% normally found in the coronary sinus). Regional ischemic myocardial dyskinesis was not corrected, presumably because regional blood flow was not enhanced (it probably decreased). It was surmised that coronary sinus occlusion traps an increased blood volume within the coronary veins at elevated pressures, unless coronary veno-venous connections are effective in providing an adequate circulation as well as drainage via Thebessians. Displacement of blood from nonoccluded to occluded myocardial zone vessels was also postulated. Gregg and DeWald did not believe that simple coronary sinus occlusion was potentially useful in treating coronary artery occlusions.

Extending the duration of these experiments, Thornton and Gregg [7] found that the early circulatory redistribution with coronary occlusion was only temporary. Substantial collaterals developed with longer coronary artery occlusions, and the authors found that after longer periods of coronary sinus occlusion − as compared to early post occlusion observations − blood taken from distal to the LAD occlusion was again of normal arterial character. Coronary sinus ligation was seen to be followed almost immediately by substantial epicardial coronary vein engorgement, and a general enlargement of vascular channels. There was a substantial mortality associated with coronary sinus ligation. However, the surviving animals with LAD and coronary sinus occlusion exhibited a reduction in perfusion defects and in infarction.

With the occluded coronary artery unavailable for a retrograde bleed, a substantial resistance may be expected to retrograde coronary venous flow via the capillaries. Following acute LAD occlusion, peripheral backflow measured in one particular experiment was of the order of 1 cc before, and immediately rose to an average of around 3 cc after coronary sinus occlusion. The bleed flow from distal to the LAD occlusion then increased substantially with time form the coronary sinus occlusion. As indicated, the blood from distal to LAD occlusion always analyzed as very low in oxygen content after coronary sinus occlusion, compared being to essentially arterial in the absence of coronary sinus occlusion. Most of the backflow was found to originate from the other two coronary systems, apparently providing some coronary veno-venous collateral pathway to the ischemic zone. Yet, regional ischemic myocardial function was generally found not to be improved as a result of coronary sinus occlusion. It seems that to provide a sufficient nutritional retrograde perfusion with coronary sinus occlusion, the peripheral coronary artery would require an alternate low resistance conduit for drainage, through Thebessians or other vessels.

Dr. Eckstein, a pioneer associated with the Beck surgical arterialization retroperfusion treatment of coronary artery disease, reevaluated the acute effects of coronary sinus pressure elevation [8]. The coronary sinus was cannulated or occluded in dogs to achieve a mean coronary venous blood pressure of 40−50 mmHg, which was deemed "safe". The left circumflex coronary artery was then ligated, and subsequent effects were observed. Sixty to seventy percent of the dogs survived for 1 h, as compared to a 70% mortality with untreated left circumflex coronary artery ligation. With supplementary arterialization of the coronary sinus, survival in the dogs studied was found to be 100%. The desired 50 mmHg pressure in the coronary sinus was sometimes difficult to achieve, because of major coronary veno-venous shunting directly at the coronary sinus ostium.

A special effort was made to evaluate the effects of the coronary sinus pressure elevation on the amount and nature of blood collected from distal to the occluded coronary artery, and on peripheral coronary pressures. When the occluded left circumflex coronary artery was opened to atmosphere via a low resistance cannula, 0.5−0.8 cc/min of arterial blood could be collected, and this blood originated from the alternate coronary arteries (via collaterals). Pressure distal to the coronary artery occlusion was elevated by partial or complete coronary vein occlusion. The collected backflow was consequently increased, and the blood became venous in character. When the coronary sinus pressure was increased from 30 to 90 mmHg, the backflow rose from 6 to 11.6 cc/min, and oxygen content of the collected blood decreased. The question was whether this finding was proof of a retrograde capillary perfusion from the coronary sinus, or could there be retroflow via veno-arterial shunts? That was not at all certain, but the authors did seem to be able to show that the venous blood accumulating in the peripheral stump of the left circumflex coronary artery originated in the coronary sinus. Great cardiac vein occlusion resulted in a much lesser increase in peripheral left circumflex backflow and pressure. One was left to wonder what the actual circulatory situation is in the absence of backflow from the peripheral left circumflex coronary artery.

With regard to drainage pathways from the occluded coronary arteries, hypotheses addressed both the retrograde supply and backflow data. In his study, Eckstein used an aorta-to-coronary sinus shunt, for purposes of coronary vein arterialization, with concommittant pressure vein elevation. This produced a greater peripheral effect distal to the left circumflex coronary artery occlusion than was the case with corresponding coronary sinus pressure elevation alone. The difference was explained by the higher diastolic coronary venous pressure levels associated with aortic blood shunting. Nevertheless, within as little as 3 h, all these effects were found substantially blunted. Hemodynamic measurements indicated only mild changes. Elevation of coronary sinus pressure by itself in the normal heart caused a mild regional left ventricular ischemia, attributed to reduction in coronary inflow. Subsequently, marked myocardial ischemia supervened when the arterial inflow was very significantly lowered by coronary artery occlusion. However, the ischemia due to complete left circumflex coronary artery occlusion was found to be partially alleviated by the coronary sinus occlusion, and also by alternate intracoronary vein pressure elevation. The myocardial vascular bed became expanded. The amount of oxygen uptake by the myocardium due to coronary sinus arterialization was 14−25% of the normal state. While the effects were interesting and sometimes significant, Eckstein et al. concluded that, following coronary artery ligation, acute coronary sinus occlusion per se produced only limited benefits, which would be insufficient to reverse a severe ischemic regional dysfunction.

In reconsidering all the diverse and sometimes contradictory early coronary sinus occlusion studies, several questions need to be asked. Firstly, was the reported coronary sinus occlusion always at the ostium, encompassing all of its branches, or else occasionally "somewhere" within the coronary sinus or even in one of its communicating coronary veins? Secondly, to what extent were the effects of the coronary sinus occlusion on blood flow and pressures influenced by the variable, i.e., more or less pronounced, coronary veno-venous shunting? Third, was the coronary sinus occlusion brief, or of a duration associated with vascular engorgement or edema? Fourth, was the coronary artery stenosed or totally occluded, what was the site of the obstruction and what was the relative timing of coronary vein vs coronary artery occlusion? Fifth, how adequate were the measurements of the effects on the ischemia? The peripheral coronary artery-to-ambient bleed clearly depends on the distal artery pressure, and it alone does not provide direct interpretation of the microcirculatory perfusion. Similarly, interpretation of changes in oxygen saturation of blood distal to a coronary artery obstruction during retrograde intervention may be fraught with difficulties, since one often does not know the source, mode, and extent of retrogradely induced circulation in the ischemic zone. One must also certainly differentiate the effects of coronary sinus occlusion experienced in a noncontracting heart vs effects in a beating heart, where systolic squeezing of myocardial and elevated diastolic wall stress can further restrict the retrograde coronary venous delivery to the ischemic zone capillaries.

The diverse early investigations led to controversial conclusions. It is not at all clear how to quantitate the extent and influence of a pressure-induced enlargement of the coronary venous blood volume, and potential deleterious effects appeared to outweigh the few observed favorable indications. In spite of this, some interest persisted for obvious reasons. If clinical non-phased coronary venous obstruction alone could prove beneficial in a particular setting of acute myocardial ischemia, simplicity of the method alone would lead to its reconsideration. In terms of current interventions, for example, should a several minute non-phased coronary sinus occlusion be safe and efficacious during one to several minutes of coronary artery occlusion, it might qualify as a potential support (e.g., during PTCA balloon inflation). While very short coronary sinus occlusions could well be safe, persisting coronary vein occlusion (i.e., beyond a few minutes to several hours) cannot be considered safe, and their use alone for clinical treatment may be counterproductive.

**Further studies and interpretation of coronary vein occlusion**

In the mid-1960s, Gensini and DiGiorgi [9, 10] became very interested in retrograde methods. They performed not only a series of retrograde measurements, but also provided direct contrast venography observations of the coronary venous and veno-venous shunt system, in closed-chest animals and also in the human. Occluded coronary vein pressure measurements (e.g., Fig. 1) elucidated the retrograde anatomic pathways. They noted that when the coronary sinus was obstructed at its ostium, only the anterior cardiac vein, and to some extent the elusive Thebesian vessels, were available to accommodate the entire coronary venous outflow.

Muers and Sleight [11] observed and investigated principally the existence of reflexive effects of coronary sinus occlusion. There was some conjecture about potential stimulation of pressure receptors located within the walls of the coronary sinus. These authors, too, pointed out the marked distension of coronary veins whenever the coronary sinus pressure was appreciably increased, and they also pointed to lymphatic changes. Enzymatically assessed myocardial pathology after

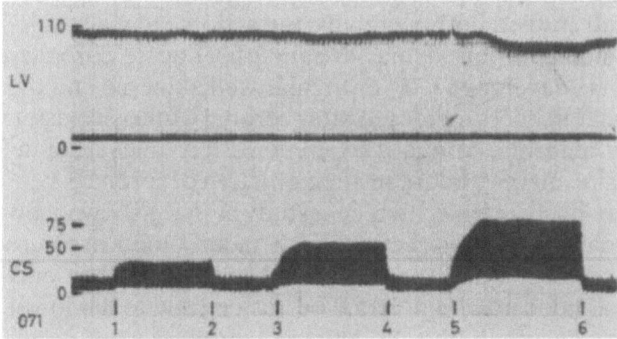

**Fig. 1.** Changes in the level of the coronary venous occluded pressure according to the level of the coronary venous obstruction. LV: left ventricular pressure; CS: coronary venous occluded pressure; 1–2: occlusion at anterior interventricular vein; 3–4: occlusion at left marginalvein; 5–6: occlusion at posterior interventricular vein. (By permission, from: DiGiorgi S, Gencini GG (1965) The coronary venous pressure: its morphology, origin and use as an expression of phasic coronary flow. Cardiologia 46:337–347.)

cardiac venous and lymphatic obstruction was studied in dogs by Pick et al. [12]. They concluded that interference with the outflow of both venous blood and lymph during coronary inflow can indeed lead to significant ischemic myocardial pathology.

Since the 1970s, many experimental and modeling investigations addressed the previously mentioned fundamental issues of coronary flow-pressure relations (without or with coronary sinus occlusion), in normal states, during pharmacologic interventions, and in the presence of ischemia. Bellamy et al. [13] found, for example, that changes in the coronary pressure-flow relation occurred within seconds of a coronary sinus occlusion, raising the coronary vascular critical closing pressure. Rouleau and White investigated more recently [14] the effects of coronary sinus pressure elevation on coronary blood flow distribution in dogs. While Rouleau's results agree with those of Bellamy, their coronary pressure data placed in question the simple single "waterfall" model, in favor of a more complex regional mechanism. As coronary sinus pressure was successively increased, the subendo-cardial-to-subepicardial blood flow ratio increased, primarily due to a reduction of the subepicardial flow with decreasing arterial input-to-coronary venous pressure differentials. At a relatively high coronary sinus pressure (e.g., 84/22 mmHg), the measured circumflex artery blood flow was found to be significantly reduced during both systole and diastole. The effects of coronary sinus occlusion, maximal vasodila-tation coronary flow, resistance, and Pzf were recently investigated in swine during LAD pressure variation [15]. At an LAD pressure of 30, 40, and 50 mmHg, coro-nary sinus occlusion reduced blood flow from 53 ml/min to 24 ml/min, from 79 to 49 ml/min, and from 105 to 74 ml/min, respectively. The mean LAD pressure at which flow stopped (Pzf) was increased by coronary sinus occlusion from $10 \pm 2$ to $20 \pm 4$ mmHg, while pressure-flow slopes (resistance) remained unchanged. The authors implicated primarily the induced extravascular pressure affecting the venous system.

Studies in the early 1980s applied modern instrumentation in an attempt to characterize myocardial injury during experimental ligation of coronary veins. Thus, Jang et al. [16] ligated the coronary sinus in open-chest goats in the absence of coronary artery occlusion and studied both regional contraction and wall thick-ness, the latter providing an indication of edema. Within 2 h of acute coronary sinus occlusion, 86% of the goats developed left ventricular wall akinesis, and the rest had hypokinesis. Mean wall thickness increased after coronary sinus occlusion by 82%. This will indicate that in normal coronary artery perfusion zones substan-tial edema can be expected as a result of retrograde augmentation of coronary ven-ous blood pressures [17]. A previously referred to recent study in sheep showed few if any effects of partial coronary sinus obstruction, with mean blood pressure increased from $2 \pm 1$ mmHg to $19 \pm 1$ and then to $31 \pm 1$ mmHg [1]. The only change noted was an increased endocardial/epicardial perfusion ratio at the lower level of coronary sinus pressure augmentation. In contrast, coronary sinus occlu-sion in closed-chest dogs resulted in significant ECG-ST segment elevations (simi-lar to changes observed with coronary artery occlusion) and caused hemorrhagic myocardial necrosis, primarily in the epicardial layer [17]. Figure 2 indicates the observed changes in serum enzymes in the closed chest dogs, while Fig. 3 shows changes in left ventricular pressure, pulmonary artery pressure, and LAD coronary

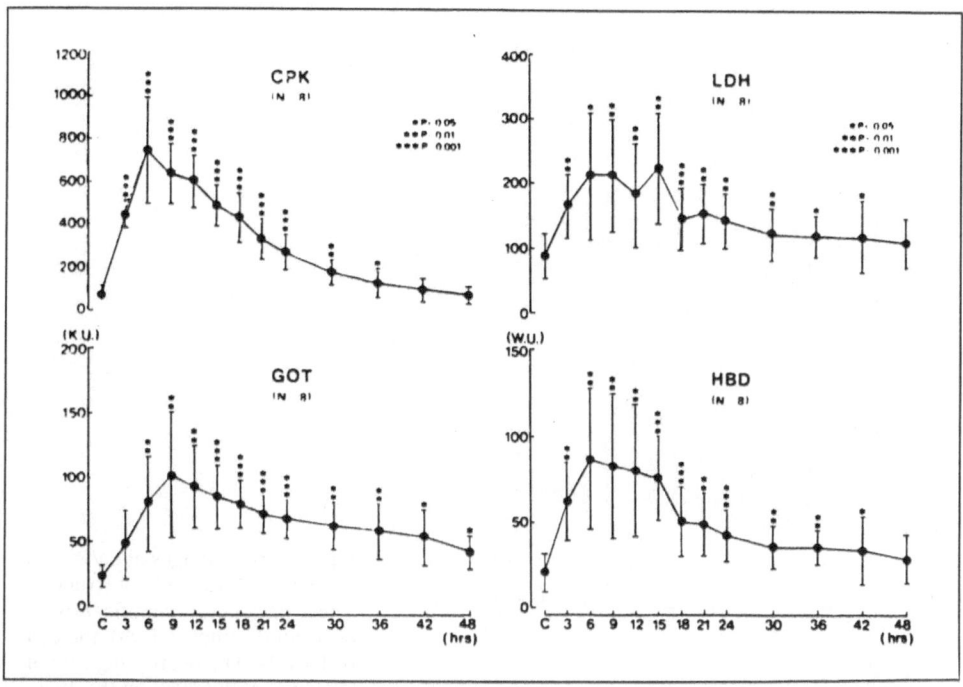

**Fig. 2.** Serial changes of serum enzymes after the CS obstruction. The activities of all enzymes significantly increased compared with the control levels, and reached their peaks at 6 to 9 hours after the CS obstruction. CS: coronary sinus. (By permission, from: Miguhara K et al. (1988) Experimental study of acute coronary sinus thrombosis: clinical references to coronary sinus thrombosis and coronary venography. Jap Circ I 52:44−62.)

blood flow obtained in another open-chest series. Another recent article reports on a study of ventricular fibrillation thresholds in anesthetized dogs following coronary sinus obstruction [18]. The interesting and perhaps revealing observation was that up to a developed coronary sinus pressure of about 40 mmHg, there was no decrease in coronary blood flow (compensatory coronary vasodilatation?) and ventricular fibrillation threshold actually increased, whereas at higher coronary sinus pressures (e.g., 60 mmHg) both coronary blood flow and the ventricular fibrillation threshold decreased markedly. Thus, apart from the factor of coronary sinus occlusion duration, lower coronary venous blood pressures might produce benefits, e.g. through redistribution and vasodilatation. At higher coronary sinus pressure levels one should be concerned about deleterious effects, possibly associated with congestion due to significantly increased intravascular, extracellular or intracellular fluid volumes.

### Applications of non-phased coronary sinus occlusions

One specific proposal to apply persistent coronary sinus occlusion concerns retrograde cardioplegic support in the presence of significantly obstructed coronary

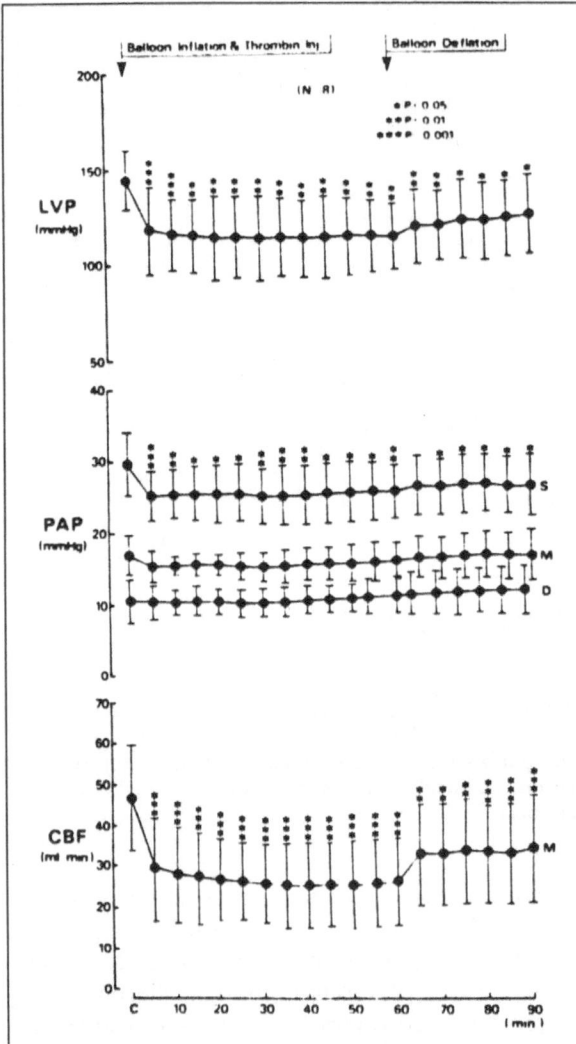

**Fig. 3.** Serial changes of LVP, PAP and CBF after the CS obstruction. These parameters significantly decreased compared with the control levels. The decreasing rate of CBF was about 40% on the average. The rebound of the blood flow was observed immediately after the balloon deflation. The same phenomena were observed both in the LVP and PAP. CS: coronary sinus; LVP: left ventricular pressure; PAP: pulmonary arterial pressure; CBF: coronary arterial blood flow. (By permission, from: Migahara K et al. (1988) Experimental study of acute coronary sinus thrombosis: clinical references to coronary sinus thrombosis and coronary venography. Jap Circ J 52:33–62.)

arteries. Schwarke et al. [19] studied myocardial metabolism of ischemic and normal zones during coronary sinus occlusion in dogs. The LAD was ligated and the coronary system was first antegradely perfused by a cardioplegic solution. It was evident that the zone distal to the LAD occlusion was insufficiently perfused and poorly cooled antegradely. When coronary sinus pressure was then elevated to 40 mmHg, the myocardial temperature fell in both occluded and nonoccluded zones, and the blood flow in the tissue distal to the LAD occlusion increased by more than 100%. Intracoronary sodium concentration measurements indicated that the coronary capillary bed perfusion distal of the LAD occlusion had improved. Conversely, measurements of pH, sodium and potassium metabolisms, all revealed that absence of the retrogradely induced coronary sinus pressure elevation was associated with marked ischemia in the region beyond the obstructed LAD coronary artery.

A diagnostic application of coronary sinus occlusion should also be mentioned here. Jacobs and Faxon [20, 21] studied coronary sinus occlusions in dogs, as well as in the catheterization laboratory, from a view point of possibly correlating coronary vein pressure vs left ventricular performance. Previous experimental and catheterization work had indicated a similarity between occluded coronary venous pressure patterns and those measured in the left ventricle. Jacobs [20] found in the dog a 30% reduction in circumflex coronary artery inflow due to coronary sinus occlusion, while release led to a 73% hyperemic artery response above baseline. In both control states and adenosine vasodilated states, occluded coronary sinus pressure correlated with left ventricular systolic blood pressure. The rate of rise of coronary sinus occlusion pressure was directly proportional to the preocclusion sinus flow. The magnitude of diastolic coronary sinus occlusion pressure was in between the aortic and left ventricular diastolic pressure levels.

Faxon [21] studied 27 patients during routine cardiac catheterization. A 7F catheter balloon was placed and inflated in the coronary sinus. Left ventricular end-diastolic pressure was not significantly different from end-diastolic sinus occlusion pressure. In contrast, systolic coronary sinus occlusion pressure was significantly lower than left ventricular systolic pressure, and was unrelated to right-sided heart pressures. Thus, in contrast with results in the animal, human end-diastolic coronary sinus occlusion pressure closely paralleled left ventricular end-diastolic pressure, which is of obvious potential clinical interest, but such a correlation did not apply in systole. The study also suggested the existence of hemodynamically important Thebesian vessel connections that may facilitate direct drainage of a significant amount of blood from the coronary vessels into the cardiac chambers.

## Conclusion

In dealing with non-phased coronary venous occlusions, one should beware of deleterious effects (e.g., reduced inflow and signs of ischemia) in "normally" perfused myocardium. Except for very brief occlusions (perhaps at most, several minutes), hazardous vascular injury and myocardial edema may be encountered when coronary sinus pressure is very high (say, beyond 50 mmHg). Considering currently available improved measurement techniques, it would seem reasonable to reexamine the effectiveness and safety of a short period of moderate pressure retroinfusion into an occluded coronary sinus, particularly when the duration of the concommittant coronary arterial occlusion is of the order of one or at most a few minutes. The various mechanisms anticipated in such a setting may include a retrogradely forced shunting of circulation from the nonoccluded myocardial zone into the ischemic region, and a significantly potentiated washout via Thebesian vessels. Longer term (persisting) coronary sinus occlusion is to be avoided in the beating heart, even though it appears applicable in the arrested heart during antegrade or low pressure retrograde cardioplegic/hypothermic infusion, in the presence of coronary artery obstructions.

# References

1. Ward KE, Fisher DJ, Michael L (1988) Elevated coronary sinus pressure does not alter myocardial blood flow or left ventricular contractile function in mature sheep. J Thorac Cardiovasc Surg 95:511−515
2. Hammond GL, Davies AL, Austen WG Jr (1967) Retrograde coronary sinus perfusion: a method of myocardia protection in the dog during left coronary artery occlusion. Ann Surg 39:166
3. Thatcher C, Serra FB, Lajos TZ, Montes M, Siegel JH (1979) Coronary venous hypertension: a potentiator of myocardial ischemic injury. J Surg Res 26:45−57
4. Gross L, Blum L, Silverman G (1937) Experimental attempts to increase the blood supply to the dog's heart by means of coronary sinus occlusion. J Exp Med 65:91
5. Ungerleider H, Kerkof A, Fahr D (1937) Venous pressure as a factor in determining collateral circulation in the heart. Proc Exp Biol Med 34:703−704
6. Gregg DE, DeWald D (1938) Immediate effects of coronary sinus ligation on dynamics of coronary circulation. Proc Soc Exp Biol Bed 39:202
7. Thornton JJ, Greg DE (1939) Effect of chronic cardiac venous occlusion on coronary arterial and cardiac venous hemodynamics. Am J Physiol 128:179
8. Eckstein RW, Hornberger JC, Sano T (1953) Acute effects of elevation of coronary sinus pressure. Circulation 7:422
9. Gencini GG, DiGiorgi S, Murad-Netto S (1963) Coronary venous occluded pressure. Arch Surg 86:72−80
10. DiGiorgi S, Gencini GG (1965) The coronary venous pressure: its morphology, origin and use as an expression of phasic coronary flow. Cardiologia 46:337−347
11. Muers MF, Sleight P (1972) The reflex cardiovascular depression caused by occlusion of the coronary sinus in the dog. J Physiol 221:259−282
12. Pick R, Miller AJ, Glick G (1974) Myocardial pathology after cardiac venous and lymph flow obstruction in the dog. Am Heart J 87(5):627−632
13. Bellamy RF, Lowensohn SH, Ehrlich W, Baer WR (1980) Effects of coronary sinus occlusion on coronary pressure flow relations. Am J Physiol 239:H57−H64
14. Rouleau JR, White M (1985) Effects of coronary sinus pressure elevation on coronary blood flow distribution in dogs with normal preload. Can J Physiol Pharmacol 63:787−797
15. Pantely GA, Bristow JD, Ladley HD, Anselone CG (1988) Effect of coronary sinus occlusion on coronary flow, resistance and zero flow pressure during maximum vasodilatation in swine. Cardiovasc Res 22:79−86
16. Jang GC, Bansal R, Mitchell WA, Grube G (1982) Characterization of myocardial injury by sector-scan in acute experimental ligation of coronar veins. Invest Radiol [Abstracts] 5:401
17. Miyahara K, Satoh F, Sakamoto H (1988) Experimental study of acute coronary sinus thrombosis: clinical references to coronary sinus thrombosis and coronary venography. Jap Circ J 52:44−62
18. Kralios AC, Nappi JN, Tsagaris TJ, Kralios FA, Kuida H (1988) Paradoxical increase of ventricular fibrillation threshold in response to coronary sinus obstruction. Am Heart J 115:334−340
19. Schwarke K, Meyer ED, Picht J, Pill P, Saggau S, Schulz B, Spath J (1985) Ischemic myocardial metabolism with and without increased coronary sinus pressure. An experimental study. Z Kardiol [Abstract] 39:95 (12th Ann. Mtg in Bad Bauheim)
20. Jacobs AK, Faxon DP, Apstein CS, Coats WD, Gattsman SB, Ryan TJ (1984) The hemodynamic consequences of coronary sinus occlusion. In: Mohl W, Wolner E, Golgar D (eds) The coronary sinus. Steinkopff, Darmstadt, pp 430−437
21. Faxon DP, Jacobs AK, Kellett MA, McSweeney SM, Coats WD, Ryan TJ (1985) Coronary occlusion pressure and its relation to intracardiac pressure. Am J Cardiol 56:457

# Pressure-controlled intermittent coronary sinus occlusion (PICSO)

W. Mohl and F. Neumann

## The concept

Based on experience with Beck's operation [1] it became evident that permanent occlusion of the coronary sinus is detrimental to the myocardium. In 1976, when J. Moll, a cardiac surgeon from Poland, gave a lecture at the Second Surgical Clinic of the University of Vienna on the application of the Beck II procedure, he reported on definitive improvements in some of the patients. Although the audience remained sceptical, certain obvious discrepancies stimulated one of the authors (WM) to initiate further research in this field. This research was based on an impression that the reported beneficial effects (e.g., as pain relief) were primarily due to retrogradely altered microcirculatory pressure and flow dynamics, rather than being caused by the active retroperfusion or arterialization of the coronary veins. Judging from the research developments during the following decade, this hypothesis still seems viable and remains the rationale for clinical application of PICSO for treatment of myocardial jeopardy.

## The pathophysiology of PICSO

As can be seen from Fig. 1, PICSO induces a cyclic variation of the venous outflow impedance, which significantly modifies the phasic coronary sinus pressures. During the occlusion of the coronary sinus (CS), the venous pressure increases gradually until the systolic pressure peaks reach a plateau. Subsequently, when the CS

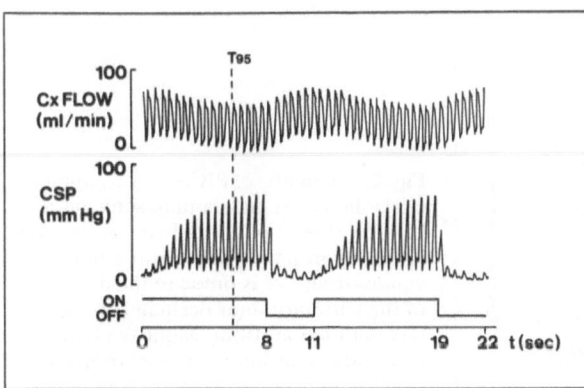

Fig. 1. Coronary sinus pressure (CSP) and coronary flow in the CX artery during PICSO. Note how the cyclic pressure variation is reflected in arterial inflow. T95 = 95% rise time of systolic coronary sinus occluded pressure.

145

occlusion is released, venous blood drains freely into the right atrium. In accordance with this sequential pattern, there occurs a forced redistribution of coronary venous blood toward underperfused myocardial zones during the coronary sinus occlusion (accompanied by an exchange of substrates) followed by an active washout of accumulated metabolites. The redistributive effect of PICSO has been impressively demonstrated by Papp et al. [27] by means of infrared thermography of regionally ischemic myocardium. Furthermore, blood density measurements performed by our group [23] have shown a cyclic variation of venous hematocrit, paralleling the occlusion/release pattern of PICSO, and indicating redistribution of venous blood as well as plasma skimming into the microcirculation, followed by enhanced washout during the CS release. Similar conclusions were reached by Ciuffo et al. [3] and Jacobs et al. [11], both of whom showed (by means of Xenon-133 injections) that PICSO effectively increases myocardial perfusion in the ischemic zone and enhances washout of cellular edema.

Neely et al. [26] and Murry et al. [24] found a relationship between cell integrity, viability, functional recovery, and catabolite accumulation during ischemia and reperfusion. This fundamental physiologic finding may serve as an explanation of the beneficial effects observed with the PICSO technique. The residual oxygen tension in the coronary venous circulation may also contribute to the observed improvement in myocardial function. In early experiments, our group [17] found an increased oxygen saturation of the myoglobin molecule in the center of the ischemic area during CS occlusion. We therefore speculated that redistribution of coronary venous flow changed the perfusion distances between the cells in the microcirculation and myocardial cells, thereby improving myocardial metabolism.

## The PICSO technique

Since our concept of PICSO is based on the assumption that safety and effectiveness can only be achieved by physiologically adjusting the periodic CS occlusions to the dynamics of CS pressure (see Fig. 2), we have developed a computer-aided control system for occlusion/release adjustment. Using a double-lumen balloon catheter connected to a pneumatic pump device, the coronary sinus is periodically

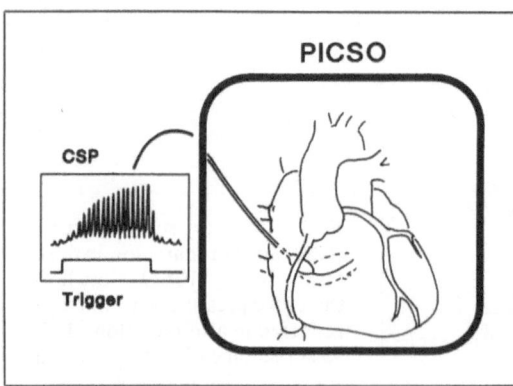

**Fig. 2.** Schematic of PICSO. A computer-controlled pneumatic pump activates a balloon at the tip of a catheter positioned into the coronary sinus. Obstruction of venous drainage is timed to the dynamics of the coronary sinus occluded pressure. (By permission, from: Mohl W (1987) Coronary sinus interventions: from concept to clinics. J Cardiac Surg 2:467–493.)

occluded, while CS pressure is continuously monitored and thus available for subsequent A/D conversion, on-line data analysis, and occlusion/release phasing control. A mathematical function is fitted as an envelope to systolic CS pressure peaks, from which one calculates quantities for both analysis of the pressure dynamics (e.g., plateau or slope of the fitted curve), and for feedback control (see also chapter XI of this book). Presently, the estimated time required for the CS occluded pressure to reach 90% of its (calculated) plateau (= 90% rise time) is used for PICSO's phasic occlusion regulation. Release time regulation is based on heart rate, allowing arterial inflow to reach its maximum during the CS release phase, just before the coronary sinus is once again occluded. When the calculated values for occlusion and release times exceed certain limits (i.e., when the algorithm fails because of artifacts, missing data, etc.), physiologically reasonable default values for pump control are used instead.

## Experimental series and clinical trials

In the following, we will review and briefly discuss a number of experimental and clinical studies, showing the recent developments and the present status of research in intermittent coronary sinus occlusion (ICSO) and pressure-controlled ICSO (= PICSO). We will describe investigations of the effect of (P)ICSO on infarct size development, on myocardial left ventricular function, and on myocardial metabolism, under a variety of experimental and clinical conditions. With a few exceptions, the studies on infarct size limitation have convincingly demonstrated the potential of (P)ICSO to increase myocardial salvage during coronary artery occlusion, as well as during the reperfusion period. Although the studies on ventricular function have been performed under various conditions and using different methods of left ventricular function assessment (e.g., regional or global ischemia; animal preparations in dogs or pigs; human trials) most of them have shown favorable effects of (P)ICSO on myocardial function during ischemic episodes. To the extent that myocardial metabolism has been investigated, the results seem to support the hypothesis of significantly enhanced washout of metabolic substrates and subsequent reduction of ischemic injury. In none of the studies were negative effects of (P)ICSO on global hemodynamics reported. Clinical studies (serial trials as well as particular case studies) demonstrated that (P)ICSO is a safe technique, and a feasible method of myocardial protection.

## A. Experimental studies

### A.1. Effects of (P)ICSO on infarct size and myocardial salvage

In a first experimental realization of the PICSO concept, i.e., using coronary sinus pressure readings as a guide for occlusion and release timing of the coronary sinus blockade, Mohl et al. [18] were able to demonstrate that the amount of necrosis could be significantly reduced when PICSO was applied during evolving myocardial infarction (c.f. Table 1). The left anterior descending (LAD) coronary artery

147

**Table 1.** Infarct size after (P)ICSO treatment in LAD occlusion and reperfusion studies

| | Infarct size %<br>(P)ICSO (vs control) | | p |
|---|---|---|---|
| Mohl et al. [18] | 56 ± 7 | (99 ± 3) | < 0.001 |
| Zalewski et al. [33] | 84 ± 5 | (100 ± 5) | n.s. |
| Ciuffo et al. I [3] | 15 ± 6 | (57 ± 8) | < 0.001 |
| Ciuffo et al. II [3] | 19 ± 7 | (49 ± 9) | < 0.02 |
| Jacobs et al. [9] | 17 ± 4 | (33 ± 4) | < 0.025 |
| Jacobs et al. [10] | 29 ± 6 | (45 ± 2) | < 0.02 |
| Diltz et al. [4][a] | 30 ± 7 | (33 ± 3) | n.s. |
| Guerci et al. [7] | 30 ± 8 | (75 ± 4) | < 0.01 |

Infarct size is expressed as percentage of myocardium at risk. Values are means and SEM.
Note that area at risk was determined using different techniques in the studies cited above: For in vivo delineation of the myocardium at risk of necrosis, labeled microspheres were injected as bolus into the left atrium before any therapeutic intervention. Post mortem identification of the hypoperfused areas is performed either autoradiographically or after atrial injection of monastral blue dye.
[a] SEM is a pooled estimate from the original data of risk mass and infarct mass.

was permanently ligated in 31 dogs. In the treatment group (n = 16) PICSO was initiated 15 min after LAD occlusion and continued for 6 h up until the end of the experiment, employing mean CS occlusion times of 15 ± 3 s and release times of 4 ± 2 s. With the myocardial area at risk found similar in both groups by radio-nuclide microsphere measurements shortly after LAD ligation, post mortem infarct size was significantly smaller in the PICSO treated group than in the control group. We found the salvage occurring predominantly in the subepicardial region which is supposed to be affected prior to endocardial layers when any retrograde intervention is applied. In these experiments we also found indications of a PICSO-induced redistribution mechanism. When pressure measurements from the occluded artery were observed, distal LAD pressure readings resembled the pattern of the CS pressure induced by the periodic outflow obstruction of PICSO.

In an extensive study of the efficacy of differing coronary sinus interventions (without or with retrograde infusion and arterialization), Zalewski et al. [33] found no reduction of infarct size in the ICSO treated dogs, as compared with LAD occlusion controls (Table 1). Intermittent occlusion of the great cardiac vein (GCV) (without arterialization) was applied for 6 h starting at 3 min after complete LAD occlusion in nine dogs. Constant occlusion/release timing of 15/4 s was used to alternately cause the great cardiac vein occluded pressure to reach a "safe" plateau, and then to restore the great cardiac vein pressure to low baseline values. Since CS pressure was not continuously monitored and evaluated in this study, the GCV occlusion/release phasing might have been inefficiently applied. The great cardiac vein may not have been totally occluded, and the systolic occluded pressure not sufficiently elevated. Since washout and, hence, the efficacy of PICSO is related to mean systolic occluded coronary venous pressure (see Fig. 3), insufficient occlusion might have contributed to the negative findings of the study.

In 1984, Ciuffo et al. [3] corroborated the positive results of early studies (i.e., significant infarct size limitation, [6, 31], using ICSO during 3 h coronary artery occlusion in 15 dogs (vs 13 control dogs), starting at 30 min after LAD ligation.

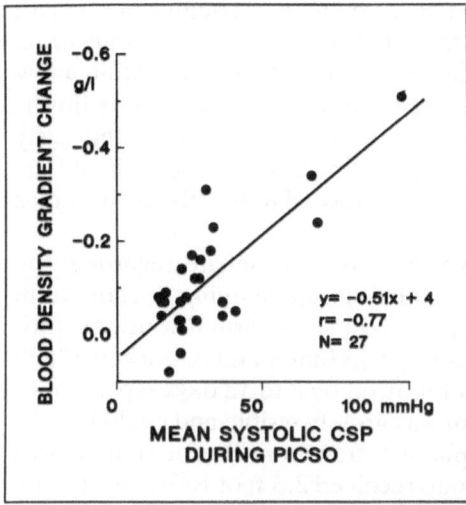

Fig. 3. Relationship between washout, measured as the arterio-venous density gradient change, and coronary sinus pressure. (By permission, from: Mohl W (1987) Coronary sinus interventions: from concept to clinic. J Cardiac Surg 2:467−493.)

Constant occlusion/release cycling was set to 15/7 s. As can be seen from Table 1 (Ciuffo I), the percentage of infarct size after 3 h of coronary artery occlusion in both groups is considerably less than after a 6-h-period of LAD infarction, the infarct limiting effect of ICSO, however, is equally well pronounced. Ciuffo also investigated the effects of ICSO (applied as above) after 2 to 4 h reperfusion following 3 h of ischemia in 13 dogs (vs nine controls). At the end of the reperfusion period the infarct/risk ratio in the intervention group was significantly smaller than in the control group (Table 1, Ciuffo II).

Similar results were obtained in an occlusion/reperfusion model of ischemia by Jacobs et al. [9] in 1985, applying PICSO during infarction and reperfusion. A 3-h-period of LAD occlusion was followed by 3 h of reperfusion. In 11 dogs randomly assigned to the treatment group, with PICSO initiated 30 min after the LAD had been completely ligated, a significantly smaller infarct/risk ratio was observed than in the control group (11 dogs receiving reperfusion without PICSO), c.f. Table 1.

In order to investigate the potential of PICSO specifically during reperfusion of infarcted myocardium, Jacobs et al. [10] designed a study demonstrating that PICSO is able to potentiate the salvage of ischemic myocardium by enhancing the washout of cellular edema. Out of a total of 24 dogs receiving 3 h LAD occlusion followed by 3 h of reperfusion, 12 dogs were subjected to the PICSO treatment during reperfusion, starting just prior to release of the LAD ligation. From their significant results (see Table 1) the authors concluded that PICSO may reduce or prevent at least part of the damage attributable to the reperfusion process.

In a 4-h-occlusion/6-h-reperfusion model of ischemia applied in 14 dogs, Diltz et al. [4] were not able to find any beneficial effect of ICSO on infarct mass. In the treatment group, ICSO was administered with a constant occlusion/release cycle of 30/30 s during LAD occlusion. There was no difference between ICSO-treated and untreated animals (c.f. Table 1), indicating that the effect of recannalization could not be further enhanced by ICSO. Considering the particular occlusion/release cycle of ICSO, these results are self-explaining. Prolonged CS occlusion, i.e.,

149

longer than it takes the systolic CS values to reach a plateau, is counterproductive, whereas 30 s of CS release is much longer than it takes the coronary sinus to be emptied from venous blood. Complete drainage is assumed to occur within a few seconds after CS release, before the hyperemic response of coronary artery flow is observed (see Fig. 1). Enhanced washout is accomplished only during this early phase of rapid emptying of the coronary sinus [23]. From these results the difference between time dependent ICSO and physiologically adaptive PICSO becomes evident.

Not only the negative results cited above, but also the debate regarding the assessment of infarct size by triphenyltetrazolium cloride staining in short-term studies (less than several days) gave rise to persistent skepticism toward the anti-ischemic efficacy of ICSO. Therefore, in a study by Guerci and coworkers [7] 22 dogs were subjected to 3 h of LAD ligation followed by 8 to 12 days reperfusion, after which period old infarcted myocardium was grossly visible and could be reliably delineated from viable tissue when triphenyltetrazolium cloride staining was used. Dogs randomized to the treatment group received 2.5 h of ICSO, beginning at 30 min after coronary artery ligation and performed at a constant occlusion/release cycling of 10/5 s. Again, the effect of ICSO on infarct size limitation could be shown clearly (Table 1), although it has to be mentioned that all ICSO treated dogs had extensive thrombosis of the coronary sinus.

## A.2. Effects of (P)ICSO on myocardial function

Of the studies to be discussed in the following, in those by Heimisch et al. [8] and Toggart et al. [32] sonomicrometry in open-chest models was used to establish the effects of ICSO or PICSO on myocardial function. Simon and Jacobs [29] and Lazar et al. [15, 16] have used echocardiography in open-chest models, whereas, Mohl et al. [19] have used both methods in open- and closed-chest dogs, respectively.

In this latter study, Mohl et al. used sonomicrometry in six dogs to assess the effect of PICSO on ischemic segment shortening after severe stenosis (i.e., reduction of LAD flow to $7.6 \pm 7.6\%$ of baseline flow) and after complete LAD occlusion. Two pairs of tubular piezoelectric transducers were implemented subendocardially for continuous measurement of regional dimensions in the ischemic (LAD) and non-ischemic (circumflex artery, CX) zone, respectively. Measurements of percent segment shortening were obtained at baseline ($14.0 \pm 2.4\%$), after stenosing the LAD ($5.5 \pm 1.2\%$), and after its complete occlusion ($-0.1 \pm 2.1\%$) and at the end of a $10 \pm 3$ min period of PICSO application during LAD stenosis and occlusion, respectively ($8.9 \pm 2.6\%$ with LAD stenosis, and $2.3 \pm 1.2\%$ with LAD occlusion). Measurements were repeated $5 \pm 2$ min after discontinuing PICSO. In both settings (stenosis and occlusion) an increase of segment shortening during the PICSO treatment could be observed in the ischemic region, although statistical significance was not reached. The improvement persisted during 5 min after termination of PICSO.

In the same study, the authors performed standardized 2D-echocardiographic measurements of left ventricular function in 20 dogs at baseline and at 30, 90, and

150

180 min after LAD occlusion. Sectional as well as segmental (using eight segments of short-axis cross-sectional views) systolic fractional area change (FAC) were employed as indexes of contractile function. Based on the open-chest experiments, PICSO cycling in the study group was performed with an occlusion/release ratio of $9.9 \pm 1.5/1.8 \pm 0.3$ s throughout a 2.5 h period of LAD coronary artery occlusion, beginning at 30 min after LAD ligation. Global ventricular function remained relatively stable (around 25% FAC) during LAD occlusion in the PICSO treated group, whereas in the control group it deteriorated from $16.3 \pm 2.7\%$ at 30 min LAD occlusion to $10.0 \pm 3.3\%$ after 3 h of occlusion. Also, regional FAC of ischemic segments decreased in the control group from $12.6 \pm 6.1$ to $4.1 \pm 6.9\%$ during LAD occlusion, whereas in the PICSO treated dogs regional FAC increased from $-0.4 \pm 10.1\%$ at the beginning to $14.4 \pm 4.4\%$ at the end of the LAD occlusion period. A general, sometimes even significant decrease of contractile function was observed in remote (non-ischemic) segments in all dogs. While the assessment of regional wall motion using sonomicrometry allowed determination of the treatment effect directly in the center of the ischemic zone; by means of echocardiography the overall effects on the total myocardial area could be evaluated. In both cases a beneficial effect of PICSO was established.

In a similar setting, Heimisch et al. [8] used sonomicrometry in seven open-chest dogs to investigate the effects of ICSO in normally perfused myocardium as well as during critical LAD stenosis and occlusion. Additionally, two disc-shaped transducers were implanted for measuring ventricular wall thickness of the ischemic LAD area. ICSO, with an average period of balloon occlusion of $25 \pm 3$ s and release of $5 \pm 2$ s was performed for 5−10 min under control conditions, until recordings were stabilized for data collection. Subsequently, the LAD was narrowed until the amplitude of the ischemic segment shortening was diminished by approximately 60%. After stabilization, ICSO was performed for 5−10 min as previously, upon which the LAD was released. The same procedure was repeated with LAD occlusion. ICSO application in the non-impaired myocardium led to a moderate depression of ventricular function which expressed itself in a $4.1 \pm 0.9\%$ increase of wall thickness and a considerable (14% and 19% in the LAD and CX segments, respectively) reduction of systolic shortening. The resulting decrease of stroke volume by 12% and a concomitant increase in heart rate by 6% led to a reduction of cardiac output to $6.2 \pm 2.1\%$. Applying ICSO during LAD stenosis could significantly improve LAD segmental shortening fraction from ischemic $5.6 \pm 1.4\%$ to treated $9.3 \pm 2.9\%$, reaching 66% of the prestenosis value. Similarly favorable results of ICSO application were obtained during LAD occlusion with a significant increase of LAD segmental shortening fraction from ischemic $0.9 \pm 1.7\%$ to treated $2.6 \pm 1.1\%$, restoring 18.4% of the control value.

Toggart et al. [32] studied the efficacy of ICSO in preserving global and regional function during acute LAD occlusion in an experimental setting without significant arterial collateral vessels using 17 swine heart preparations undergoing extracorporeal coronary perfusion. Global left ventricular function was assessed by shifts in left ventricular end-diastolic pressure (LVEDP) and left ventricular dp/dt. Regional function was evaluated by means of transmurally placed ultrasonic crystals. Measurements were obtained shortly before and after LAD occlusion and at 15 min intervals for 1 h. ICSO with a constant occlusion/release cycling of 15/5 s was started after randomization and performed throughout the whole period of LAD occlusion in the treatment group. There were no statistically significant differences between the ICSO treated animals and control animals in terms of global ventricular function. Ischemic bed wall thickening was not improved during ICSO. However, LVEDP tended to fall in the treatment group after it had initially risen in both groups upon LAD ligation, but remained elevated permanently in the control group. Also, peak dp/dt, after its initial fall upon LAD occlusion, tended to rise in the treatment group at 30 and 45 min after LAD occlusion, whereas it remained depressed in the control group. These results seem to be affected by a preparation in which arterial inflow was kept constant and, thus, leading to an overall engorgement of the circulation. As seen from Fig. 1, reduction of arterial inflow during CS occlusion is repaid during its release, a phenomenon which was suppressed in Toggart's study, thus leading to significantly increased diastolic CS occluded pressure readings, as compared to those usually observed.

In a 3-h-occlusion/3-h-reperfusion model, Simon and Jacobs [29] tried to determine the effect of PICSO on left ventricular function in 22 open-chest dogs. PICSO was initiated at 30 min after LAD occlusion and continued until the end of the reperfusion period in the treatment group. Non-PICSO control animals had reperfusion alone. Systolic wall thickening of ischemic segments as determined from 2D-echocardiograms of three different cross-sections (with eight segments each), was found to significantly improve in PICSO-treated dogs as compared to control animals, regaining about 50% and 25% of the pre-occlusion values, respectively.

Lazar et al. [15] undertook an experimental study in five pigs to determine the effects of PICSO on the normal myocardium as well as its effectiveness in altering perfusion damage after a period of ischemic arrest. All animals were placed on cardiopulmonary bypass (CPB) and control measurements of left ventricular stroke work index (LVSWI), end-diastolic pressure (LVEDP), and end-diastolic volume (LVEDV) were performed after 20 min. In normal hearts (five pigs), PICSO was then carried out for 1 h with an occlusion/release ratio of 10/4 s, and the measurements were repeated. In 14 pigs the aorta was crossclamped for 2 h and the LAD artery occluded. Upon release CPB was continued for 1 h more at the end of which post-ischemic measurements were performed. In the treatment group (n = 7) PICSO was applied during the period of ischemic arrest. No hemodynamic or global functional changes during PICSO were found in normal hearts. After ischemic arrest, neither of the two groups exhibited a significant change in the left ventricular pressure-volume relation (LVEDP vs LVEDV) from prior control measurements. LVSWI at given LVEDP value, however, was significantly higher in the PICSO group than in controls (e.g., $0.87 \pm 0.07$ vs $0.61 \pm 0.05$ g $\cdot$ m/kg at LVEDP = 10 mmHg; $p < 0.01$) and resembled pretreatment measurements. Ejection fraction as well returned to control levels ($50 \pm 2\%$) in the treated group, whereas in the untreated animals it did not ($33 \pm 6\%$; $p < 0.01$).

In a recent study, performed to investigate the potential of PICSO in supporting the distribution of antegrade cardioplegia, Lazar et al. [16] not only established the improvement of cardioplegic delivery by means of PICSO, but also found post-ischemic functional indices practically restored to baseline level compared to significantly lower values in control animals. Twenty pigs were prepared as previously. In the treatment group, PICSO was applied during the period of ischemic arrest. At post-ischemic measurements, at LVEDP values of 60 and 70 ml, stoke work index was significantly higher in the PICSO treated group and reached more than 80% (vs about 50% in untreated pigs) of the control measurements at the beginning of CPB. The post-ischemic pressure-volume curves (LVEDP vs LVEDV) were significantly different between the groups at LVEDP ranging from 10 to 12 mmHg, indicating a difference in left ventricular compliance (LVEDV = $67 \pm 6$ ml in the PICSO group vs $53 \pm 4$ ml in controls, at 12 mmHg LVEDP; $p < 0.02$). Moreover, within this LVEDP range, the pressure-volume curve of the PICSO treated group was similar to the curve obtained from control measurements before treatment.

## A.3. Effects of (P)ICSO on myocardial metabolism

In a number of studies, most of which were primarily performed to investigate the effect of (P)ICSO on myocardial function experimentally of clinically, it was also attempted to establish the influence of (P)ICSO on metabolic processes in the myocardium. Since the original design of these studies did not emphasize the metabolic aspect, results proved to be heterogeneous and difficult to interpret.

In the experimental ICSO study by Toggart et al. ([32], see above), oxygen saturation measurements were obtained from the left main coronary artery, from the LAD artery distal to the ligature, and from the coronary sinus. Main coronary arterial oxygen saturation remained unchanged (at about 100%) during LAD occlusion and was similar in both groups. Oxygen-saturation in the coronary sinus as well did not change during LAD occlusion, but tended to be lower in treated animals than in controls (roughly between 30% and 45%), with statistical significance reached immediately after LAD ligation and at 45 min postocclusion. Distal LAD oxygen saturation in both groups fell from about 100% to approximately 30% upon LAD ligation. A further decrease (to less than 20%) which reflects improving oxygen utilization was observed in the treatment group, as compared to a slight increase, i.e., a compensation to more than 30% in the control group. The between-group differences reached significance at 30, 45, and 60 min after LAD ligation.

In the above mentioned study of Lazar et al. [16], myocardial tissue pH was measured in the distribution of the distal LAD artery and the CX artery at baseline as well as during ischemic arrest and reperfusion. pH values were standardized according to LV temperature, and the pH measurements were similar (6.8−6.9) in both groups at baseline. The decrease of tissue pH values during ischemia was significantly less pronounced in the PICSO treated group than in control animals ($6.58 \pm 0.09$, PICSO, vs $6.31 \pm 0.09$, controls; $p < 0.05$). This trend continued during reperfusion, with the pH values in the treated group being slightly above baseline at 60 min of reperfusion, and in the control group remaining below control measurements. As it is well known that high pH values correspond to high ATP levels, depressed pH values can be interpreted as evidence of stunned myocardium.

In the earlier study [15] using the same protocol, Lazar et al. obtained similarly favorable results regarding the effect of PICSO on myocardial metabolism. At the end of the reperfusion period after 1 h PICSO treatment following 2 h ischemic arrest, treated hearts had significantly higher pH values in the distal LAD than controls ($6.99 \pm 0.06$ vs $6.67 \pm 0.05$; $p < 0.01$). Again, postischemic ventricular function parameters (LVSWI vs LVEDP, and ejection fraction) were significantly higher after the PICSO treatment, regaining the levels of pretreatment measurements (see above).

In a recent study in 22 swine ([5], see also [2]) of the effects of PICSO on myocardial ischemia under reduced flow (80% LAD stenosis) conditions and pacing at a constant heart rate (130–145 bpm), measurements of hemodynamics, regional blood flow, oxygen, lactate, and nucleotide metabolism were obtained at baseline, 30 min after coronary artery stenosis, at 30 and 60 min of the subsequent PICSO treatment period, and 30 min after discontinuation of PICSO, and at comparable times in the untreated control animals. In 10 animals PICSO was begun at 30 min of LAD stenosis and continued for 1 h with occlusion times of 7.5 s to 9 s and a constant release time of 4 s. Although between-group differences cannot easily be discussed since two different control groups have been used, there appears to be a slightly more pronounced decrease of oxygen content in the anterior interventricular vein during the PICSO treatment than was observed in untreated animals. Myocardial oxygen extraction and consumption were comparable in all groups. Considering lactate utilization, a general reversal from consumption to production was observed upon LAD stenosis. However, in the treatment group, lactate production was significantly reduced during the PICSO treatment, whereas in controls, it tended to remain relatively unchanged.

## B. Clinical studies

### B.1. PICSO application in open heart surgery

In a first clinical trial performed by Mohl et al. in 1985/1986 [22, 20], PICSO was evaluated in 30 patients undergoing bypass surgery. PICSO was applied in 15 patients for 1 h during the early reperfusion period, using average PICSO phasic occlusion and release times of $6.67 \pm 1.97$ and $4.33 \pm 0.44$ s, respectively. Analyzing the effects of PICSO on myocardial function, regional wall motion as systolic fractional area change (FAC) was assessed by 2D-echocardiography, moderately hypokinetic segments were found to be preserved significantly better in PICSO-treated patients as compared to control patients ($-1.3 \pm 2.4$ vs $-9.1 \pm 2.6\%$ FAC; $p < 0.04$). Although not statistically significant, the same trend was found for normal and severely hypokinetic segments.

As to the effects of PICSO on myocardial metabolism, a positive relationship between cumulative CK release and coronary sinus occluded pressure of $r = 0.94$ ($p < 0.006$) could be established. As expected, three months after operation, functional classification was similarly favorable in both groups. In this study, the importance of continuous monitoring of CS pressure clearly became evident (see Figs. 4a and b). Sudden changes and slow variations of CS pressure parameters have to be detected by continuous monitoring and used for feedback control of PICSO.

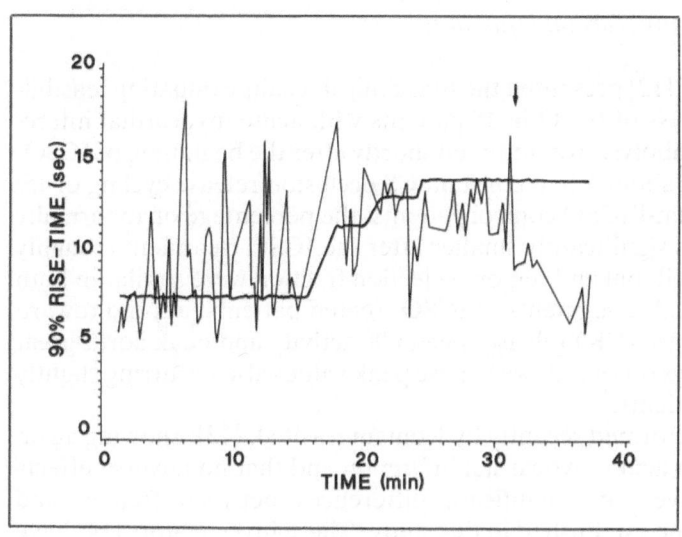

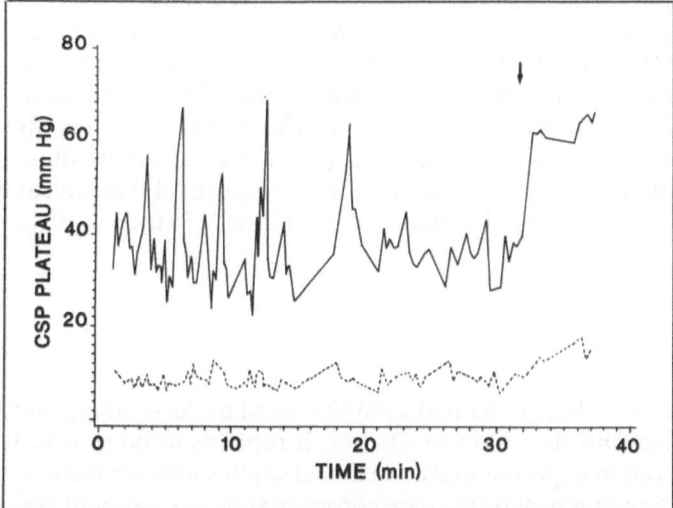

**Fig. 4.** Coronary sinus pressure (CSP) during PICSO. a) Time course of the 90% rise time (= time that takes peak pressure to reach 90% of its plateau) of systolic occluded CSP. Bold line indicates actual CS occlusion time. Note the sudden decrease of the 90% rise time after the bypass grafts were opened (arrow); b) Time course of CSP plateaus for systolic (solid line) and diastolic pressure (broken line). Note the sudden increase of the systolic plateau during opening of the bypass grafts (arrow), but only a subtle change in diastolic pressure. Data obtained in one patient during bypass surgery. (By permission, from: Mohl W et al. (1988) Clinical evaluation of pressure-controlled intermittent coronary sinus occlusion: randomized trial during coronary artery surgery. Ann Thorac Surg 46:192–201.)

## B.2. ICSO application in myocardial infarction

In 1987, Komamura et al. [12] presented the first clinical results evaluating feasibility, safety and effectiveness of ICSO in 15 patients with acute myocardial infarction. Intracoronary thrombolysis was initiated shortly after the beginning of ICSO, which was performed for about 1 h at constant CS occlusion/release cycling of 15/5 s. As compared to 22 non-PICSO control patients, the percentage of abnormally contracting segments was significantly smaller after the ICSO treatment (roughly 20% vs 35%; $p < 0.05$). Global and regional ejection fraction were similar in both groups. In recannalized LAD segments of ICSO treated patients, a trend toward greater total creatine kinase (CK) release, peak CK activity and peak aorta-great cardiac vein difference was observed, with these peak values also occurring slightly earlier than in control patients.

A similar study was performed recently by Komamura et al. [13], showing again the feasibility of ICSO in acute myocardial infarction and that no adverse effects were detectable. However, no significant differences between treated and untreated patients could be established in this study. These investigators [14] have also applied ICSO during PTCA in 10 patients with severe LAD stenosis, in order to evaluate its effects on brief myocardial ischemia. ICSO with constant occlusion/release timing of 10/5 s was performed for 3 min before and after PTCA, and during the occlusion of the LAD artery with the PTCA balloon. Hemodynamics and coronary arterial flow velocitzy remained stable during the ICSO treatment, whereas arterio-venous $O_2$-difference and lactate extraction ratio significantly increased during ICSO application before PTCA (from $8.2 \pm 2.9$ ml/dl to 9.0, d.h. $9.0 \pm 2.4$ ml/dl, and from $11 \pm 27\%$ to $22 \pm 20\%$, respectively) as well as after PTCA (from $8.9 \pm 1.7$ ml/dl to $9.5 \pm 1.1$ ml/dl and from $27 \pm 22\%$ to $31 \pm 19\%$, respectively).

## Clinical implications

Coronary sinus interventions challenge the pathophysiological understanding and knowledge of the researcher and the clinician. As yet, it remains to be resolved whether the benefits observed in experimental and clinical studies warrant the current efforts made of providing practical retrograde coronary venous treatment systems. However, limitations in primary interventions such as lytic therapy have become evident, and the present knowledge of reperfusion injury points to further development of new concepts, e.g., staged reperfusion or supplementation of reperfusion. Thus, a recent experimental study of Nakagawa et al. [25] appears to suggest a perfect model of a clinical setting for application of coronary sinus interventions and suggests the potential of PICSO as a valuable supplement of the lytic therapy. The authors have tested the effectiveness of PICSO in combination with low-dose intravenous administration of urokinase in dogs (n = 8), as compared to high-dose (n = 6) and low-dose i.v. drug (n = 10) administration alone, and also to direct intracoronary low-dose infusion (n = 8). Lytic therapy and PICSO were started 1 h after angiographic confirmation of complete coronary artery occlusion. PICSO was applied with mean CS occlusion times of $12 \pm 3$ s and release times of

5 ± 1 s. The success rate of reperfusion in the treatment group was significantly higher (six of eight) than in the high-dose group (one of 10) and almost as high as with intracoronary infusion (seven of eight). The incidence of ventricular premature contraction was significantly lower (4 ± 3/min) in the PICSO treated group than with intracoronary infusion (38 ± 28/min), but similar to high-dose (3/min) and low-dose administration without PICSO (6 ± 3/min). Studies performed by Ropchan et al. [28] and Simon et al. [30] also indicate the superiority of combining arterial blood retroperfusion with PICSO in salvaging ischemic myocardium. Our own intraoperative trial suggests the potential of PICSO to limit reperfusion injury.

PICSO fits very well into these new interventional cardiology concepts. Redistribution of coronary venous flow may be compared with staged reperfusion, and venous blood represents a most potent physiologically available buffer system. Adding enzymes to the reperfusate is one option of myocardial protection, while enhancing the washout of catabolites is another. Together with the possiblity of redistributing not only venous blood but also other substrates, the PICSO technique seems to encompass all components required for relief of myocardial jeopardy. We believe that whenever conventional therapy (e.g., lytic therapy) requires supplement, PICSO might prove to be the assist therapy of choice during severe myocardial ischemia.

## References

1. Beck CS (1949) Revascularization of the heart. Surgery 26:82–88
2. Capone RJ, Fedele F, Most A, Gerwitz H (1986) Pressure controlled intermittent coronary sinus occlusion (PICSO) improves myocardial ischemia in swine. In: Mohl W et al. (eds) Clinics of CSI. Steinkopff, Darmstadt, pp 333–334
3. Ciuffo AA, Guerci AD, Halperin G, Bulkley G, Casale A, Weisfeldt ML (1984) Intermittent obstruction of the coronary sinus following coronary ligation in dogs reduces ischemic necrosis and increases myocardial perfusion. In: Mohl W et al. (eds) The coronary sinus. Steinkopff, Darmstadt, pp 454–464
4. Diltz EA, Mames RN, Lee JW, Underwood TR, Mishra A, Vanheyningen CA, Niclas JM (1985) Intermittent coronary sinus occlusion does not reduce infarct size or ischemic dysfunction in an occlusion/reperfusion model (abstract). Circulation [Suppl III] 72:120
5. Fedele FA, Capone RJ, Most AS, Gerwitz H (1988) Effect of pressure-controlled intermittent coronary sinus occlusion on pacing-induced myocardial ischemia in domestic swine. Circulation 77:1403–1413
6. Gross L, Blum L, Silverman G (1936) Experimental attempts to increase the blood supply to the dog's heart by means of coronary sinus occlusion. J Exp Med 65:91–106
7. Guerci AD, Ciuffo AA, DiPaula AF, Weisfeldt ML (1987) Intermittent coronary sinus occlusion in dogs: reduction of infarct size 10 days after reperfusion. JACC 9:1075–1081
8. Heimisch W, Mohl W, Mendler N, Hagl S (1984) Intermittent coronary sinus occlusion: effects on regional function of the normal and ischemic myocardium. In: Mohl W et al. (eds) The coronary sinus. Steinkopff, Darmstadt, pp 465–472
9. Jacobs AK, Faxon DP, Coats WD, Mohl W, Apstein CS, Schick EC, Ryan TJ (1985) Pressure-controlled intermittent coronary sinus occlusion (PICSO) during reperfusion markedly reduces infarct size (abstract). Clin Res 33:197A
10. Jacobs AK, Faxon DP, Coats WD, Mohl W, Ryan TJ (1986) The effect of pressure controlled intermittent coronary sinus occlusion during reperfusion. In: Mohl et al. (eds) Clinics of CSI. Steinkopff, Darmstadt, pp 345–347
11. Jacobs AK, Rothendler J, Faxon D, Minihan A, Coats W, Simon P, Ryan T (1988) Enhancement of ischemic zone perfusion by intermittent coronary sinus occlusion (abstract) 3rd Intl Symposium on Myocardial Protection via the Coronary Sinus, Boston

12. Komamura K, Kodama K, Nanto S, Hirayama A, Koretsune Y, Hori M, Inoue M (1987) Human experience with intermittent coronary sinus occlusion in acute myocardial infarction: feasibility, safety and effectiveness (abstract). The 8th Intl Conference of the CSDS, Osaka

13. Komamura K, Mishima M, Hirayama A, Kodama K (1988) Human experience with intermittent coronary sinus occlusion in acute myocardial infarction (abstract). 3rd Intl Symposium on Myocardial Protection via the Coronary Sinus, Boston

14. Komamura K, Nanto S, Yamamoto K, Kodama K (1988) Time-controlled intermittent coronary sinus occlusion during LAD coronary angioplasty in man (abstract). 3rd Intl Symposium on Myocardial Protection via the Coronary Sinus, Boston

15. Lazar HL, Rajaii A, Roberts AJ (1988) Reversal of reperfusion injury after ischemic arrest with pressure-controlled intermittent coronary sinus occlusion. J Thorac Cardiovasc Surg 95:637−642

16. Lazar HL, Khoury T, Rivers S (1988) Improved distribution of cardioplegia with pressure-controlled intermittent coronary sinus occlusion. Ann Thorac Surg 46:202−207

17. Mohl W, Gueggi M, Haberzeth K, Losert U, Pachinger O, Schabert A, Borek U, Wolner E, Kessler M (1980) Effects of intermittent coronary sinus occlusion (ICSO) on tissue parameters after ligation of LAD. Bibl Anat 20:517−521

18. Mohl W, Glogar DH, Mayr H, Losert U, Sochor H, Pachinger O, Kaindl F, Wolner E (1984) Reduction of infarct size induced by pressure-controlled intermittent coronary sinus occlusion. Am J Cardiol 53:923−928

19. Mohl W, Punzengruber C, Moser M, Kenner T, Heimisch W, Haendchen R, Meerbaum S, Maurer G, Corday E (1985) Effects of pressure-controlled intermittent coronary sinus occlusion on regional ischemic myocardial function. JACC 5:939−947

20. Mohl W, Simon P, Neumann F, Punzengruber C, Schreiner W, Schuster J, Spiss C, Müller M, Tüchy G, Cisar A, Wenzel R, Czaky-Palawichini T (1986) PICSO-Workshop. In: Mohl W et al. (eds) Clinics of CSI. Steinkopff, Darmstadt, pp 363−377

21. Mohl W (1987) Coronary sinus interventions: from concept to clinics. J Cardiac Surg 2:467−493

22. Mohl W, Simon P, Neumann F, Schreiner W, Punzengruber C (1988) Clinical evaluation of pressure-controlled intermittent coronary sinus occlusion: randomized trial during coronary artery surgery. Ann Thorac Surg 46:192−201

23. Moser M, Mohl W, Kenner T (1984) The arteriovenous density gradient as an index for myocardial function. In: Mohl W et al. (eds) The coronary sinus. Steinkopff, Darmstadt, pp 497−507

24. Murry CE, Jennings RB, Reimer KA (1986) Preconditioning with ischemia: a delay of lethal cell injury in ischemic myocardium. Circulation 74:1124−1136

25. Nakagawa Y, Ikeoka K, Tateishi J, Kawasima S, Fujitani K, Iwasaki T (1988) The effect of coronary thrombolysis by intermittent coronary sinus occlusion (ICSO) in combination with intravenous administration of urokinase (UK) (abstract). 3rd Intl Symposium on Myocardial Protection via the Coronary Sinus, Boston

26. Neeley JR, Grotyohann LW (1984) Role of glycolytic products in damage to ischemic myocardium. Dissociation of adenosine triphosphate levels and recovery of function of reperfused ischemic hearts. Circ Res 55:816−824

27. Papp L, Kékesi V, Osváth B, Juhász-Nagy A, Szabó Z (1986) The efficiency of coronary sinus occlusion and coronary bypass in the ischemic dog heart. In: Mohl W et al. (eds) Clinics of CSI. Steinkopff, Darmstadt, pp 339−344

28. Ropchan G, Wilson G, Cruz J, Feindel C (1988) Nonsynchronized coronary sinus retroperfusion to salvage ischemic myocardium. 3rd Intl Symposium on Myocardial Protection via the Coronary Sinus, Boston

29. Simon P, Jacobs AK (1987) The effects of PICSO on left ventricular function during reperfusion. International Working Group on Coronary Sinus Interventions. Newsletter 1:8

30. Simon P, Jacobs AK, Hogfeldt V, Owen A, Faxon D, Ryan T (1988) Infarct size reduction using coronary sinus techniques: is arterial blood necessary? (abstract) 3rd Intl Symposium on Myocardial Protection via the Coronary Sinus, Boston

31. Smith G, Denning J, Eleff M, Echstein R (1952) Further studies on the effect of arterial venous fistulas and elevation of sinus pressure on mortality rates following acute coronary occlusions. Circulation 5:262−266

32. Toggart EJ, Nellis SH, Liedtke AJ (1987) The efficacy of intermittent coronary sinus occlusion in the absence of coronary artery collaterals. Circulation 76:667–677
33. Zalewski A, Goldberg S, Slysh S, Maroko PR (1985) Myocardial protection via coronary sinus interventions: superior effects of arterialization compared with intermittent occlusion. Circulation 71:1215–1223

# Surgical modes of retroperfusion

S. Meerbaum

Retroperfusion and arterialization of the coronary veins generally entails a dual process: 1) the retrograde infusion and arterialization via the coronary veins, and 2) obstruction of the coronary sinus, to preclude retroperfusate shunting back to the low pressure right atrium. The objective has always been to effectively retro-deliver circulation and nutrition to myocardium deprived of its normal coronary artery supply.

The first systematic and enthusiastic efforts aimed at a surgical mode of retroperfusion were those of the Beck era in the 1940s and 1950s. These efforts have been amply described and discussed elsewhere [1−5], and a schematic of Beck's operation is shown in Fig. 1. Basically, the objective of Beck and his contemporaries was to treat patients with coronary artery disease by 1) placing a bypass graft from the aorta to the coronary sinus, and 2) obstructing or constricting the coronary sinus so as to facilitate effective retrograde perfusion of the heart. Such a procedure consti-tutes, of course, a forced modification of the myocardial circulation. Among the consequences observed were occasionally distinct benefits in terms of reduced ischemia and apparently preserved myocardial viability. Such benefits were par-ticularly effective when they extended up to the time when a significant compensa-tory coronary arterial collateral supply replaced the effects of retroperfusion and provided permanent correction of the condition. Unfortunately, longer term observations frequently indicated closing-off of the aorta-to-coronary vein graft, and revealed deleterious effects of retroperfusion (i.e., vascular injury, myocardial edema, hemorrhages). The reported human applications were preceded by sub-

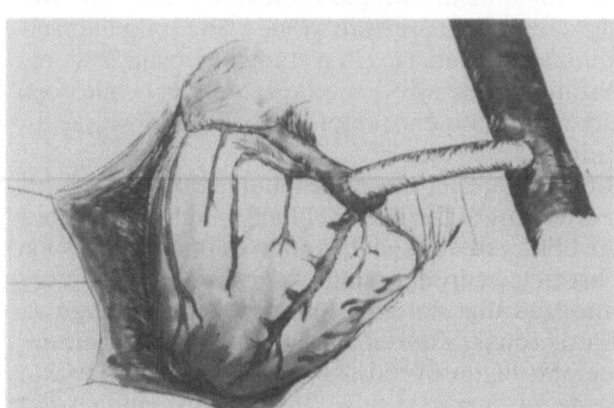

**Fig. 1.** Completed Beck II operation. Vein shunt aorta to coronary sinus. Note arteriali-zation of venous system as artist's concept. Courtesy of Dr. Charles P. Bailey. (By per-mission, from: Harken DE (1986) To Beck and back. In: Mohl W, Faxon D, Wolner E (eds) Clinics of CSI. Stein-kopff. Darmstadt, pp 5−12.)

stantial experimental investigations in dogs, simulating the chronic retroperfusion setting. More basic physiologic studies also employed limited periods of arterial blood retroperfusion during coronary occlusions.

It was noted in retrogradely treated acute coronary occlusions that some hearts continued to beat regularly for substantial periods (e.g., 24 h), whereas there was no such survival of acutely ischemic dogs without the retroperfusion. Also encouraging were laboratory investigations of hearts with dye injections into the coronary sinus (however, at an elevated pressure of 100 mmHg): the left myocardium was noted to exhibit "signs of complete injection of the capillaries" [6]. As they refined their methods of aorta-to-coronary sinus anastomosis, some of the Beck era investigators predicted optimistically that surgical retroperfusion treatment of coronary artery disease would become a useful technique for reduction of mortality and limiting infarction. Yet, these seasoned investigators also recognized fundamental problems, some of which became evident during the experimental and limited human applications.

One of the puzzling problems articulated was how to determine and achieve an optimal retrograde delivery to the jeopardized myocardium. Too much retrograde flow was found to be harmful, causing edema and hemorrhages, whereas too little retroinfusate was just not effective. Another general question concerned the manner in which a satisfactory coronary venous drainage could possibly be maintained while the coronary veins were being retroinfused and the normal coronary sinus outflow was severely restricted. The existence, variability and significance of coronary veno-venous shunts was beginning to be understood, and it was well recognized that further studies were needed to resolve questions regarding retroperfusion effects and effectiveness.

**Supporting investigations**

Based on a report by MacAllister et al. [7], the conclusion was reached that the coronary sinus should be obstructed but not completely ligated, either prior to or simultaneous with, or subsequent to mobilizing the vascular connection between the aorta and the coronary sinus. It was felt that, were the coronary sinus to be completely occluded, too much blood would be introduced through the graft into the coronary venous tree, and that the consequent pressure would lead to vascular and myocardial injury, as well as to impaired drainage. In a study involving 30 dogs, Eckstein [8] proved that arterialization effectively protects acutely ischemic dogs from ventricular fibrillation, which is normally encountered shortly following acute circumflex coronary artery occlusion.

Examining the surgically instituted retroperfusion technique, Bailey et al. [9] concluded that some (but only a fraction) of the arterial blood could be delivered to the myocardial capillaries. The observed marked dilatation of the cardiac veins during retroperfusion into the obstructed coronary sinus was considered a "safety mechanism". It was also hypothesized that aorta-to-coronary vein grafting can stimulate the development of intercoronary arterial anastomoses. In fact, in several experiments lasting one week, the degree of coronary arterial collateralization became sufficient so as to nullify the retrograde supply. Finally, it was emphasized by some that even a few supplemental milliliters of oxygenated blood retro-delivered to jeopardized regional myocardium could make the difference between tissue viability or necrosis.

Perhaps the most authoritative experimental evaluation of Beck's retroperfusion treatment was presented by Eckstein et al. [10–11] and Bakst et al. [12–13]. An aorta-to-coronary sinus graft and coronary sinus obstruction were simulated in dogs, and results were followed at intervals up to 1 year. Eckstein investigated the amount, source, and oxygen content of the blood collected retrogradely from the coronary artery, along with its peripheral pressure. The arterio-venous fistula flow was measured, its effect on heart weight was determined, and pathologic changes in vessels and myocardium were also studied. Basically, these studies showed that, in the early period of the revascularization treatment (up to about 5 weeks), retroperfusion could be performed with some degree of effectiveness, even in the presence of limited residual antegrade coronary artery supply. However, as previously noted, after 5 weeks of the procedure, the coronary veins would tend to spontaneously occlude, and the functional effects of retroperfusion would cease. Significant protection agains coronary artery occlusion could nevertheless persist, although the heart would then benefit primarily through development of its interarterial coronary collateral circulation. In his experimental chronic retroperfusion, Eckstein found that retroperfusion flows ranging from 400 cc/min to as high as 1500 cc/min, resulted in substantial myocardial edema.

In his study, Bakst [12–13] noted that circumflex coronary artery ligation in normal dogs was associated with a 90% mortality rate up to 6 h post occlusion, and all animals died eventually of ischemic injury. Ligation of the left anterior descending coronary artery in dogs resulted in a 60% mortality rate. When the coronary sinus was arterialized for 1–2 months prior to ligation of the circumflex coronary artery, mortality was reduced to only 20%, and dogs that did die in this retroperfusion series exhibited a spontaneous occlusion of the aorto-coronary sinus graft. On the other hand, ligation of the circumflex coronary artery in animals in which the coronary sinus had been arterialized for 6 months, produced a 66% mortality. It was therefore not clear to what extent significant protection against ventricular fibrillation could be achieved by arterialization of the coronary sinus. Loss of protection was invariably associated with retroperfusion frustrated by virtue of intimal proliferation and thrombosis developed within the coronary sinus and its branch vessels.

**Outcome and modification of surgical retroperfusion procedure**

One coronary artery disease series (186 patients) treated with coronary vein arterialization was that of the Beck et al. [14]. The two-stage operative treatment (placement of aorta-to-coronary sinus graft, followed 2 weeks later by coronary sinus obstruction) appeared somewhat beneficial in terms of relief of angina, but the mortality with the Beck II procedure was high (26%), and there remained problems in spite of only partial coronary sinus occlusion. Among the issues repeatedly mentioned were chronic graft closures, vascular trauma, myocardial edema, hemorrhages, fibrosis, and venous thrombosis. An appraisal by Feil [15] indicated even higher mortality 38%.

Another attempt at applying the Beck surgical procedure was made in less than 100 patients by Moll et al. of Poland in the period up the 1970s [16]. The overall rationale, in view of the now widely practiced coronary artery bypass surgery, was selection of those cases where insufficient coronary runoff and diffuse artherosclerosis favored the use of coronary veins, which are generally unaffected by the atherosclerotic process. Moll et al. reported a 7% mortality, an 11% incidence of postoperative myocardial infarction, and a 75% bypass graft patency 6 months after the operation. Proliferative venous bypass changes were noted in 10% of the cases. These surgeons occasionally used a modified regional retroperfusion by placing the graft between the aorta and an obstructed great cardiac vein. The selective concept retroperfuses the specific area involved by the coronary artery disease, and does not interfere with coronary venous drainage from the noninvolved zones. Substantial experimental research of regional coronary vein arterialization was carried out at about the same time [17−25], resulting in some indications and a hope that a more regionally applied surgical graft retroperfusion might alleviate the major difficulties encountered with the more global Beck procedure. Occasional cases of patient retroperfusion using the regional graft method were reported in the US, and most of these as well as the background studies were summarized and evaluated in review articles by Hochberg [26−27].

Hochberg et al. [24] used an aorta-to-great-cardiac-vein bypass preparation. The graft to the coronary veins was very close to the proximal LAD occlusion site, and thus would be optimal for retrodelivery of arterial blood toward the ischemic myocardium. The graft flow in these dogs averaged 53 ml/min. After a period ranging from 3−5 months, 10 of 14 coronary vein bypass grafts were evidently patent. Subsequent open-chest radioactive microsphere study in the dogs revealed particulary effective subendocardial delivery by retroperfusion, as compared to untreated coronary occlusions. The untreated control series with equiva-

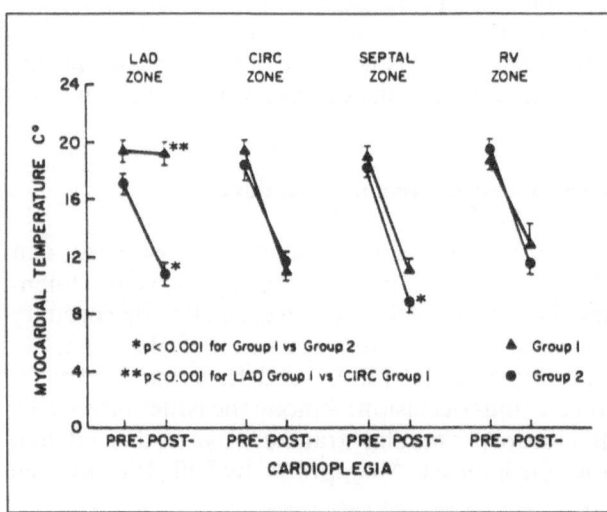

**Fig. 2.** A comparison of myocardial temperatures before and after delivery of cardioplegia through either the coronary veins or the coronary arteries, in the four zones tested. Note the lack of cooling in the distribution of the left anterior descending coronary artery (LAD) in Group 1 animals (aortic cardioplegia). Data are shown as mean ± standard deviation. (CIRC = circumflex coronary artery; RV = right ventricular.) (By permission, from: Gundry SR, Kirsh MM (1984) A comparison of retrograde cardioplegia vs antegrade cardioplegia in the presence of coronary artery obstruction. Ann Thor Surg 38(2):124−127.)

lent LAD occlusions exhibited much higher mortality. The report indicated minimal damage when compared to the global Beck procedure problems, apparently no edema or hemorrhages. This would favor surgical application of the regional coronary vein arterialization technique whenever feasible, and particular mention was made of a patient subset with diffuse atherosclerosis of a previously failed coronary artery bypass.

There were individual applications of oxygenated blood coronary vein retroperfusion during aortic valve replacement surgery [28], and also during revascularization of coronary arteries [29]. Based on these human applications, and supporting experimental studies [30–32], there remains the possibility of utilizing temporary retroperfusion support as a means to avoid intraoperative myocardial infarction. However, the evidence of benefits for these retrograde modes was either insufficient or unconvincing, and it also appears that there is little current interest in pursuing the chronic global or regional bypass retroperfusion approach. Instead, surgeons have now turned to applying the promising coronary sinus perfusion as a means for selective retrograde cardioplegia delivery. Strong rationales for an effective retrograde vs prevalent antegrade cardioplegia include: reduced interference with the surgical procedure, elimination of trauma at the aortic valve and coronary ostia, and – particularly – overcoming hazardous cardioplegic maldistributions along with nonhomogenous cooling beyond severe coronary artery obstruction.

**Retrograde cardioplegia via the coronary sinus**

Recent reviews of this attractive intraoperative retroperfusion modality were presented by Mohl and Roberts [33], Mohl [34], and Lazar [35]. Menasche [36], Chiu [37], Gundry [38], Fabiani [39], Okike [40], Wechsler [41], and others, have all contributed significantly to the appliation of cardioplegic delivery via the coronary sinus. The primary purpose was and is to provide a more uniform distribution of cardioplegia in the presence of significant coronary artery lesions, leading to more effective protection of the jeopardized myocardium.

Experimental studies such as those of Gundry et al. [38] proved that cardioplegia retrogradely administered via the coronary sinus (40–50 mmHg) assures uniform cardioplegia delivery and cooling, which preserves the jeopardized underperfused myocardium, and leads to a more rapid return of cardiac function. Figures 2 and 3 compare the temperature distribution and systolic shortening in the involved zone with antegrade vs retrograde cardioplegia. Uniform ventricular cooling in spite of coronary artery occlusions was also demonstrated in the canine by Saylam et al. [42], Bolling et al. [43], Mori et al. [44]. Figure 4 (from Masuda et al. [45]) shows the improved ATP preservation in myocardium beyond an LAD occlusion when retrograde cardioplegia was applied.

In 1982, Menasche et al. reported on application of retrograde cardioplegia in patients undergoing aortic valve surgery [36]. A 12-F balloon-tipped catheter was placed in the coronary sinus, and the perfusion pressure was limited to 40 mmHg. The retrograde technique was shown to be safe, simple to perform, beneficial when direct cannulation of coronary ostia may be traumatic, and as effective as antegradely delivered cardioplegia. There have been improvements in the catheter bal-

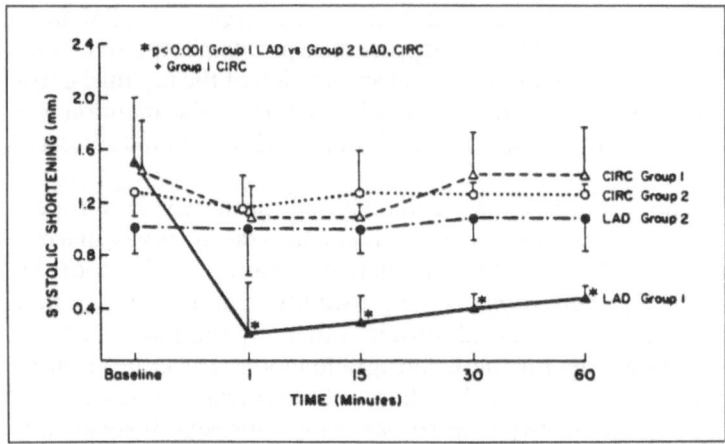

**Fig. 3.** A comparison of systolic shortening of the septal to free wall axis of major fibers in the circumflex (CIRC) distribution and the distribution of the left anterior descending (LAD) coronary artery for Group 1 (aortic cardioplegia) and Group 2 (coronary vein cardioplegia). Note the lack of recovery of shortening in the LAD region of Group 1. Data are shown as mean ± standard deviation. (By permission, from: Gundry SR, Kirsh MM (1984) A comparison of retrograde cardioplegia vs antegrade cardioplegia in the presence of coronary artery obstruction. Ann Thor Surg 38(2):124–127.)

loon design, aimed at reliable catheter positioning and better coronary sinus occlusion, without obstruction of important tributary vessels. Others have also been applying the retrograde cardioplegia [46, 47], and have indeed reported on better myocardial protection and improved left ventricular function as compared with the antegrade method.

In 1984, Fabiani [48] studied a somewhat different retrograde approach. Primarily to avoid potential coronary sinus injury, he proposed to administer cardioplegia

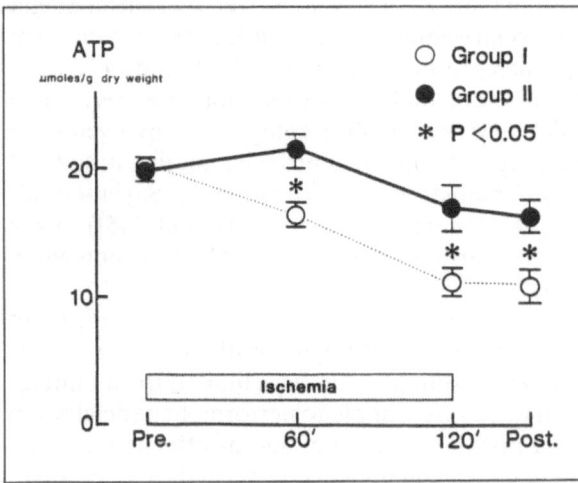

**Fig. 4.** Sequential changes of adenosine triphosphate *(ATP)* during 120 min of global ischemia and 30 min of reperfusion in the distribution of the occluded left anterior descending artery. Group I is the antegrade perfusion group and Group II, the retrograde perfusion group. * p < 0.05 compared between Groups I and II at the same time. (By permission, from: Masuda M et al. (1986) Myocardial protection in coronary occlusion by retrograde cardioplegia perfusion via the coronary sinus in dogs. J Thorac Cardiovasc Surg 92:255.)

into the right atrium via a catheter, after snaring both vena cavae and the pulmonary artery. Using atrial pressures of 60−80 mmHg, crystalloid cardioplegia solution was pumped in at a rate of 1 liter over a period of 3 min, and the cardioplegic solution was delivered into the aortic root. More recently, Fabiani [49] reported having performed the procedure in 1200 patients without complications. In particular, the distension of the right heart cavities was said to be well tolerated, and the improved cooling of the right atrium was thought to be responsible for a lower incidence of arrhythmias. Yet, other investigators [50, 51] found that coronary sinus cardioplegia provides superior myocardial protection − particularly in the right ventricle − as compared to the right atrial delivery. Post operative right ventricular dysfunction may be more common with right atrial cardioplegia.

Retrograde coronary sinus perfusion with cardioplegic solutions is now considered a useful method during surgical revascularization. In an experimental study in pigs, Horneffer et al. [52] showed optimal salvage of ischemic myocardium with the retrograde (vs the antegrade) method. Figure 5 shows a comparison of infarction with differing surgical protection regimens. In patients undergoing coronary artery bypass surgery, a randomized study by Guiraudon et al. [53] found no statistically significant differences between antegrade and retrograde cardioplegia, but surmised there would be benefits of the retrograde approach in particular patient subsets. Assessment by means of thermovision in patients with critically obstructed and occluded coronary arteries [54] demonstrated that retrograde cardioplegia did provide more uniform hypothermia. Recent studies also compare blood vs crystal-

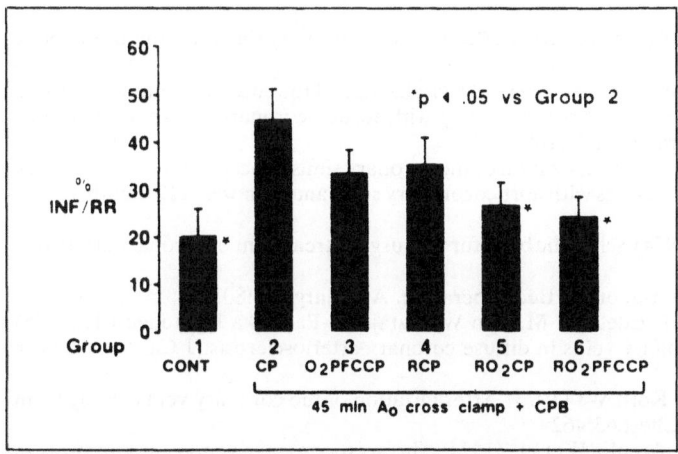

**Fig. 5.** The amount of infarcted myocardium in the six experimental groups (N = 54). (CONT = the control group, which underwent reflow after 30 minutes of coronary occlusion without global ischemic arrest; CP = a single antegrade infusion of cardioplegic solution; $O_2$PFCCP = two antegrade infusions of oxygenated perfluorocarbon cardioplegic solution; RCP = retrograde perfusion with nonoxygenated cardioplegic solution; $RO_2$CP = retrograde perfusion with oxygenated cardioplegic solution; and $RO_2$PFCCP = retrograde perfusion with oxygenated perfluorocarebon cardioplegic solution; % INF/RR = the amount of infarcted myocardium in the region at risk. $A_o$ = aortic; CPB − cardiopulmonary bypass.) (By permission, from: Horneffer PJ et al. (1986) Retrograde coronary sinus perfusion prevents infarct extension during intraoperative global ischemic arrest. Ann Thorac Surg 42:132−142.)

loid retrograde cardioplegia. These two modalities appeared to be equivalent in effectiveness [55]. Finally, Okike et al. [56] presented a new pulsatile variant of retrograde cardioplegia, which was found in dogs to provide superior cooling and ATP preservation as compared with steady retrocardioplegia.

## References

1. Beck CS (1948) Revascularization of the heart. Ann Surg 128:54
2. Beck CS, Hahn RS, Leininger DS, McAllister FF (1951) Operation for coronary artery disease. JAMA 147:1726
3. Hochberg MS, Austen WG (1980) Selective retrograde coronary venous perfusion. Ann Thorac Surg 29(6):578–588
4. Harken DE (1986) To Beck and back. In: Mohl W, Faxon D, Wolner E (eds) Clinics of CSI. Steinkopff, Darmstadt, pp 5–12
5. Meerbaum S (1984) The Beck era: a springboard for renewed research of coronary venous retroperfusion aimed at treatment of myocardial ischemia. In: Mohl W et al. (eds) The coronary sinus. Steinkopff, Darmstadt, pp 320–327
6. Roberts JT, Browne HS, Roberts G (1943) Nourishment of the myocardium by way of the coronary veins. Fed Proc 2:90
7. McAllister FF, Leininger D, Beck CS (1951) Diastolic retroperfusion of acutely ischemic myocardium. Am J Cardiol 37:558–598
8. Eckstein RW, Smith G, Eleff M et al. (1952) The effect of arterialization of the coronary sinus in dogs on mortality following acute coronary occlusion. Circulation 6:16
9. Bailey CP, Truex RC, Angulo AW et al. (1953) The anatomic (histologic) basis and efficient clinical surgical technique for the restoration of the coronary circulation. J Thorac Surg 25:143
10. Eckstein RW, Hornberger JC, Sano T (1953) Acute effects of elevation of coronary sinus pressure. Circulation 7:422
11. Eckstein RW, Leininger DS (1954) Chronic effects of aortocoronary sinus anastomosis of beck in dogs. Circ Res 2:60
12. Bakst AA, Adam A, Goldberg H et al. (1955) Arterialization of the coronary sinus in occlusive coronary artery disease: III. Coronary flow in dogs with aortico-coronary sinus anastomosis of 6 month's duration. J Thorac Surg 29:188
13. Bakst AA, Bailey CP (1956) Arterialization of the coronary sinus in occlusive coronary artery disease: IV. Coronary flow in dogs with aorticocoronary sinus anastomosis of 12 month's duration. J Thorac Surg 31:559
14. Beck CS, Leininger DC (1954) Scientific basis for the surgical treatment of coronary artery disease. JAMA 159:1264
15. Feil H (1943) Clinical appraisal of the Beck operation. Ann Surg 118:807
16. Moll JW, Dzieatkoviak AJ, Edelman M, Iljin W, Ratajczyk-Pakalska E, Stengert K (1975) Arterialization of the coronary veins in diffuse coronary arteriosclerosis. J Cardiovasc Surg 16:520
17. Arealis EG, Volder JGR, Kolff WJ (1973) Arterialization of the coronary vein coming from an ischemic area (letter). Chest 63:462
18. Andreadis P, Natsikas N, Arealis E et al. (1974) The aortocoronary venous anastomosis in experimental acute myocardial ischemia. Vasc Surg 8:45
19. Bhayana JN, Olsen DB, Byrne JP et al. (1974) Reversal of myocardial ischemia by arterialization of the coronary vein. J Thorac Cardiovasc Surg 67:125
20. Chiu CJ, Mulder DS (1975) Selective arterialization of coronary veins for diffuse coronary occlusion. J Thorac Cardiovasc Surg 70:177
21. Demos S, Brooks H, Holland R et al. (1974) Retrograde coronary venous perfusion to reverse and prevent acute myocardial ischemia. Circulation [Suppl 3:III] 49, 50:168
22. Gardner RS, Magovern GJ, Park SB et al. (1974) Arterialization of coronary veins in the treatment of myocardial ischemia. J Thorac Cardiovasc Surg 68:173
23. Hochberg MS (1977) Hemodynamic evaluation of selective arterialization of the coronary

venous system: an experimental study of myocardial perfusion using radioactive microspheres. J Thorac Cardiovasc Surg 74:774

24. Hochberg MS, Roberts WC, Morrow AG et al. (1979) Selective arterialization of the coronary venous system: an encouraging long-term flow evaluation utilizing radioactive microspheres. J Thorac Cardiovasc Surg 77:1

25. Park SB, Magovern GJ, Liebler GA et al. (1975) Direct selective myocardial revascularization by internal mammary artery-coronary vein anastomosis. J Thorac Cardiovasc Surg 69:63

26. Hochberg MS, Austen WG (1978) Selective retrograde coronary venous perfusion. Ann Thorac Surg 29:478

27. Hochberg MS, Roberts AJ, Parsonnet V et al. (1986) Selective arterialization of coronary veins: Clinical experience of 55 american heart surgeons. In: Mohl W, Faxon DP, Wolner E (eds) Clinics of CSI. Steinkopff, Darmstadt, pp 195−201

28. Gott VL, Gonzalez JL, Zuhdi MN et al. (1957) Retrograde perfusion of the coronary sinus for direct-vision aortic surgery. Surg Gynecol Obstet 104:319

29. Lillehei CW, DeWall RA, Gott VL et al. (1956) The direct-vision correction of calcific aortic stenosis by means of pump-oxygenator and retrograde coronary sinus perfusion. Dis Chest 30:123

30. Blanco G, Adam A, Fernandez A (1956) Direct experimental approach to the aortic valve. J Thorac Cardiovasc Surg 32:171

31. Hammond GL, Davies AL, Austen WG (1967) Retrograde coronary sinus perfusion: A method of myocardial protection in the dog during left coronary artery occlusion. Ann Surg 166:39

32. Solorzano J, Teitelbaum G, Chiu RCJ (1978) Retrograde coronary sinus perfusion for myocardial protection during cardiopulmonary bypass. Ann Thorac Surg 25:201

33. Mohl W, Roberts AJ (1985) Coronary sinus retroperfusion and pressure-controlled intermittent coronary sinus occlusion (PICSO) for myocardial protection. Surg Clin North Am 65(3):477−495

34. Mohl W (1987) Retrograde cardioplegia via the coronary sinus. Ann Chir Gynecol 76:61−67

35. Lazar HL (1988) Coronary sinus interventions during cardiac surgery. Ann Thorac Surg 46:475−482

36. Menasche P, Kural S, Fauchet M, Lavergne A, Commin P, Bercot M, Touchaut B, Georgipoulos G, Piwnica A (1982) Retrograde coronary sinus perfusion: a safe alternative for ensuring cardioplegic delivery in aortic valve surgery. Ann Thorac Surg 34:647

37. Chiu RC (1984) Cold cardioplegia via the retrograde coronary sinus infusion for myocardial protection. In: Mohl W, Wolner E, Glogar D (eds) The coronary sinus. Steinkopff, Darmstadt, p 275

38. Gundry SR, Kirsh MM (1984) A comparison of retrograde cardioplegia vs antegrade cardioplegia in the presence of coronary artery obstruction. Ann Thorac Surg 38(2):124−127

39. Fabiani JN, Relland J, Carpentier A (1984) Myocardial protection via the coronary sinus in cardiac surgery: comparative evaluation of two techniques. In: Mohl W, Wolner E, Glogar D (eds) The coronary sinus. Steinkopff, Darmstadt, pp 305−311

40. Okike ON, Phillips D, Hsi C et al. (1986) Pulsatile cardioplegia retroinfusion. Chest 89:487

41. Wechsler AS, Salter DR, Murphy CE, Goldstein JP, Brunsting LA, Abd-Elfattah AS (1987) Metabolic differences of retrograde cardioplegia. In: Mohl W, Faxon D, Wolner E (eds) Clinics of CSI. Steinkopff, Darmstadt

42. Saylam A, Aytac A, Andac O, Tuncor I, Aslan A (1982) Retrograde coronary sinus perfusion of cold cardioplegic solutions in the presence of coronary artery occlusions. Experimental study. J Thorac Cardiovasc Surg 30:378−382

43. Bolling SF, Flaherty JT, Bulkley BH, Gott VL, Gardener TJ (1983) Improved myocardial preservation during global ischemia by continuous retrograde coronary sinus perfusion. J Thorac Cardiovasc Surg 86:659

44. Mory F, Ivey TD, Tabayashi K, Thomas R, Misbach GA (1986) Regional myocardial protection by retrograde coronary sinus infusion of cardioplegic solution. Circulation [Suppl III] 74:111−116

45. Masuda M, Yonengak K, Shiki K et al. (1986) Myocardial protection in coronary occlusion by retrograde cardioplegic perfusion via the coronary sinus in dogs. J Thorac Cardiovasc Surg 92:255

46. Walter PJ, Kindl F, Pdzuweit T, Schaper J (1984) Metabolic and ultrastructural changes in the ischemic myocardium due to additional perfusion of the coronary sinus with "Brettschneider" cardioplegic solution. In: Mohl W, Wolner P, Glogar D (eds) The coronary sinus. Steinkopff, Darmstadt/Springer, New York, pp 284–290
47. Fundaro P, Salati M, Beretta L, Santori C (1986) Retrograde vs antegrade cardioplegia for myocardial protection in coronary artery bypass graft surgery. In: Mohl W, Faxon D, Wolner E (eds) Clinics of CSI. Steinkopff, Darmstadt/Springer, New York, pp 229–233
48. Fabiani JN, Relland J, Carpentier A (1984) Myocardial protection via the coronary sinus in cardiac surgery: comparative evaluation of two techniques. In: Mohl W, Wolner E, Glogar D (eds) The coronary sinus. Steinkopff, Darmstadt/Springer, New York, pp 305–311
49. Fabiani JN, Deloche A, Swanson J, Carpentier A (1986) Retrograde cardioplegia through the right atrium. Ann Thorac Surg 41(1):101–102
50. Gundry SR (1987) Retrograde cardioplegia. Correspondence. Ann Thorac Surg 43:121–123
51. Salter DR, Goldstein J, Abd-Elfattah A, Murphy C, Brunsting LA, Wechsler AS (1987) Ventricular function after atrial cardioplegia. Circulation [Suppl V] 76:5–129
52. Horneffer PJ, Gott VL, Gardner TJ (1986) Retrograde coronary sinus perfusion prevents infarct extension during intraoperative global ischemic arrest. Ann Thorac Surg 42:132–142
53. Guiraudon GM, Campbell CS, McLellan DG, Kostuk WJ, Purves PD, MacDonald JL, Clelland AG, Tabros NB (1986) Retrograde coronary sinus vs aortic root perfusion with cold cardioplegia: randomized study of levels of cardiac enzymes in 40 patients. Circulation [Suppl III] 74:111–105
54. Shapira N, Lemole GM, Spagna PM, Bonner FJ, Fernandez J, Morse D (1987) Antegrade and retrograde infusion of cardioplegia: assessment by thermovision. Ann Thorac Surg 43:92–97
55. Goldstein JP, Salter DR, Murphy CE, Abd-Elfattah AS, Morris III JJ, Wechsler AS (1986) The efficacy of blood vs christalloid coronary sinus cardioplegia during global myocardial ischemia. Circulation [Suppl III] 74:III-99
56. Okike ON, Phillips D, Chi C, Gore JM, Mojica WA, Weiner B, Alpert JA, Vander Salm TJ (1986) Pulsatile cardioplegia retroinfusion (Abstract). Chest 89(6):487

# Clinically oriented phased retroperfusion systems

S. Meerbaum

## Rationales and concepts

### Retroperfusion safety and effectiveness

In spite of some convincing rationales [1−4], clinical application of retroperfusion (e.g., in the setting of coronary artery obstructions) appeared at first less than promising. The contracting heart features a characteristic and significant phasic increase in systolic intramyocardial pressures or wall stress, resulting in both near cessation of arterial blood inflow and active extravascular squeezing, which normally promotes drainage of coronary venous blood. Coronary stenosis or occlusion can result in significant regional underperfusion, but some antegradely directed flow and pressure usually persists in the peripheral coronary artery. During myocardial ischemia, the "noninvolved" portion of the heart is not only vital to preservation of essential global cardiac function, but it is often depended on to provide a degree of compensation for the regional deficiency. Obstruction of the coronary venous efflux generates a series of complex interactions, some of which can be beneficial in terms of resupplying an ischemic zone, but others are fraught with potentially dangerous consequences. Thus, clearance of toxic metabolic products from ischemic myocardium may be severely hindered, exacerbating the injury in an already jeopardized cardiac tissue. Depending on coronary venous anatomy, site of coronary vein obstruction and mode of retroperfusion, perfusion in the "uninvolved" myocardium may be significantly interfered with. Pressures within coronary veins may become excessive, leading to general vessel engorgement, unacceptable vascular trauma, myocardial edema, hemorrhages, and permanent damage.

The ischemic condition to be treated can, of course, be of a greater or lesser severity, depending on the history and prevailing physiologic state, including the degree of coronary artery flow restriction and available coronary collateral supply. Competition between residual antegrade vs retrogradely produced blood pressures and flows (reflected in a variety of veno-arterial pressure differentials), is a frequent feature in clinically germane settings. The effectiveness of a retrovenous intervention is found to depend on the particular location of the coronary artery obstruction, as well as on the relative site (and mode) of the retrograde coronary venous treatment. As previously demonstrated, a significant antegrade supply, produced by major coronary collateral development, essentially prevented retrograde perfusion, and persisting coronary vein obstruction along with retroinfusion produced complications such as thrombosis and irreversible myocardial damage.

171

To take advantage of the potential clinical benefits of selective coronary venous retrodelivery to an antegradely blockaded ischemic myocardial region, a new rationale and modality of retroperfusion had to be developed.

## Temporary retrograde support

With regard to the clinical application rationale, a new consensus developed in the 1970s to concentrate on the specific objective of providing temporary retroperfusion support to regionally ischemic or infarcting myocardium. Such an adjunct treatment was mostly aimed at adequately maintaining myocardial viability and function pending correction of the underlying problem, using the established surgical revascularization or one of the emerging newer interventions such as percutaneous transluminal angioplasty (PTCA) and thrombolytic reperfusion. The complex issues of permanent or longterm retroperfusion were therefore set aside. The primary goal was to provide the most prompt and most effective coronary venous retrosupply of the jeopardized myocardium, without incurring the deleterious consequences known to be associated with persisting obstruction and excessive pressurization of the coronary venous system.

## Phased retroperfusion

Synchronized retroperfusion (SRP) was to obstruct the coronary vein and retroperfuse arterial blood only in diastole, when venous efflux is minimal and the heart is most receptive to safe coronary venous intervention. SRP was also designed to facilitate normal systolic drainage via the coronary sinus, when ventricular contraction causes a spontaneous "milking" or "squeezing out" of the coronary venous blood. The systolic drainage, as well as the arterial blood retroinfusion with simultaneously timed coronary vein occlusion, were synchronized in SRP by means of the electrocardiogram and usually accomplished by means of an intervascular balloon at the tip of the retroperfusion catheter. One variation of this concept (possibly of interest for very short retrograde intervention) is to maintain a steady retrograde blood infusion, but appropriately synchronize and diastolically occlude the coronary sinus or great cardiac vein. Conversely, the coronary bein could be scheduled to remain obstructed for short periods, while the retroinfusate flow is phased to occur only in diastole. The terms of retroperfusion and retroinfusion are really equivalent, but the former is generally associated with arterialization of the coronary veins, while retroinfusion has been largely reserved to characterize retrograde delivery of pharmacologic agents.

## Intermittent retroperfusion

An alternate approach was to simply forego the retroinfusion altogether, and merely apply a relatively rapid alternation between short non-phased coronary sinus occlusions (e.g., for 10 s), which are deemed inherently safe (because of coronary venous compliance), and intervening periods with full coronary sinus release (e.g., of 5 s duration). This technique, described and discussed in chapter VI of this book, is termed intermittent coronary sinus occlusion (ICSO) and was initially applied with fixed timing of the two phases. When the ICSO phasing is feed-back controlled in relation to measured pressures developed within the coronary veins, the more complex system was given the name PICSO. This relatively simple methodology, i.e., without any arterial blood retroinfusion, was based on the expectation that rapid redistribution and resupply of venous blood from nonischemic zones to the ischemic myocardial region (during the brief coronary sinus occlusion period) would enhance the underperfused zone's circulation, and possibly also provide some degree of nutrition. The subsequent sudden complete coronary sinus release and the generated flow acceleration were deemed to significantly enhance the washout of toxic metabolites from the ischemic myocardium.

The effects of multiple short coronary venous obstructions are not well established, and may not even be readily generalized because of the highly variable coronary venous system. The myocardial venous capacitances are not readily defined at this time, although one might presume that sufficient epicardial and myocardial vascular capacitance (including coronary veno-venous shunts) exists to safely accommodate a short "backing up" of venous blood. There is also indication that the generated coronary veno-arterial pressure gradients result in a beneficial redistribution toward the regionally ischemic myocardium. On the other hand, and as already emphasized, coronary sinus occlusion in the presence of substantial antegrade perfusion is potentially hazardous, and can also lead to an undesirable underperfusion and dysfunction in the nonischemic zone.

*Phased-intermittent system*

Ideally, and for the sake of flexibility in the face of a variable anatomy and physiology, it would appear that a more universally controllable phasic retroperfusion/retroinfusion concept should be aimed at. Such a system would permit diastolically phased retroinfusion (and/or coronary vein occlusion) in each cardiac beat or else several beats apart. The system would also permit applying an intermittent coronary venous obstruction, without concommittant retrograde infusion of arterial blood. Simple steady (yet short) occlusive coronary vein modalities may also be safe and advantageous, e.g., when performing temporary drug retroinfusions into an ischemic myocardial region. In this latter case, the pharmacologic agent is not only selectively retrodelivered to the injured zone, but may have to persist for a certain minimal period within the regional tissue to provide effective treatment.

After this introductory discussion of clinically oriented retrograde interventions, we will now describe and illustrate pertinent experimental SRP investigations. These, along with recent clinical trials discussed in chapter 11, form the basis for hopeful projections of phased coronary venous retroperfusion treatment during myocardial ischemia or evolving acute myocardial infarction.

## Initial synchronized retroperfusion study in open-chest dogs

Meerbaum et al. described in 1976 [5] an open-chest dog study in which great cardiac vein synchronized diastolic retroperfusion of arterial blood was applied during acute LAD coronary artery ligation. The arterial blood was shunted from the dog's carotid artery and pumped into the great cardiac vein through a positive displacement roller or finger pump, placed in series with an ECG synchronized pulsed bladder device. Electromagnetic probe measurements indicated significantly augmented retrograde diastolic flow pulses. The 7-F retroperfusion catheter in this study was of a simple nonocclusive single lumen configuration.

The tip of this retroperfusion cannula was thought to be generally wedged in the great cardiac vein during the diastolic retrograde arterial blood infusion phase. In some cases, however, the catheter was probably sufficiently separated from the coronary vein wall to permit some diastolic shunting of the retroinfusate toward the coronary sinus and the right atrium. Pressure measurements in the regional coronary vein during the diastolic phase shed some light on these differing situations. In systole, when left ventricular contraction normally generates the "milking-squeezing" action facilitating coronary venous drainage, the catheter was evidently generally minimally obstructive (based again on pressure observations), so that drainage around the catheter could proceed with little hindrance. In some instances, though, partial or full catheter wedging may have led to significant interference with coronary venous efflux. It was recognized that the above variable catheter positioning and action had to be overcome by a controllable and generally improved catheter design.

Catheterization problems in this initial study were believed to be responsible for the observed variability of the synchronized retroperfusion performance, which ranged from excellent to poor, and was mostly intermediate. Nonetheless, measurements of ECG-ST segment elevations, regional cardiac contraction, myocardial metabolism, distal coronary artery as well as coronary vein pressures, and flow collected retrogradely from distal to the LAD occlusion, all indicated the potential benefits of the synchronized diastolic retroperfusion technique, when it is applied during acute myocardial ischemia. As already mentioned, pressure measurements in the anterior interventricular coronary vein helped the interpretation of effective vs ineffective instances of the retrograde intervention. The degree of coronary venous diastolic pressure augmentation played a major role, as anticipated, and so did the coronary venous systolic pressure release during drainage. Pathologic examination cooroborated the significantly safer approach of synchronized retroperfusion with facilitated systolic coronary venous drainage, as compared to the previously tested non-phased retroperfusion into a persistently obstructed coronary vein.

Figure 1 shows the initial experimental preparation, including the mode of regional coronary vein catheterization, the "home-made" retrograde synchronized pumping device, the proximal LAD coronary artery ligation, and a variety of measurements. Of particular note: the instrumentation aimed at regional measurement of myocardial contraction and of regional coronary vein blood pressure. Whereas more accurate means for experimental measurement are now available, the miniaturized piezo-resistive myocardial gauges, which indicated local force/

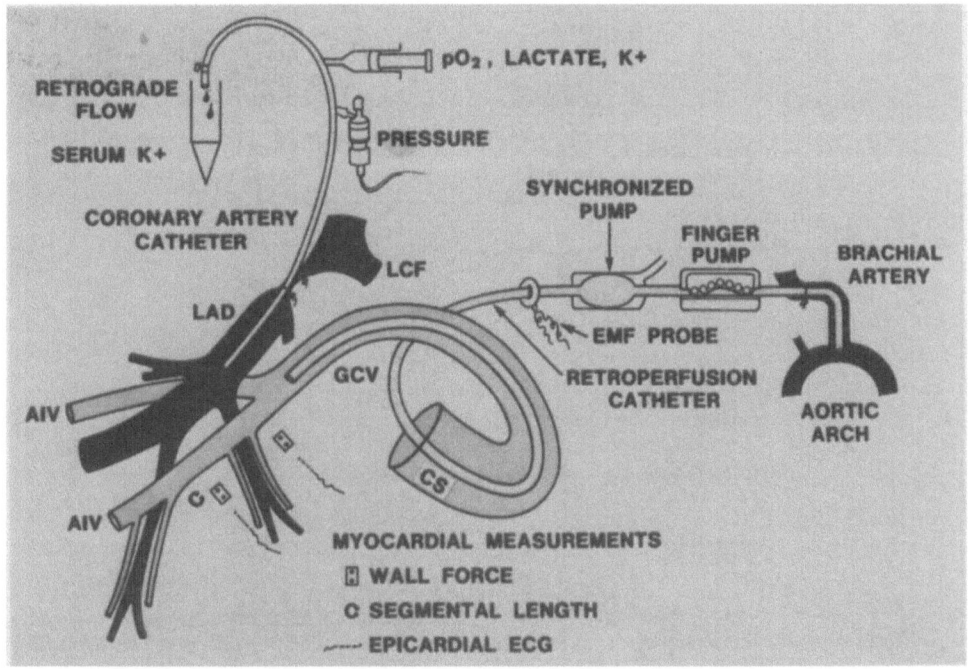

**Fig. 1.** Diastolically augmented coronary venous retroperfusion. Schematic drawing indicates the diastolic retroperfusion system and sites of myocardial measurements performed in this study. Arterial blood from the brachial artery is shunted to the regional coronary vein draining the occluded left anterior descending coronary artery (LAD) through a positive displacement finger pump and an in-series electrocardiographic synchronized pump. The latter is programmed to produce diastolic flow augmentation. The pulsed flow is delivered by means of an electromagnetic flow (EMF) probe into a nonocclusive retroperfusion catheter inserted through the coronary sinus (CS) into a coronary vein (GCV) draining the ischemic myocardial segment. Myocardial force gauges and subepicardial electrocardiographic (ECG) probes were attached within the coronary occluded segment and the adjoining "border" zone to monitor alterations in regional function during coronary occlusion and retroperfusion. Delivery of retroperfusate to the ischemic zone was assessed by measuring pressure, flow and blood PO$_2$ from the coronary artery distal to the occlusion. AIV = anterior interventricular vein; LCF = left circumflex coronary artery. (By permission, from: Meerbaum S et al. (1976) Diastolic retroperfusion of acutely ischemic myocardium. Am J Cardiol 37:588−598.)

stress, segmental length and wall thickness, provided an adequate evaluation of regional contractile function.

The special measurements of blood pressure in the anterior interventricular coronary vein reflected first a general reduction due to lowered blood flow during untreated LAD occlusion, and then demonstrated the phasic alternations between an SRP-induced retrograde diastolic augmentation and systolic pressure release facilitating coronary venous drainage. As in other contemporary research studies, pressures distal to the LAD occlusion were measured and blood was collected and sampled from the peripheral LAD "stump". Ideally, the new type retroperfusion was to promote prompt and effective diastolic retrograde delivery of circulation

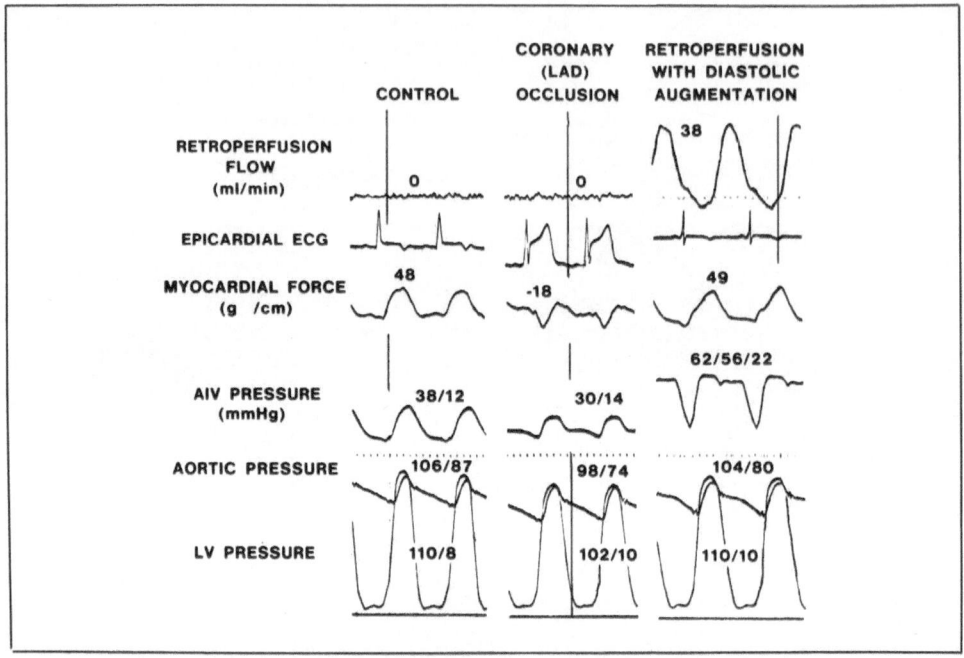

**Fig. 2.** Effects of diastolic augmented retroperfusion after 30 minutes of occlusion of the left anterior descending coronary artery (LAD) in one of the experiments in which retroperfusion effectively improved the ischemic zone contractile dysfunction. The diastolic anterior interventricular venous (AIV) pressure was increased by the synchronized pumping to 56 mmHg, thus providing augmented retroperfusion in diastole. Programmed phasing of the coronary venous pressure also facilitated systolic drainage from the regional coronary vein around the nonocclusive catheter into the coronary sinus. After coronary occlusion, ischemic zone myocardial force measurements indicated pronounced segmental dyskinesia coupled with epicardial S-T segment elevations. Retroperfusion caused a return to preocclusion regional mechanics and reversal of S-T elevation in the ischemic zone. Small changes in aortic and left ventricular (LV) pressure were observed during the coronary occlusion and subsequent reperfusion periods. ECG = electrocardiogram. (By permission, from: Meerbaum S et al. (1976) Diastolic retroperfusion of acutely ischemic myocardium. Am J Cardiol 37:588−598.)

and arterialization into the capillaries subserving the jeopardized LAD-occluded acutely ischemic myocardium. As can be seen in Figs. 2 and 3, this objective was partly accomplished. Thus, although some systolic coronary venous obstruction undoubtedly persisted in this initial experimental protocol due to imperfect catheter design, a significant diastolic regional coronary vein augmentation was clearly achieved, and this was associated with improved regional contraction, as well as with signs of decreased ischemic injury (Fig. 4).

The specific protocol of this experimental study consisted of a 15 min untreated LAD occlusion followed by 45 min of retroperfusion with maintained LAD occlusion. The purpose was to first investigate the effectiveness of a synchronized retroperfusion in the setting of a totally reversible acute ischemia. In contrast with many studies up to that time (early 1970s), when interventions for infarct salvage were

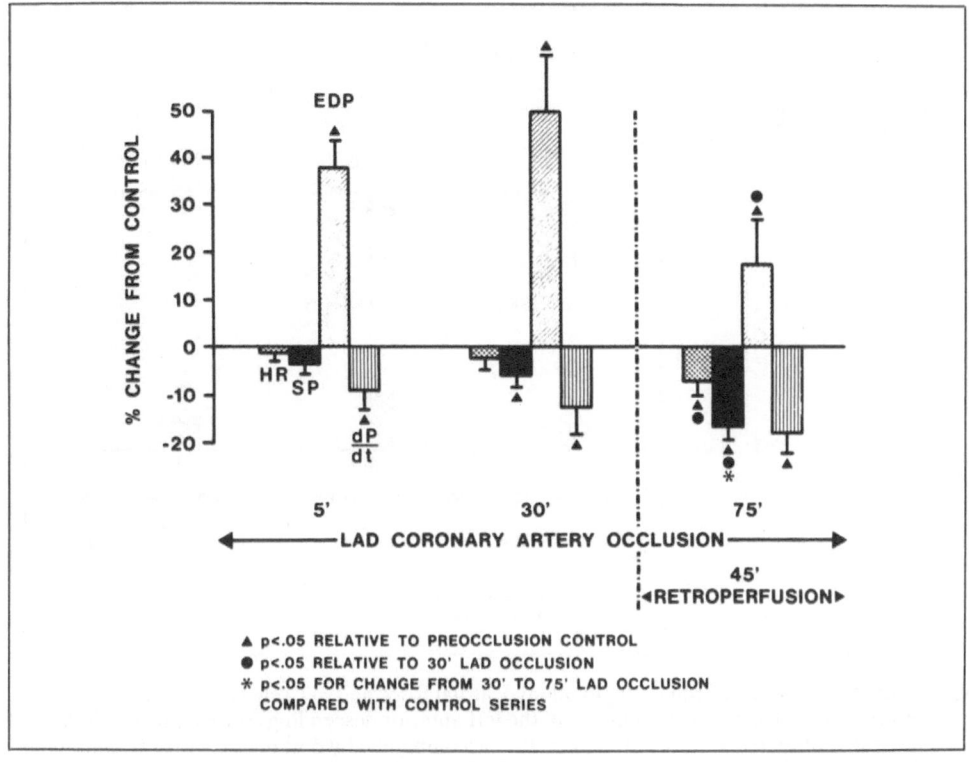

**Fig. 3.** Hemodynamic effect of retroperfusion after occlusion of the left anterior descending coronary artery (LAD) in series B (16 dogs). Bar chart presents statistical data on heart rate (HR), systolic blood pressure (SP), left ventricular end-diastolic pressure (EDP), and maximal isovolumetric rise of left ventricular pressure (dP/dt) in response to coronary occlusion and retroperfusion. Forty-five minutes of retroperfusion after 30 minutes of occlusion resulted in a further reduction in systolic pressure that was significant relative to data in the untreated series at 75 minutes of occlusion. (By permission, from: Meerbaum S et al. (1976) Diastolic retroperfusion of acutely ischemic myocardium. Am J Cardiol 37: 588–598.)

often started before or simultaneously with a coronary artery occlusion, this investigation applied retroperfusion in the presence of already established major consequences of profound acute ischemia, characterized by severe regional myocardial dysfunction. The duration of the experimental SRP was too short to derive conclusions as to infarct size, chronic effects or ultimate safety of the procedure. However, substantial prior studies used simple steady (nonsynchronized) shunting of arterial blood into an obstructed great cardiac vein (e.g., wedged retroperfusion cannulae). As one might expect with this simple shunting, there was evidence of increased coronary venous pressure during systolic contraction as a result of the nonphased coronary vein occlusion, while electromagnetic flowmeter measurements showed that most of the nonpulsed retroperfusion flow occurred in diastole. With such steady shunting to the great cardiac vein, even longer retroperfusion durations and higher retroinfusion flows (as compared to those used in this SRP

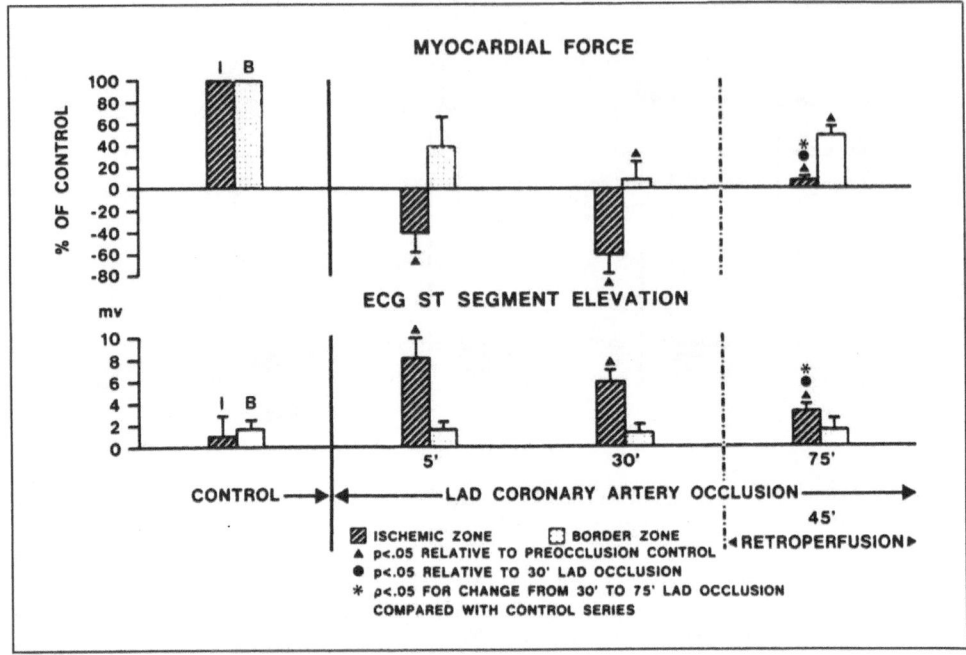

**Fig. 4.** Effects of retroperfusion on regional myocardial force and S-T segment elevation in series B (16 dogs). After 30 minutes of occlusion of the left anterior descending coronary artery (LAD), there was a loss of ischemic region (I) contraction accompanied by dyskinesia. Forty-five minutes of diastolic retroperfusion (with persisting occlusion) caused a significant partial return of regional myocardial force. An improvement in force is also noted in the border zone (B). Retroperfusion resulted in a significant reduction in the ischemic region epicardial S-T elevation but order zone S-T segments remained unaltered. ECG = electrocardiographic. (By permission, from: Meerbaum S et al. (1976) Diastolic retroperfusion of acutely ischemic myocardium. Am J Cardiol 37:588–598.)

study) failed to produce the improvements in regional cardiac function and signs of reversal of ischemia generally experienced with SRP. Similarly, when such a shunt retroperfusion was potentiated by a nonsynchronized finger pump, the resulting positive flow displacement still did not result in benefits comparable to those achieved with SRP. Based on the admittedly few and relatively short experiments, the SRP modality also appeared to clearly diminish the vascular trauma, coronary vein engorgement, myocardial edema and hemorrhages, frequently encountered with nonsynchronized modalities featuring obstruction of the normal coronary venous drainage.

It needs to be pointed out in reviewing this first SRP study that, while improvements of ischemic derangements were significant (and indeed in some cases very marked, with practically full reversal of acute ischemic injury), statistical data reflected a substantial variability of treatment effectiveness and indicated incomplete return of regional function. It was hypothesized, based on the above observations, that the variably effective obstruction of regional coronary vein caused some of diastolically retroinfused arterial blood to be shunted back toward the coronary sinus and away from the retrograde ischemic zone perfusion. However, critical evaluation of the data of this study leads to a conclusion that, even under optimally synchronized and controlled phasing of both retroinfusion and coronary vein occlusions, treatment effectiveness may remain only partial. Furthermore, there remained a host of questions about SRP effects if the experimental protocol is modified, e.g., in terms of duration of the coronary artery occlusion and length of the SRP treatment. Finally, it was clearly desirable to proceed toward an improved pump-catheter system.

## Initial SRP study in closed-chest animals

A second important SRP study (reported by Farcot et al. in 1978) was performed in closed-chest dogs [6]. Although a double lumen catheter could have been used, an autoinflatable balloon-tipped single lume catheter was devised and used in conjunction with the ECG-synchronized bladder pump, so as to provide diastolic retroinfusion of arterial blood during coronary vein catheter balloon inflation, (Fig. 5). Systolic coronary venous drainage around the SRP catheter was facilitated during systole as the retroperfusion balloon was rapidly collapsed, (Fig. 6). An example of SRP-induced augmented diastolic pressure and maintained systolic blood pressure in the regional coronary vein is shown in Fig. 7. The protocol was

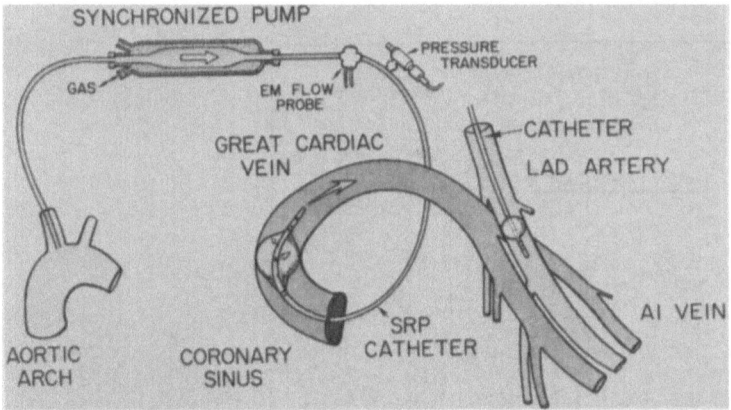

**Fig. 5.** Schematic diagram of synchronized retroperfusion experimental system. Arterial blood is shunted from the brachial artery into the great cardiac vein and the regional anterior interventricular (AI) coronary vein that adjoins the left anterior descending (LAD) coronary artery. The latter is occluded by means of an intracoronary balloon catheter to create a zone of acute ischemia to be treated by retroperfusion. Arterial blood is pumped by means of an electrocardiograph-synchronized gasactuated bladder pump that propels blood in retrograde manner during diastole through a special autoinflatable balloon catheter into the coronary vein. The retroperfusion flow rate and blood pressure are monitored by means of an electromagnetic flowmeter (EM FLOW) probe and a pressure transducer. SRP = synchronized retroperfusion. (By permission, from: Farcot JC et al. (1978) Synchronized retroperfusion of coronary veins for circulatory support of jeopardized ischemic myocardium. Am J Cardiol 41:1191–1201.)

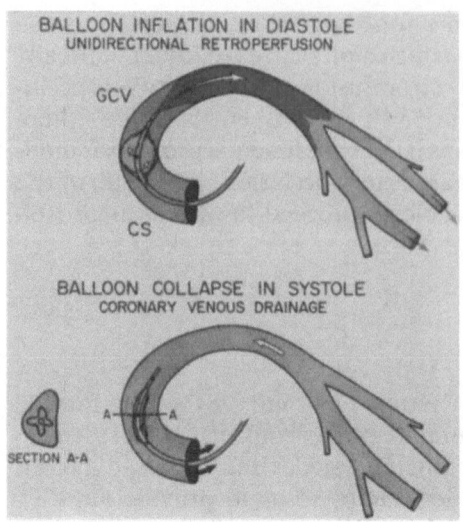

**Fig. 6.** Phasic operation of synchronized retroperfusion. During cardiac diastole, arterial blood is pumped through the catheter placed within a regional coronary vein such as the great cardiac vein (GCV). This also inflates the catheter balloon, ensuring unidirectionality of the retrograde flow. During systole, the electrocardiographic synchronized pump stops forward flow, causing a sharp decrease in pressure within the catheter. This leads to rapid balloon collapse, which facilitates coronary venous drainage. Note the characteristic mode of balloon folding in the form of a star around the body of the catheter (SECTION A-A). CS = coronary sinus. (By permission, from: Farcot IC et al. (1978) Synchronized retroperfusion of coronary veins for circulatory support of jeopardized ischemic myocardium. Am J Cardiol 41:1191–1201.)

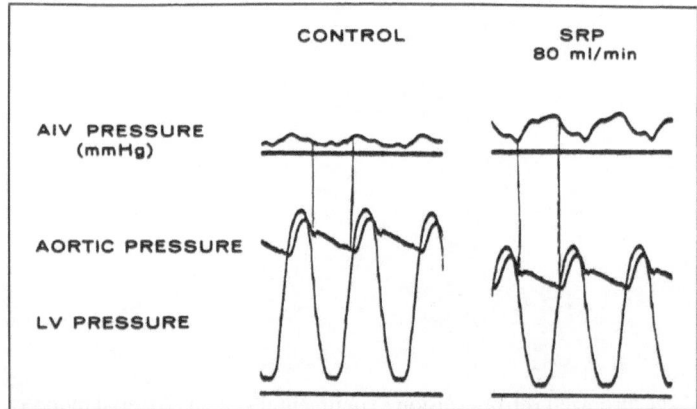

**Fig. 7.** Regional coronary venous pressure dynamics during synchronized retroperfusion. Illustration of the effect of synchronized retroperfusion (SRP) on the pressure measured within the anterior intraventricular vein (AIV) that adjoins the occluded left anterior descending coronary artery. Application of synchronized retroperfusion causes a rise in the diastolic level of the anterior intraventricular vein pressure, enhancing retrograde delivery of arterial blood into the ischemic myocardium. In contrast, synchronized retroperfusion does not significantly alter the systolic phase of the venous pressure, indicating that near normal systolic drainage is produced with the new technique. LV = left ventricular. (By permission, from: Farcot IC et al. (1978) Synchronized retroperfusion of coronary veins for circulatory support of jeopardized ischemic myocardium. Am J Cardiol 41:1191–1201.)

180

also modified, with SRP via the great cardiac vein initiated 1 h after the acute LAD coronary artery occlusion, and the treatment was extended over another 3 h of LAD occlusion. In addition to hemodynamics, metabolism and cardiac function were measured. Infarct size was determined in this SRP series and compared to infarcts in an untreated control group of closed-chest dogs.

The LAD occlusion was accomplished with an intracoronary catheter balloon. Hemodynamics and global function were found improved by SRP, while myocardial metabolism was enhanced and infarct size significantly lowered. Most of the results (Figs. 8, 9, and Table 1) corroborated the promising features of SRP, but performance improvements also reflected only partial effectiveness. This study was still limited in that it included infarct size data on only a small number of dogs, and insufficient measurements of global and regional cardiac function. The need was also evident to further improve the system, if it was to be clinically applied.

There followed a series of important SRP validation studies. Thus, Berdeaux [7] studied regional myocardial blood flow during both untreated and retroperfusion treated experimental myocardial ischemia. The LAD coronary artery occlusion was of 2 h duration, with a 1-h untreated control state followed by retroperfusion treatment. Using a complex system of radioactive microsphere measurements at several sites of the circulation, Berdeaux determined that ischemic zone myocardial perfusion could be increased with SRP to as much as about 50% of the normal level. In addition, he found that the endocardial-to-epicardial myocardial blood flow ratio within the jeopardized region could be significantly increased by SRP, i.e., from a mean of 0.46 during the pre-SRP ischemia to 0.64 with SRP. The selective enhancement of the endo-epi ratio confirmed a previous finding by Hochberg [8].

Smith [9] and Geary [10] employed baboons for their studies of SRP. Great cardiac vein SRP (40−50 ml/min) was instituted 15 min or 1 h after LAD occlusion and continued up to 4 h of the LAD occlusion, following which antegrade reperfusion was caused by releasing the coronary artery occlusion. The animals were followed for 24 h. As compared to untreated control occluded baboons, in whom there was a 31% left ventricular infarction, synchronized retroperfusion after 15 min LAD occlusion reduced infarction to as little as 4.8% of the left ventricular mass. SRP reduction of underperfused risk zone necrosis after 1 h LAD occlusion was also reported, from 94% down to 57%. Gundry [11] investigated diastolic retroperfusion, in both normal and hypertrophied dogs. He occluded the LAD coronary artery for 40 min and began synchronized retroperfusion at 10 min post-occlusion. A 30-min SRP treatment restored 37% of systolic shortening, compared to no restoration in control dogs. He also noted normalization by retroperfusion of ischemic changes in heart rate, cardiac output, aortic pressure, dp/dt and left ventricular size, as compared to no improvements noted in control LAD occlusions. Gundry found recovery in 10 of 13 treated dogs following reperfusion, while only 2 of 13 untreated dogs survived. He thought that SRP might provide a useful support in acute myocardial ischemia, particularly prior to emergency coronary artery bypass.

Farcot [12] compared SRP to reperfusion in dogs when SRP was instituted at 10 min during a 180 min coronary occlusion. Again, SRP exhibited favorable improvement in transmural blood flow and redistribution toward the endocar-

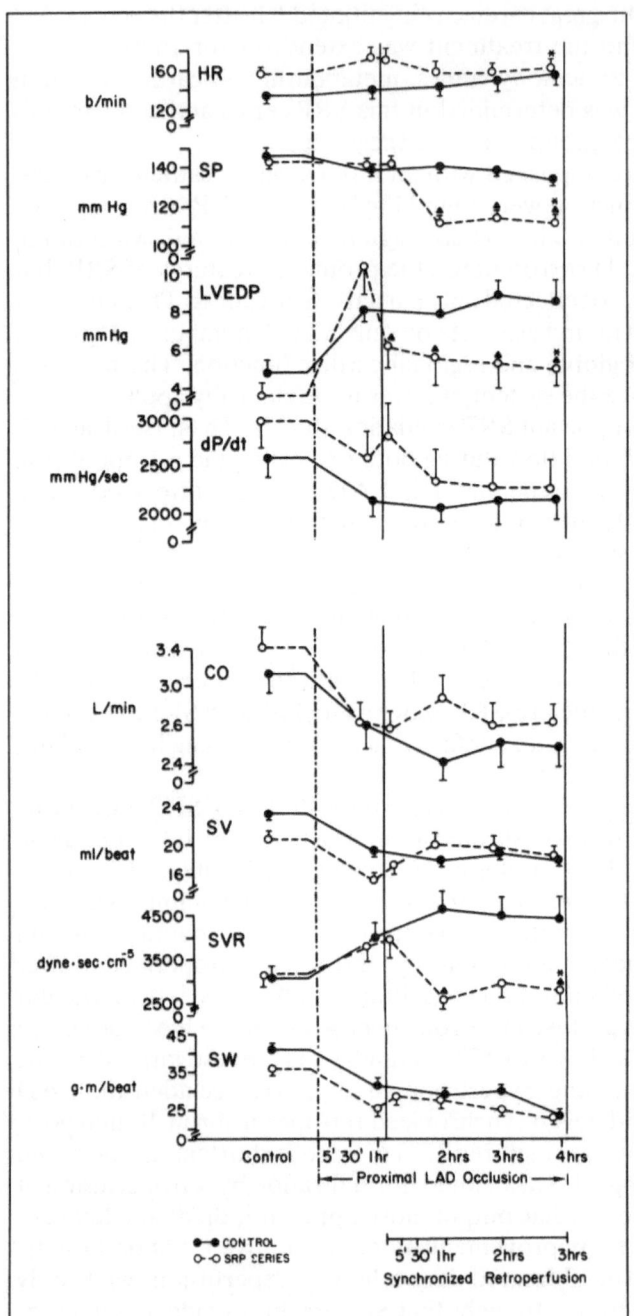

dium, with simultaneous significant improvements in regional function. However, these improvements were substantially less than those with an equivalent early antegrade reperfusion.

Yamazaki et al. [13] provided detailed data on regional and global cardiac function during SRP initiated 30 min postocclusion and carried on up to 6 h of the proximal LAD occlusion in dogs. Evidence was presented that significant beneficial SRP effects develop rapidly, within 5 min or less (Fig. 10). In a similar protocol, Drury et al. [14] applied the SRP system in a similar protocol (pre-clinical) 6 h coronary artery occlusion study. Effective infarct salvage and improved function were corroborated, along with evidence that SRP caused no significant vascular injury, myocardial edema, hemorrhages, or other potential damage including blood hemolysis (Fig. 11). Clinical SRP applications were at first based on this study's system (Fig. 12, 13) and data evaluation. More recent clinical trials are reviewed in chapter 11 of this book.

## Hypothermic SRP

Meerbaum et al. [15] initiated investigations of moderate regional hypothermic synchronized retroperfusion. The superimposed hypothermia was explored as a means to further enhance the SRP effectiveness in treating myocardial ischemia. These studies, continued by Haendchen et al. [16], reported on closed-chest dog studies with 3 and 6 h LAD coronary artery occlusion, with hypothermic SRP initiated 30 min post-occlusion. The arterial blood was cooled to 20 °C at the pump, and then delivered retrogradely in diastole to the acute ischemic myocardium. Regional myocardial temperatures were decreased by just a few degrees (to avoid potentially hazardous myocardial gradients), but heart rate was found to be diminished on the average by 30 beats/min, and rate-pressure product as well as systemic vascular resistance were significantly reduced. It appeared that the infarct size reduction with hypothermic SRP exceeded that of normothermic SRP. Two-dimensional echo measurements revealed striking improvements in both ischemic and nonischemic zone function, (Fig. 14). In an important study, Haendchen et al. [17] extended the hypothermic SRP studies to encompass 7 days' reperfusion following a 3-h LAD occlusion, with a 2.5 h SRP treatment begun 30 min post-occlu-

◄——————————————————————————————

**Fig. 8.** Statistical summary and comparison of hemodynamics in dogs with synchronized retroperfusion versus untreated series. Synchronized retroperfusion (SRP) maintained cardiac output (CO) and increased left ventricular stroke volume (SV), whereas these values decreased further in the untreated series. Synchronized retroperfusion resulted in sharply decreased systemic vascular resistance (SVR) in contrast with increased resistance during untreated occlusion of the left anterior descending coronary artery (LAD). There was no significant difference in left ventricular stroke work alterations. Δ and * represent significance as defined in the footnote to Table 1. b = beats; dP/dt = rate of rise of first derivate of left ventricular pressure; HR = heart rate; L = liter; LVEDP = left ventricular end-diastolic pressure; SW = stroke work. (By permission, from: Farcot IC et al. (1978) Synchronized retroperfusion of coronary veins for circulatory support of jeopardized ischemic myocardium. Am J Cardiol 41:1191–1201.)

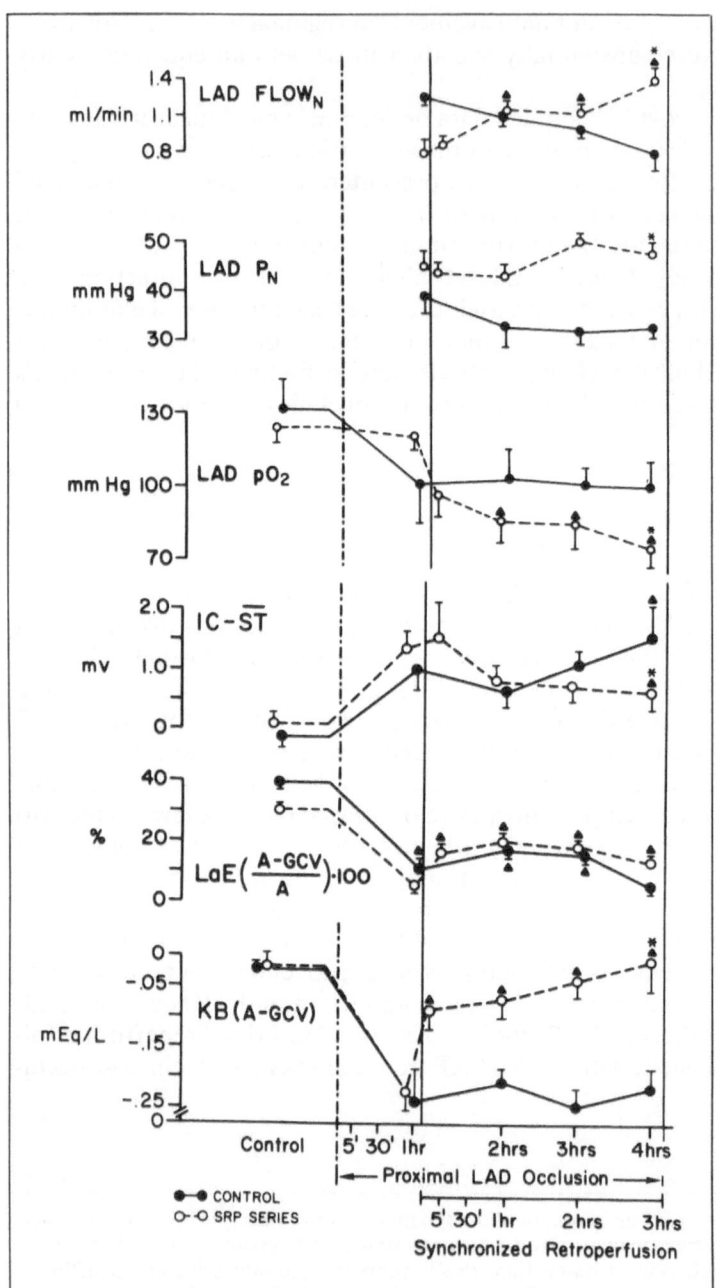

**Table 1.** Nitro-blue tetrazollum delineation of ischemic myocardium at 4 h of proximal left anterior descending coronary arterial occlusion.

| | Control Group | | | Treated Group[a] | |
|---|---|---|---|---|---|
| Dog. no. | Weight LV (g) | Ischemic Zone (% of LV) | Dog no. | Weight LV (g) | Ischemic Zone (% of LV) |
| C-1 | 176.6 | 16.3 | T-1 | 194.3 | 4.5 |
| C-2 | 141.2 | 13.5 | T-2 | 145.3 | 1.3 |
| C-3 | 197.7 | 4.2 | T-3 | 141.6 | 0 |
| C-4 | 207.8 | 33.8 | T-4 | 188 | 0.85 |
| C-5 | 170 | 13.35 | T-5 | 199.8 | 10.2 |

[a] Three h of synchronous retroperfusion after 1 h of occlusion.
LV = left ventricle. (By permission, from: Farcot et al. (1978) Synchronized retroperfusion of coronary veins for circulatory support of jeopardized ischemic myocardium. Am J Cardiol 41: pp 1191−1701.)

sion. Comparing results with untreated but similarly reperfused control occlusion dogs, use of SRP resulted in significant improvements in post-reperfusion cardiac function and its rate of recovery, while major infarct salvage improvements were also demonstrated (Fig. 15). An interesting observation indicated that retroperfusion instituted prior to the reperfusion greatly diminished the early post-reperfusion regional wall thickness (index of edema), a derangement frequently noted immediately after untreated reflow (Fig. 16).

## Coronary venous retroinfusion

Great interest was frequently expressed in retrogradely supplying drugs for treatment of ischemic myocardium [18]. Thus (without using the synchronization fea-

◀——————————————————————————

**Fig. 9.** Measurements of ischemic zone perfusion, injury and metabolic derangements. Flow and pressure (P) measurements distal to the occluded left anterior descending coronary artery (LAD) were normalized to prevailing blood pressure at 1 hour of coronary occlusion. Coronary pressure and flow increased with synchronized retroperfusion but dropped or remained unaltered in the untreated occlusion. Synchronized retroperfusion (SRP) decreased the $PO_2$ distal to the coronary occlusion, yet no change was apparent in the untreated series. Intracoronary S-T segment elevations (IC-ST) distal to the coronary occlusion rose sharply in the first hour of occlusion. Subsequently, synchronized retroperfusion caused a significant reduction in S-T elevation (measured in eight dogs), whereas there was a further rise in the untreated series. These findings may indicate that retroperfusion reduced the degree of myocardial injury in the ischemic zone. Regional lactate extraction (LaE) was slightly improved by synchronized retroperfusion. The potassium balance (KB) returned essentially to control levels after 3 hours of synchronized retroperfusion, in contrast with persisting potassium loss in the ischemic zone in the untreated series. Δ and * represent significance as defined in the footnote in Table 1. A = coronary artery; A-GCV = coronary artery-great cardiac vein gradient; L = liters; N = normalized. (By permission, from: Farcot IC et al. (1978) Synchronized retroperfusion of coronary veins for circulatory support of jeopardized ischemic myocardium. Am J Cardiol 41:1191−1201.)

185

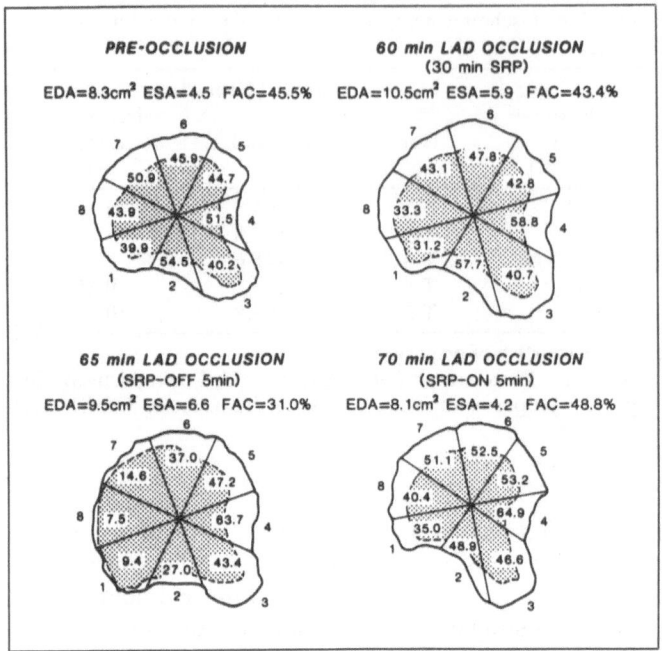

**Fig. 10.** Segmental wall motion analysis of short-axis endocardial outlines at the low left ventricular level, showing the rapid effect of synchronized retroperfusion (SRP). **Continuous lines and dashed lines** indicate left ventricular endocardium at enddiastole and end-systole, respectively. Analysis was performed before **(upper right)** and after **(lower left)** a 5-minute interruption of retroperfusion and 5 minutes after resumption of retroperfusion **(lower right)**.EDA = end-diastolic area; ESA = end-systolic area; FAC = systolic fractional area change. (By permission, from: Yamazaki S et al. (1985) Synchronized coronary venous retroperfusion: prompt improvement of left ventricular function in experimental myocardial ischemia. J Am Col Cardiol 5:655.)

ture), Kordenat [19] applied in dogs with thrombotic LAD coronary artery obstruction a retroinfusion treatment of the acutely ischemic myocardium, employing methosergide and dipyridamole. As compared to equivalent control occlusion dogs in whom stroke volume and cardiac output decreased while peripheral resistance increased. Kordenat's abstract presentation reported that the drug retroinfusions resulted in significant improvements during the evolving myocardial infarction, but details of retroinfusion mode and performance were not described.

Povzhitkov [20, 21] combined retroperfusion with retrograde coronary venous occlusion PGE₁ or mannitol. For example, during proximal LAD occlusion, PGE, extended the retroperfusion-induced functional improvements to the apical region of the left ventricle, where acute ischemic injury often tended to persist with unaided SRP. Retroinfusion resulted in significant infarct salvage, and selective use of cardioactive drugs appeared to be a potentially worthwhile direction. Berdeaux [22], on the other hand, demonstrated with SRP (retroinfusion at coronary sinus site near Marshall vein) that supplemental verapamil (2 µg/kg/min) or nitro-

186

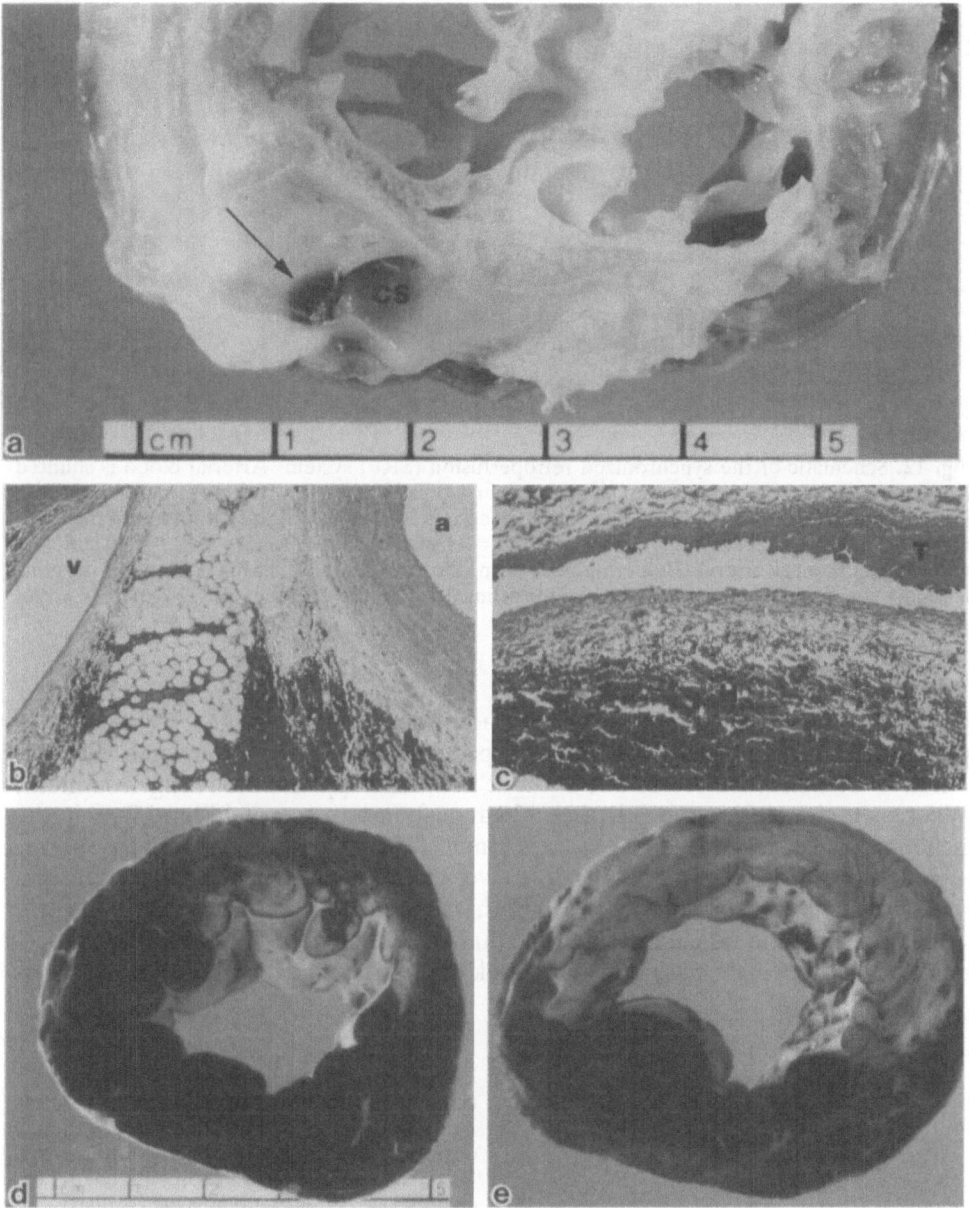

**Fig. 11.** Pathologic sections. **a,** Photograph of the right atrium in a treated dog demonstrating bruising of the atrial wall **(arrow)** near the orifice of the coronary sinus (cs). **b,** Photomicrograph of the left anterior descending coronary artery (a) and anterior interventricular vein (v) in a treated dog demonstrating hemorrhage (H) in the epicardial fat surrounding the vessels (hematoxylin-eosin stain, original magnification × 16, reduced by 38%). **c,** Photomicrograph of the great cardiac vein in a treated dog demonstrating a nonocclusive thrombus (T) and perivascular hemorrhage (H) (hematoxylin-eosin stain, original magnification × 16, reduced by 38%). **d, e,** Myocardial slabs stained with triphenyltetrazolium chloride from a treated **(d)** and a control **(e)** dog demonstrating the extent of myocardial necrosis (unstained segments). (By permission, from: Drury JK et al. (1985) Synchronized diastolic coronary venous retroperfusion: results of a preclinical safety and efficacy study. J Am Col Cardiol 6:328.)

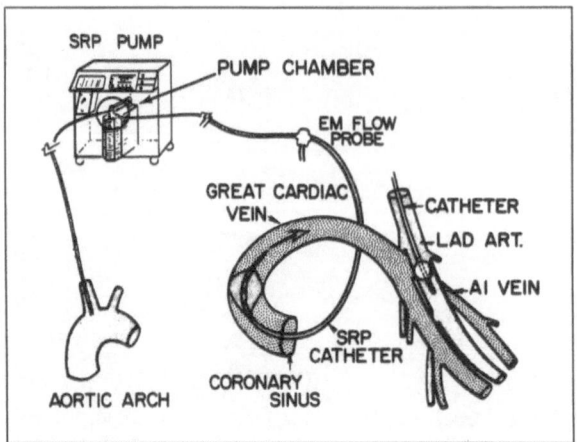

**Fig. 12.** Schematic of the synchronized retroperfusion (SRP) system. Arterial blood is shunted from a brachial artery to the pump chamber. During diastole the upward movement of a piston onto the pump chamber displaces blood through an autoinflatable balloon catheter into the coronary sinus. AI = anterior interventricular; EM = electromagnetic; LAD ART. = left anterior descending coronary artery. (By permission, from: Drury JK et al. (1985) Synchronized diastolic coronary venous retroperfusion: results of a preclinical safety and efficacy study. J Am Col Cardiol 6:328.)

glycerin (0.7 µg/kg/min) did not provide additional improvements. This study was performed in open-chest dogs with 3 h proximal LAD occlusion, and SRP was initiated 19 min post occlusion, without or with supplemental drug treatment. Whereas regional function (ultrasound gauges) and myocardial blood flow in the ischemic zone and endo/epi flow ratio (measured by radioactive microspheres) increased significantly with SRP alone, verapamil actually caused a decrease in function and myocardial blood flow in both ischemic and nonischemic zones. Nitroglycerin resulted in no changes, except that the beneficial SRP-induced enhancement of endo/epi ratio was abolished. Lack of improvements in this SRP retroinfu-

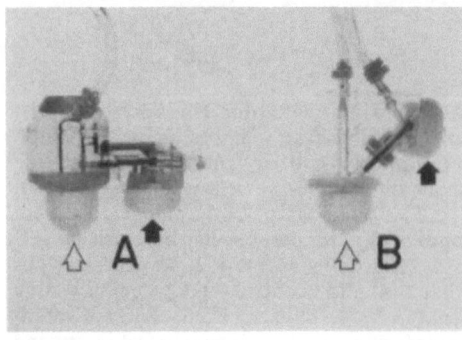

**Fig. 13.** Original (A) and modified (B) pump chamber assemblies. Arterial blood first enters a reservoir chamber **(black arrows)** and then, during systole, a pump chamber **(white arrows).** Upward movement of a piston onto the pump chamber displaces blood into the coronary sinus. A one-way valve between the reservoir and pump chambers prevents backward flow. The only difference between the two models is that the pathway from the reservoir into the pump chamber is more tortuous in model A than in model B. (By permission, from: Drury JK et al. (1985) Synchronized diastolic coronary venous retroperfusion: results of a preclinical safety and efficacy study. J Am Col Cardiol 6:328.)

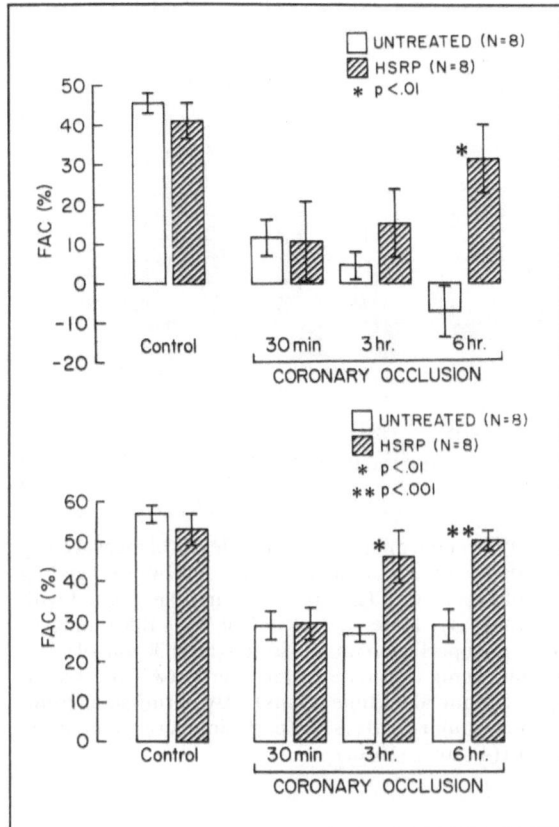

**Fig. 14.** (top) Two-dimensional echocardiographic analysis of low left ventricular section (much below the left anterior descending coronary artery occlusion site) ischemic segment, with segmental function expressed as percent systolic fractional are change (FAC). Without treatment, coronary occlusion caused severe dysfunction that progressed to frank dyskinesis. In contrast, hypothermic synchronized retroperfusion (HSRP) resulted in a sharp rise in contraction, reaching a level only moderately below the control state. (bottom) Echocardiographic quantitation of percent systolic FAC in a remote (only moderately involved) segment of a low left ventricular short-axis section. Hypocontraction during coronary occlusion was unaltered without treatment but reversed with HSRP. (By permission, from: Meerbaum S et al. (1982) Hypothermic coronary venous phased retroperfusion: a closed chest treatment of acute regional myocardial ischemia. Circulation 65:1435–1445.)

sion study may have been simply due to the vasodilators having little supplemental effectiveness in the setting of ischemia-induced coronary vasodilation. Or else, the position of the SRP catheter tip may have been too remote from the LAD occlusion site, thus reducing retroperfusion effects.

Yet another experimental retroinfusion study was aimed at retrograde delivery of antiarrhythmic agents. Karagueuzian et al. [23] retroinfused procaine amide into the great cardiac vein of dogs with 3–12 days LAD coronary artery occlusion. Whereas intravenous treatment proved effective in only 11% of the tachycardia episodes and the remainder proved refractory to such treatment, retrograde delivery of procaine amide terminated the induced sustained ventricular tachyarrhythmias in 69% of cases. It was demonstrated, by means of myocardial tissue content measurements that the antiarrhythmic drug was selectively delivered into the infarcted zone, yielding a 10-fold concentration compared to that in the remote noninfarcted region. Thus, a potentially interesting modality of antiarrhythmic treatment was proposed for critical circumstances. Otsu et al. [24] corroborated such retroinfusion benefits in another analogous study, applying lidocaine via the coronary veins during their obstruction. Retrograde administration resulted in sig-

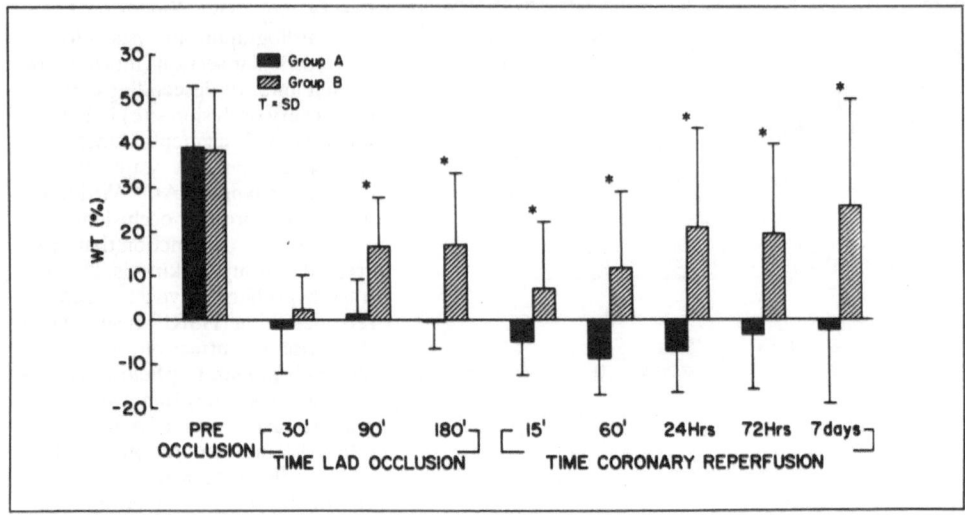

**Fig. 15.** Ischemic zone systolic wall thickening (WT) at the midpapillary muscles level measured by two-dimensional echocardiography in control dogs with 3 h of untreated left anterior descending (LAD) coronary artery occlusion and 7 days of reperfusion (group A) and in dogs treated with hypothermic synchronized retroperfusion (group B). Note restoration of wall thickening to physiologic levels in group B when hypothermic retroperfusion was instituted (after 30 min of coronary occlusion) and no improvement in wall thickening with reperfusion in group A (SD = standard deviation; * p < 0.05 relative to group A at equivalent time points). (By permission, from: Haendchen RV et al. (1983) Prevention of ischemic injury and early reperfusion derangements by hypothermic retroperfusion. J Am Col Cardiol 1(4):1067−1080.)

nificantly increased myocardial concentrations of the antiarrhythmic agent, which was most effective in terminating ventricular tachycardia.

Recent SRP retroinfusion research indicating further benefits also encompassed calcium antagonists such as diltiazem [25], and retrograde administration of superoxide dismutase. The latter drug was shown to have the capacity of minimizing derangements and injury often associated with early postreperfusion periods following acute coronary artery occlusion. Hatori et al. [26] demonstrated in animals that retrogradely delivered oxygen radical scavengers such as superoxide dismutase reduced reperfusion injury and functional stunning even more effectively than when the drug was administered antegradely. The reason is believed to be the general difficulty of antegradely delivering agents into jeopardized myocardium distal to a coronary artery occlusion.

### Retrogradely induced lysis of coronary artery thrombus

Previously mentioned peripheral coronary artery measurements of oxygen content following coronary artery occlusion alone and then with retroperfusion, indicated that the retrograde interventions were generally associated with a change of the blood beyond the occlusion from arterial to venous character. It was frequently

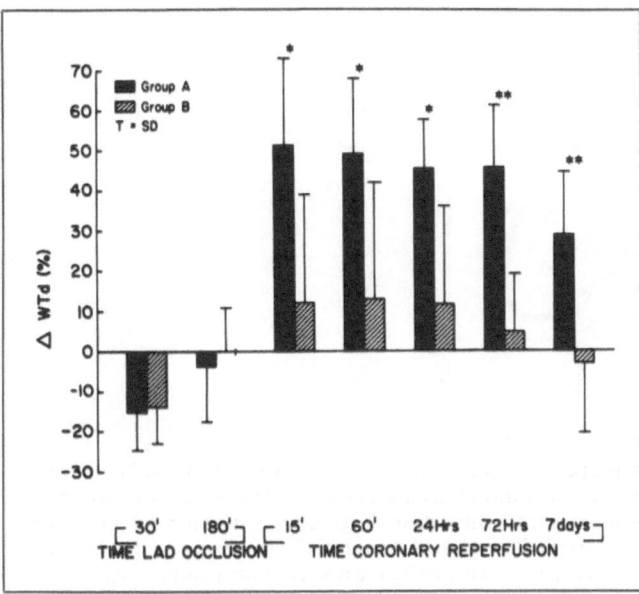

**Fig. 16.** Percent change in ischemic zone end-diastolic wall thickness from preocclusion control values (ΔWTd), measured by two-dimensional echocardiography at the midpapillary muscle level. Diastolic wall thickness decreased by approximately 15% in both groups at 30 min of left anterior descending coronary occlusion. No significant differences in diastolic wall thickness were observed between dogs in group A (untreated 3 h of left anterior descending coronary artery occlusion followed by 7 days of reperfusion) and group B (hypothermic synchronized retroperfusion treatment from 30 to 180 min of the occlusion period, followed by 7 days of reperfusion) during the occlusion period. However, early after reperfusion, dogs in group A exhibited marked increase in diastolic wall thickness which persisted throughout the 7-day period, as opposed to dogs in group B in which only minor changes were observed after reperfusion. (* p < 0.01, ** p < 0.001 relative to group B). Abbreviations as in Fig. 15. (By permission, from: Haendchen RV et al. (1983) Prevention of ischemic injury and early reperfusion derangements by hypothermic retroperfusion. J Am Col Cardiol 1(4):1067−1080.)

unclear whether this change reflected effective retroperfusion and ischemic tissue oxygen resupply and extraction, or else a forced veno-arterial shunting of blood from the coronary veins (or some combination of the two mechanisms). Regardless of the precise explanation, a variety of in vivo and in vitro experiments provided convincing evidence that retroinfusion (e.g., via the great cardiac vein) could deliver an agent beyond the regional myocardial microcirculation into the corresponding obstructed coronary artery (e.g., the LAD). The effectiveness of such delivery was unknown, some experience indicating that under favorable circumstances, about 25% of the retroinfusate might be retrodelivered into the regional coronary artery. These observations led to an attempt by Meerbaum et al. [27] to retroinfuse streptokinase for the purpose of lysing a coronary artery thrombus.

In the closed-chest dog preparation (Fig. 17), the arterial clot was generated by an intravascular thrombogenic copper coil, resulting in verified full coronary artery occlusion within a period of anywhere from 10−60 min. Retroinfusion of strepto-

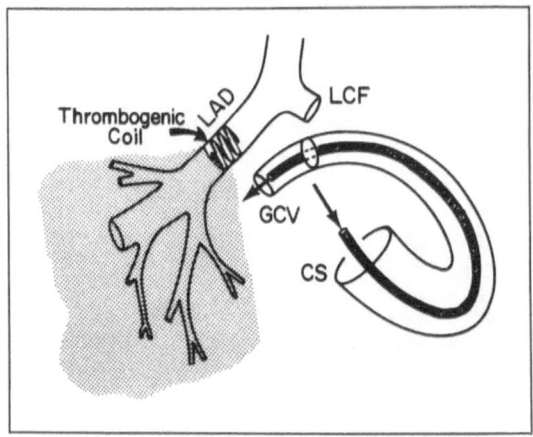

**Fig. 17.** Schematic of experimental preparation for study of coronary venous streptokinase retro-infusion treatment of induced coronary artery thrombus. Left anterior descending coronary artery (LAD) thrombosis was induced with an intravascular thrombogenic coil. The thrombolytic agent was retroinfused into a temporarily obstructed great cardiac vein (GCV) or else in conjunction with synchronized retroperfusion support of the regionally ischemic myocardium pending achievement of coronary artery reperfusion. Retrogradely induced thrombolysis was demonstrated in dogs and, in principle, a similar coronary sinus (CS) based lytic agent retroinfusion could also be envisioned in the setting of a thrombotic left circumflex coronary artery (LCF) occlusion. (By permission, from: Meerbaum S (1986) Coronary venous retroperfusion delivery of treatments to ischemic myocardium. Herz 11(1):41−54.)

kinase was initiated 90 min after LAD coil insertion (on the average about 60 min post LAD occlusion), and was carried out in two modalities: 1) alternating brief retroinfusions with intervening periods of untreated occlusion, or 2) as a supplemental drug infusion into the SRP pumping circuit. While there were no significant differences in mean results with these two retroinfusion approaches, less variability was encountered when using the SRP route. The retrograde coronary vein strepto-kinase infusion dosage was 2000 IU/min. Lysis of the coronary artery clot and partial as well as full reperfusion were examined by repeat coronary angiography. Figure 18 indicates a sequence of typical observations in the dog. Figure 19 shows the statistical results, with retrograde infusion compared with equivalent dosage intravenous administration. The streptokinase-SRP method provided significantly more rapid lysis of the coronary artery thrombus (initial signs of reperfusion at a mean of about 25 min, full reperfusion at 50 min after LAD occlusion), as compared to the intravenous streptokinase (about 125 min). This favorable result was obtained even though retroinfusion was obviously also associated with shunting to the systemic circulation. A striking feature of the thrombolytic retroinfusion was an apparent significant reduction in ischemic injury, and evidence (by 2-DE study) of more rapid post-reperfusion return of regional cardiac function. No reperfusion arrhythmias were encountered. More recently, a similar experimental SRP retroin-fusion study corroborated effects of retrogradely administered streptokinase [28]. Results were equivalent, with some changes in dosage actually indicating the possi-bility of even more rapid lysing of the coronary artery thrombus by means of strep-

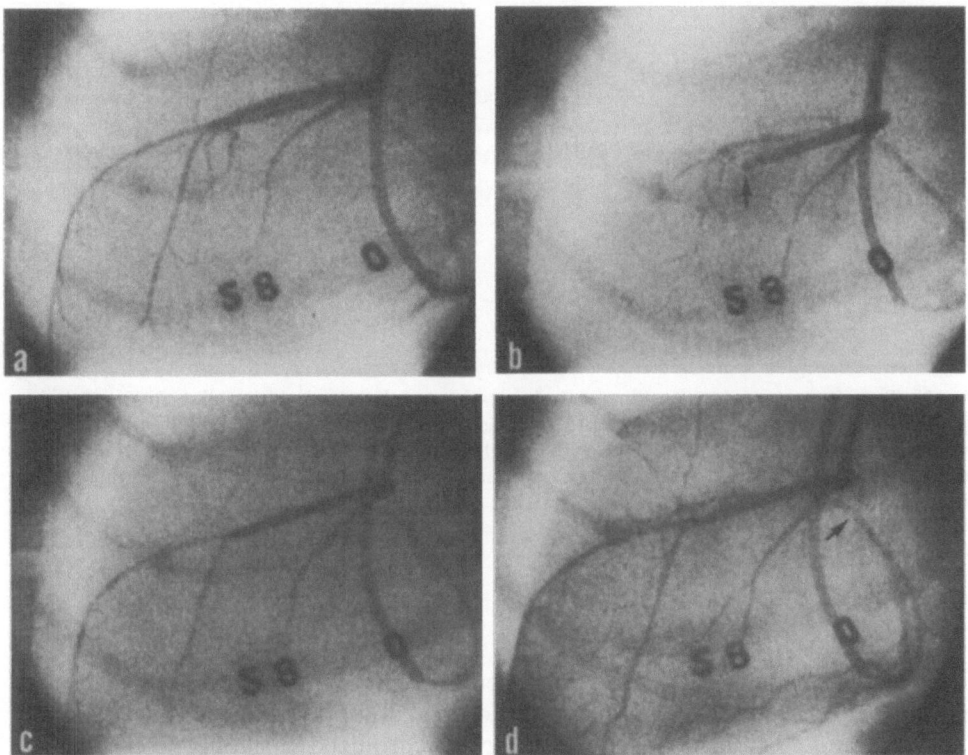

**Fig. 18.** Angiographic documentation of an experimental sequence of coronary artery events before and after streptokinase retroperfusion-induced thrombolysis in one closed-chest dog: **(a)** selective coronary angiography in the control state, showing the left anterior descending coronary artery and its branches; **(b)** total thrombotic obstruction of the left anterior descending artery (see **arrow**) beyond the small copper coil placed at a relatively proximal site; **(c)** initial left anterior descending coronary reflow after streptokinase treatment byx way of coronary venous retroperfusion; and **(d)** fully reestablished coronary flow noted 50 min after start of the streptokinase retroperfusion via great cardiac vein (see **arrows**). Similar coronary artery thrombolysis was achieved in all dogs treated with retrograde coronary venous streptokinase infusion. (By permission, from: Meerbaum S et al. (1983) Retrograde lysis of coronary artery thrombus by coronary venous streptokinase administration. J Am Col Cardiol I:1262–1267.)

tokinase retroinfusion. A further investigation also examined the effectiveness of retrograde injection of the tPA thrombolytic agent for treatment of coronary artery thrombosis [29].

## Technological aspects of SRP

Current synchronized retroperfusion systems and protocols are undergoing modification, so that it is not possible to be certain of the eventual system components and mode of application. However, certain rationales and ongoing technology should be briefly noted.

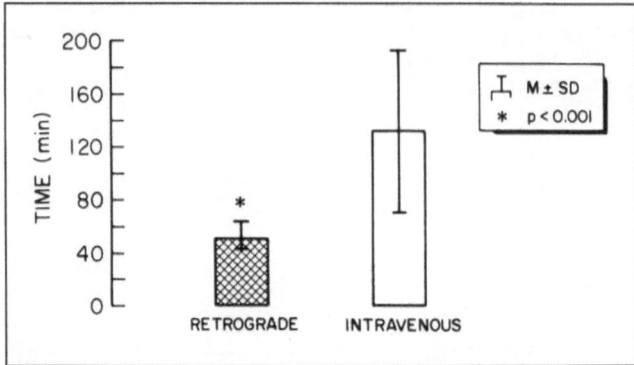

**Fig. 19.** Time from start of streptokinase administration to full thrombolysis. All retrograde administration measurements were gathered to emphasize the relatively small variability and significantly shorter time to lysis as compared with the intravenous method. (By permission, from: Meerbaum S et al. (1983) Retrograde lysis of coronary artery thrombus by coronary venous streptokinase administration. J Am Col Cardiol I:1262–1267.)

### Pumping system

During many of the reported experimental investigations, the extracorporeal pumping device consisted of variously homemade or specially produced simple fluid-actuated bladder pumps, providing synchronized diastolic propulsion of blood into the retroperfusion catheter, and interruption of this blood flow in systole. A more complex, professionally manufactured, programmed piston type pump was designed for clinical application by Retroperfusion Systems, Inc. (RSI, Costa Mesa, California). It consists of a cassette with inlet and pumping chambers, separated by a valve. The piston which pumps blood from an arterial source to the great cardiac veins is actuated by an electronic microprocessor equipped console, and there is another valve at the pump outlet. RSI has been modifying their pumping system, primarily to allow retrograde delivery of increased volumes, apparently required for effective clinical retroperfusion, and also to provide a more flexible programming for a variety of phasic/intermittent retroinfusions. The FDA approved RSI protocols for specific clinical SRP applications, including support during PTCA and circulatory assistance during unstable angina.

### Retroperfusion catheter

After employing homemade and manufactured (e.g., USCI, Billerica, New York) simple autoinflatable single lumen SRP balloon catheters during most of the reported experimental investigations, some recent clinical evaluations used improved double lumen autoinflatable balloon catheters. The latter featured an extra lumen specially dedicated for coronary vein pressure measurements, which is considered important for safety an operational control purposes. To assure total systolic deflation of the autoinflatable balloon, a valve was also placed at the cathe-

194

ter lumen tip, thus facilitating the important systolic drainage phase of SRP. Alternately, other recent clinical trials applied a triple lumen 8-F SRP catheter in which the pumped flow is delivered through the center lumen, a second lumen serves for separate controlled fluid inflation of the oval balloon (near the catheter tip), and the third lumen is again dedicated for coronary vein pressure measurements. A fixed volume of gas from a pressurized source is used to inflate the catheter balloon to a diameter of the order of 10 mm. In light of anatomic variability and need for specific catheter tip balloon positioning, it is anticipated that other catheter sizes and design configurations will be produced.

*Coronary vein catheterization*

There is substantial clinical experience with coronary sinus catheterization [30−33], and anatomic factors are taken into consideration (see chapters 4 and 5). Various approaches to the right atrium and then into the coronary sinus and great cardiac vein were discussed recently, along with pertinent catheter maneuvers [34]. Regarding catheter insertion points, the internal jugular vein or the left subclavium vein catheterization have been employed, but the brachial venous approach is also considered. It is urged to use contrast venography to assist decisions as to catheter placement, and also to keep in mind the more delicate nature of the thin coronary venous vessels. It appears that rapid (within 3 min) catheterization of the great cardiac vein can be normally achieved in 90% of attempts. More difficulty is said to be encountered in patients with right atrial dilatation. In one recent PTCA application of SRP [35], 11 out of 12 patients could be catheterized within 3 min and secure catheter positioning was achieved in the coronary sinus near the right atrium (six patients) or else deep in the coronary sinus approaching the great cardiac vein (five patients).

*Assessment of SRP effectiveness*

Monitoring during SRP protocols would normally include assessment of frequency of pain, electrocardiographic changes, blood pressure, enzymes and other blood analysis. It is also desirable to evaluate SRP effects on myocardial function and perfusion, and evaluation of regional tissue viability is of distinct interest. Two-dimensional echocardiography can be applied to provide useful measurements of regional and global left ventricular function, as is currently demonstrated in the PTCA-SRP protocols. As has been previously pointed out, contrast venography provides important information on the specific details of the coronary venous system. In addition, it now appears that special echo contrast agents featuring microbubbles of the order of 3−5 microns in diameter, make it possible to apply myocardial contrast 2-DE (see chapter 10) for assessment of changes in myocardial perfusion defects. Other methodologies and advances in monitoring and sequential measurement of myocardial perfusion (and possibly metabolism) can be anticipated in the future. If the convenient echocardiographic technique can be validated in conjunction with a potentiating agent such as papaverine, myocardial perfusion reserve and tissue viability might also be inferred, when SRP or other coronary venous interventions are used to support the jeopardized ischemic or reperfused myocardium.

## Conclusion

SRP has been experimentally shown to have potential in treating acute myocardial ischemia due to coronary artery occlusion. Clinical trials are under way (see next chapter), using upgraded SRP systems. In general, safety and feasibility seem to be well established through studies of Gore et al. [36], Berland et al. [35], Kar et al. [37], and other as yet unpublished human trials. Much current effort appears to center on treatment of ischemia associated with single-vessel occlusions during PTCA balloon inflations. Clinical SRP effectiveness in the setting which was extensively studied and found successful in animals, i.e., during evolving acute myocardial infarction, has not as yet been adequately demonstrated. There are clearly important differences between the experiments and clinical applications. One known factor is the different coronary venous anatomy, which may make it more difficult in the human to develop sufficient coronary venous pressures unless one uses greater than anticipated retroinfusion flows. Catheters, pumps, and protocols must all be reexamined in terms of reliability and optimal SRP application. Nonetheless, it is expected that the current clinical efforts will succeed in pinpointing patient subsets and conditions in which SRP or modified SRP treatment can be employed as a useful adjunct support in the area of interventional cardiology. The International Group on Coronary Sinus Interventions [38], through its biannual symposia and quarterly newsletters, provides a useful link between the various investigators and a forum for review of scientific and clinical advances.

## References

1. Meerbaum S (1986) Coronary venous retroperfusion delivery of treatment to ischemic myocardium. Herz 11(1):41−54
2. Meerbaum S (1986) Synchronized retroperfusion. Developments and current concepts. In: Mohl W, Faxon D, Wolner E (eds) Clinics of CSI. Steinkopff, Darmstadt/Springer, New York, pp 253−258
3. Corday E, Kar S, Drury JK, Ryden L, O'Byrne GT, Hajduczki I, Tadokoro H, Chang BL (1988) Coronary venous retroperfusion for support of ischemic myocardium. CVR & R:50−53
4. Mohl W, Roberts AJ (1985) Coronary sinus retroperfusion and pressure-controlled intermittent coronary sinus occlusion (PICSO) for myocardial protection. Surg Clin North Am 65(3):477−485
5. Meerbaum S, Lang TW, Osher JV, Hashimoto K, Lewis GW, Feldstein C, Corday E (1976) Diastolic retroperfusion of acutely ischemic myocardium. Am J Cardiol 37:588−598
6. Farcot JC, Meerbaum S, Lang TW, Kaplan L, Corday E (1978) Synchronized retroperfusion of coronary veins for circulatory support of jeopardized ischemic myocardium. Am J Cardiol 41:1191−1201
7. Berdeaux A, Farcot JC, Boudarias JP, Barry M, Bardet J, Jiudicelli JF (1981) Effects of diastolic synchronized retroperfusion on regional coronary blood flow in experimental myocardial ischemia. Am J Cardiol 47:1033−1040
8. Hochberg MS, Roberts WC, Morrow AJ, Austen WG (1979) Selective arterialization of the

coronary venous system: encouraging long term flow evaluation utilizing radioactive microspheres. J Thor Cardiavasc Surg 77:1

9. Smith GT, Geary GG, Blanchard W, McNamara II (1981) Reduction in infarct size by synchronized selective coronary venous retroperfusion of arterialized blood. Am J Cardiol 48:1064−1070

10. Geary GG, Smith GT, Suehiro GT, Zeman C, Siu B, McNamara JJ (1982) Quantitative assessment of infarct size reduction by coronary venous retroperfusion in baboons. Am J Cardiol 50:1424−1430

11. Gundry SR (1982) Modification of myocardial ischemia in normal and hypertrophied hearts, utilizing diastolic retroperfusion of the coronary vein. J Thor Cardiovasc Surg 83:659

12. Farcot JC, Berdeaux A, Giudicelli JF, Vilaine JP, Boudarias JP (1983) Diastolic synchronized retroperfusion vs reperfusion: Effects on regional left ventricular function and myocardial blood flow during acute coronary occlusion in dogs. Am J Cardiol 51:1414−1421

13. Yamazaki S, Drury JK, Meerbaum S, Corday E (1985) Synchronized coronary venous retroperfusion: prompt improvement of left ventricular function in experimental myocardial ischemia. J Am Col Cardiol 5:655

14. Drury JK, Yamazaki S, Fishbein M et al. (1985) Synchronized diastolic coronary venous retroperfusion: results of a preclinical safety and efficacy study. J Am Col Cardiol 6:328

15. Meerbaum S, Haendchen RV, Corday E, Fishbein M, Ritt J, Lang TW, Uchiyama T, Aosaki N. Broffman J (1982) Hypothermic coronary venous phased retroperfusion: a closed chest treatment of acute regional myocardial ischemia. Circulation 65:1435−1445

16. Haendchen RV, Corday E, Meerbaum S (1982) Hypothermic synchronized retroperfusion of the coronary veins for the treatment of acutely ischemic myocardium. Compr Ther 8(12):7−15

17. Haendchen RV, Corday E, Meerbaum S, Povzhitkov M, Ritt J, Fishbein MC (1983) Prevention of ischemic injury and early reperfusion derangements by hypothermic retroperfusion. Am Col Cardiol 1(4):1067−1080

18. Corday E, Meerbaum S, Drury JK (1986) The coronary sinus: an alternate channel for administration of arterial blood and pharmacologic agents for protection and treatment of acute cardiac ischemia. J Am Col Cardiol 7:711

19. Kordenat RK (1979) Retroperfusion of the ischemic myocardium with methysergide and dipyridamole. VII Int. Congr Throm Haem, p 6−100

20. Povzhitkov M, Haendchen RV, Meerbaum S, Fishbein M, Ritt J, Corday E (1982) Protective effect of coronary venous prostaglandin E, retroperfusion during acute myocardial ischemia. Am J Cardiol 49:1017

21. Povzhitkov M, Haendchen RV, Meerbaum S, Fishbein M, Ritt J, Corday E (1982) Mannitol coronary venous retroperfusion: improvement in ischemic left ventricular function in acute occlusion. Clin Res 30:17

22. Berdeaux A, Farcot JC, Giudicelli JF, Boudarias JP (1984) Failure of regional vasodilator drug to potentiate the retroperfusion beneficial effect in ischemic myocardium in dogs. In: Mohl W, Wolner E, Glogar D (eds) The coronary sinus. Steinkopff, Darmstadt, pp 354−359

23. Karagueuzian HS, Ohta M, Drury JK, Fishbein MC, Corday E, Meerbaum S, Mandel WJ, Peter T (1984) Coronary venous retroinfusion of procaine amide in the management of inducible ventricular tachyarrhythmias in conscious dogs during chronic myocardial infarction. In: Mohl W, Wolner E, Glogar D (eds) The coronary sinus. Steinkopff, Darmstadt, pp 385−391

24. Otsu F, Carew TE, Maroko PR (1985) Myocardial concentration and antiarrhythmic effects of lidacaine administered via coronary veins. Col Cardiol 5(2):467

25. Tadokoro H, Miyazaki A, Satomura K et al. (1988) Profound infarct size reduction with coronary venous retroinfusion of diltiazen in pigs (Abstract). J Am Col Cardiol 11:65A

26. Hatori N, Drury JK, Satomura K et al. (1987) Reduction of infarct size with retrograde but not antegrade administration of superoxide dismutase in pigs (Abstract). Circulation 76 [Suppl IV]:200

27. Meerbaum S, Lang TW, Povzhitkov M, Haendchen R, Uchiyama T, Broffman J, Corday E (1983) Retrograde lysis of coronary artery thrombus by coronary venous streptokinase administration. J Am Col Cardiol I:1262−1267

28. Miyazaki A, Drury JK, Hatori N, Corday E (1987) Improved clot lysis and infarct size reduction with coronary venous compared with intravenous infusion of streptokinase (Abstract). JACC 9(2):82A

29. Miyazaki A et al. (unpublished) Retrograde lysis of coronary artery thrombus by coronary venous tPA administration
30. Ganz W, Tamura K, Marcus HS et al. (1971) Measurement of coronary artery sinus blood flow by continuous thermodilution in man. Circulation 44:181–195
31. Baim BS, Rothman MT, Harrison DC (1980) Improved catheter for regional coronary sinus blood flow in metabolic studies. Am J Cardiol 46:997–1000
32. Bradley AD, Baim BS (1985) Measurement of coronary blood flow in man: Methods and implications for clinical practice. In: Shroeder JS (ed) Invasive cardiology. Davis Co., Philadelphia
33. Dehmer GJ, Schmitz JM, Malloy CR (1986) A new technique for cannulation of the coronary sinus from the femoral vein. Cathet Cardiovasc Diagn 12:427–429
34. Baim DS (1987) Techniques for coronary sinus catheterization. International working group on coronary sinus interventions, newsletter 1(3):10–11
35. Berland J, Farcot JC, Barrier A et al. (1988) ECG and 2-D echo assessment of myocardial protection achieved by synchronized coronary retroperfusion during LAD angioplasty (Abstract). J Am Col Cardiol 11:132A
36. Gore JM, Weiner BH, Benotti JR et al. (1986) Preliminary experience with synchronized coronary sinus retroperfusion in humans. Circulation 74:381
37. Kar S, Drury JK, Eigler N et al. (1988) Amelioration of ischemia during PTCA with diatolic coronary venous retroperfusion (Abstract). J Am Col Cardiol 11:64A
38. International working group on coronary sinus interventions. In: Mohl W (ed) (Sec Gen) 2nd Surgical Clinic, University of Vienna, Austria

# Clinical experience with diastolic coronary venous retroperfusion

J. K. Drury

Clinical trials to determine the safety and efficacy of diastolic coronary venous retroperfusion have recently been initiated, more than 10 years after successful introduction of the technique in the animal laboratory. To date, diastolic retroperfusion has been employed in two clinical situations: in patients undergoing percutaneous transluminal coronary angioplasty, and in patients with medically refractory unstable angina. The number of patients who have been studied is relatively small. Nevertheless, the preliminary results have been encouraging and suggest that diastolic retroperfusion may provide an effective alternate blood supply for temporary support of patients with acute myocardial ischemia.

## The technique

Diastolic coronary venous retroperfusion is achieved by shunting fully heparinized autologous arterial blood (usually from a femoral artery) through an ECG synchronized electromechanical pump via a catheter into the coronary veins (Fig. 1). A coronary sinus catheter is inserted percutaneously under fluoroscopic control through the right internal jugular vein (other central venous access sites may be used) into the coronary sinus under fluoroscopic control and advanced to the great cardiac vein. The clinical retroperfusion catheter currently available (Retroperfusion Systems, Inc.) has two lumens, one for blood delivery and one for pressure

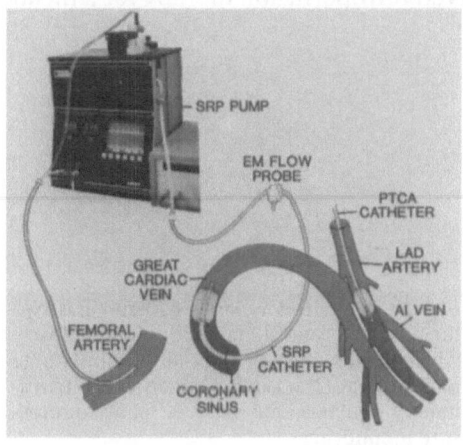

**Fig. 1.** Oxygenated arterial blood is shunted from the femoral artery through an ECG synchronized pump (SRP PUMP) via a coronary sinus through a catheter into the great cardiac vein.

199

monitoring (Fig. 2). The blood delivery lumen communicates at the distal end of the catheter with a balloon which inflates automatically during systole, compartmentalizing the coronary venous system, thereby preventing blood loss into the right atrium and vacilitating unidirectional flow to the myocardium. Blood flow is interrupted in systole and a vacuum is applied to rapidly deflate the balloon and permit normal coronary venous drainage. The catheter pressure lumen is connected to a fluid-filled transducer to provide continuous coronary venous pressure monitoring. The safe upper limits of coronary venous pressure have been determined in numerous animal studies to be 40 mmHg (mean) and 60 mmHg (peak) pressure [1, 2]. Excessive coronary venous pressures resulting in myocardial edema and hemorrhage may occur: 1) when retrograde blood flows are too high, or else; 2) when the coronary venous catheter is too large for the coronary sinus or if the catheter balloon does not deflate completely. Retrograde blood flow is adjusted in each case so as to provide a maximal blood flow without exceeding the maximum allowable coronary venous pressures. If there is evidence of a distinct mismatch between the catheter and the coronary sinus as evidenced by elevated coronary venous pressures prior to initiation of retroperfusion therapy, the catheter is removed and replaced with a catheter of smaller French size or with the same size catheter fitted with a smaller balloon. Catheters are available ranging in size from 7 to 8.5-F with balloon diameters of 5, 8, and 10 mm.

## Treatment of unstable angina

### Feasibility

Synchronized coronary venous retroperfusion therapy was studied in five patients with unstable angina refractory to conventional maximum medical therapy [3]. The procedure was successfully implemented in the five consecutive patients in which it was attempted. The mean time for catheter insertion was 5 min (range: 30 s to 10 min) and the mean time taken to implement the entire procedure was 40 min (range: 25–60 min). The mean coronary venous retrograde flow was 103 ml/min (range: 55–135 ml/min). The patients received retroperfusion therapy for a mean

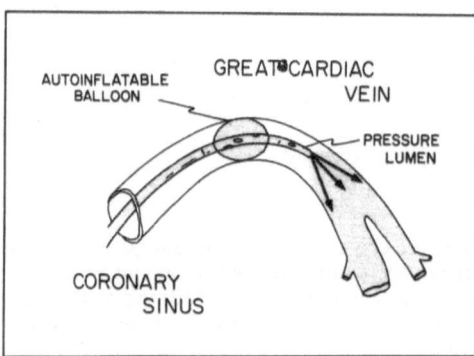

**Fig. 2.** The coronary sinus catheter has two lumens: 1) a central lumen for blood delivery which communicates with an autoinflatable balloon, and 2) a second lumen which terminates 4 mm proximal to the distal tip for pressure monitoring.

**Table 1.** Clinical conditions pre- and during retroperfusion.

|  | Pre-Retroperfusion | Druing Retroperfusion |
|---|---|---|
| Anginal episodes (per h) | $0.71 \pm 0.4$ | $0.01 \pm 0.01^*$ |
| Intravenous morphine (mg/h) | $0.74 \pm 0.26$ | $0.17 \pm 0.19$[a] |
| Sublingual nitroglycerine | $0.23 \pm 0.13$ | $0.0 \pm 0$ [b] |

[a] $p < .05$; [b] $p < .02$
Adapted from Gore, (2)

of 28.2 h (range: 12−50 h). The clinical condition 36 h pre-retroperfusion was compared to the clinical condition during retroperfusion therapy, each patient serving as his own control.

*Efficacy*

The endpoints used to determine efficacy were: 1) frequency of chest pain; 2) sublingual nitroglycerine use; and 3) narcotic analgesic use. As shown in Table 1, there was significant improvement in anginal symptoms, as well as a statistically significant reduction in nitroglicerine and intravenous morphine use.

*Safety*

There were no adverse hemodynamic effects with retroperfusion, nor was there evidence of significant red blood cell or platelet destruction as assessed by serial hematologic measurements (Table 2). All patients underwent coronary arteriography as part of the protocol and there was evidence of significant multivessel coronary artery disease in four of the five patients. The remaining patient had a 99% LAD coronary artery stenosis. Three patients subsequently underwent coronary artery bypass graft surgery, at which time the external surface of the heart was inspected. In two of the three patients there was no evidence of trauma to the heart from the retroperfusion therapy. In the remaining patient there was an ecchymosis on the anterior surface of the heart, probably secondary to elevated pressures in the regional coronary veins however, coronary venous pressure monitoring was not

**Table 2.** Hematologic parameters during retroperfusion.

|  | Before Retroperfusion | During Retroperfusion (12−24 h) |
|---|---|---|
| HB | 12.0 | 11.2 |
| PLT | 155 | 127 |
| Plasma Hb | 4.1 | 5.0 |

Hb = Hemoglobin (g/dl); Plt = platelet count (thousands/mm$^3$)
Plasma Hb = plasma free hemoglobin (mg/dl; normal 0.0−7.0)
Adapted from Gore, (2)

performed. The other two patients could not be surgically revascularized because of diffuse distal vessel disease, and they died. An autopsy was performed on one of these two patients, and careful inspection of the venous system revealed a small hematoma on the posterior surface of the heart near the coronary sinus, and a small thrombus at the site of catheter contact near the orifice of the coronary sinus. The operative and post-mortem findings were consistent with previous animal studies [2], and did not appear to be of hemodynamic significance.

### Treatment during percutaneous transluminal coronary angioplasty

Preliminary publication of data on the effects of coronary venous diastolic retroperfusion during coronary angioplasty is now available in 28 patients from four centers [4−8]. Although three different retroperfusion systems were used in these clinical studies, all featured a similar basic design employing an autoinflatable balloon catheter and an electromechanical synchronized pumping device. However, only one of these systems (Retroperfusion Systems, Inc.) featured coronary venous pressure monitoring during the retroperfusion treatment. Despite the fact that these studies were performed independently, the protocols followed were essentially the same. All of the study patients were scheduled for elective PTCA of the left anterior descending coronary artery and in each patient, PTCA inflations were performed without (controls) and with retroperfusion. The flow rates employed in these studies were approximately 90 ml/min, ranging from 60 to 120 ml/min.

*Feasibility*

Coronary sinus catheterization was successful in 90% of patients and was completed in less than 5 min. Moreover, preparing the system and instituting retroperfusion was generally accomplished in less than 15 min.

*Efficacy*

To determine the efficacy of diastolic retroperfusion during PTCA, patients were monitored with respect to chest pain, ECG, hemodynamics and in some cases by means of two-dimensional echocardiography. In four of the five studies (involving 25 patients) [4−7] reperfusion was found to reduce: 1) the incidence, onset and severity of chest pain; 2) the magnitude of ST segment change, and; 3) the severity of regional wall motion abnormality which occurred with the angioplasty produced coronary artery occlusion. One study [8] found retroperfusion not to be effective as determined by chest pain, ECG change, and hemodynamic indices (dp/dt, and V-max), however, this study involved only three patients.

In our recent experience with 17 patients at the Cedars-Sinai Medical Center we found that retroperfusion resulted in: a reduction in the PTCA balloon occlusion induced incidence of angina from 67% (untreated inflations) to 37% (treated inflations) as well as a delay in angina onset from 45 ± 13 s (untreated inflations) to 66 ± 14 s (SRP treated inflations), p < 0.001; a decrease in ST segment depression from 3.0 ± 1.5 mm (untreated inflations) to 1.4 ± 0.8 mm (treated inflations) (p < 0.005); and a reduction in the severity of wall motion abnormality with PTCA when retroperfusion treatment was applied. Specifically, using the "center-line" method of analysis, ischemic zone wall motion was found to become dyskinetic (−0.7 ± 1.1) during PTCA balloon occlusion without treatment, however, when the inflations were treated with retroperfusion, the wall motion became only mildly hypokinetic (+0.9 ± 0.9, p < 0.002) (Fig. 3).

Although not part of the above described protocol, we had the opportunity to study the effects of retroperfusion during a prolonged coronary occlusion in one patient who developed acute left anterior descending coronary artery occlusion during an attempted PTCA of a 90% proximal LAD stenosis. The LAD occlusion resulted in severe chest pain, ST-segment elevation, complete heart block and cardiogenic shock despite ventricular pacing and catechlomaine support. Two-dimensional echocardiography demonstrated septal, apical and anterior dyskinesis with an estimated left ventricular ejection fraction of 25%. Since the coronary sinus catheter was still in place, treatment with coronary sinus retroperfusion was initiated, resulting in dramatic and rapid improvement with resolution of the chest

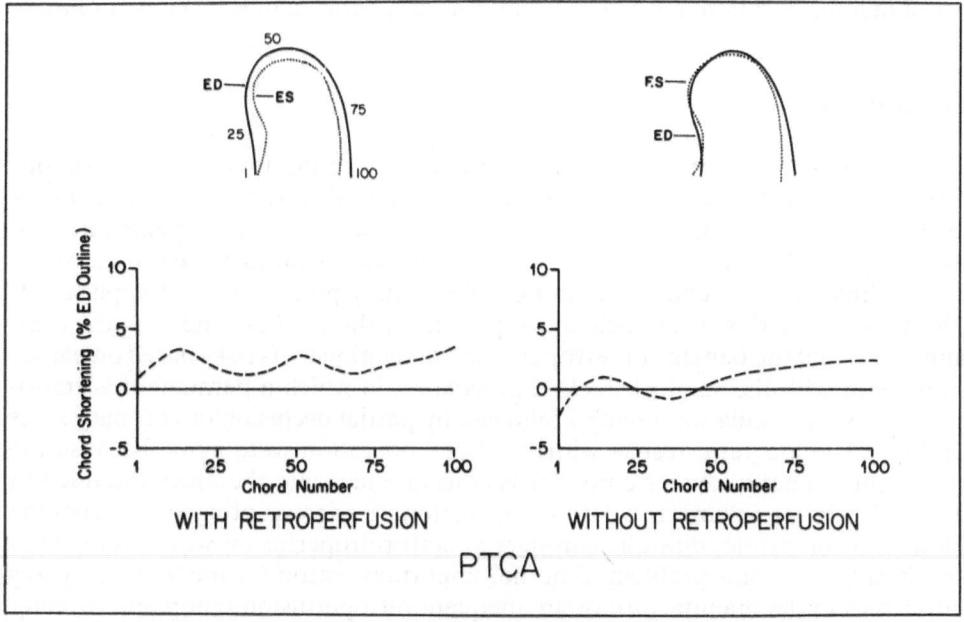

**Fig. 3.** An example of the effects of retroperfusion treatment on regional wall motion using the center line analysis method. The end systolic (ES) and end diastolic (ED) frames are entered into a computer and 100 equidistant chords are automatically drawn from ES to ED.

The echo on the left was obtained after 2 min of LAD balloon occlusion treated with retroperfusion and demonstrates normal wall motion in all chords as demonstrated by the horizontal plot on the graph. The echo on the right was obtained from the same patient after 2 min of untreated LAD balloon occlusion and demonstrated apical dyskinesia (chords 25–45).

pain, return to sinus rhythm and an increase in blood pressure to pre-LAD occlusion levels. A repeat two-dimensional (2-D) echo demonstrated significant improvement in wall motion with a left ventricular ejection fraction of approximately 50%. The patient was maintained on retroperfusion for 2.5 h while attempts were made to reestablish coronary artery patency with repeated balloon dilations. However, as this was not successful, the patient went to emergency coronary bypass graft surgery from which he had an uneventful recovery. On hospital discharge the left ventricular ejection fraction was 61% despite the prolonged coronary occlusion.

*Safety*

The only adverse effects encountered in the PTCA studies were: 1) hematomas at the coronary sinus catheter insertion site in three patients; 2) transient atrial fibrillation during catheter insertion in two patients; and 3) bruising of the coronary sinus, as noted at coronary artery bypass graft surgery in two patients. The hematomas did not require transfusion and resolved spontaneously, the atrial fibrillation converted spontaneously to sinus rhythm without treatment, and the coronary sinus bruising was not thought to be clinically significant at surgery and there were no hemodynamic consequences noted post-operatively. There was also no evidence of hemolysis or platelet destruction, however, it should be remembered that the duration of retroperfusion was very short in these PTCA studies.

**Commentary**

Clinical trials to determine the safety and efficacy of diastolic coronary venous retroperfusion have finally been initiated, more than 10 years after successful introduction of the technique in the animal laboratory, and almost 100 years after the discovery that the coronary venous system can be used to nourish the myocardium. Few techniques have undergone more careful study prior to human application. The reasons for this delay include skepticism in the medical and surgical communities about the benefits of retroperfusion in relation to its risks based on earlier experience with the surgical Beck II procedure, in which a permanent aorta-to-coronary sinus fistula was created followed by partial occlusion of coronary sinus outflow. The longterm results with this Beck operation were poor, because the treatment did not allow for coronary venous drainage and therefore often led to congestive heart failure as a late complication. By interrupting the retrograde blood flow in systole, diastolic coronary venous retroperfusion was developed to specifically avoid this problem. Another important reason for the long delay was difficulties in the manufacture of an adequate retroperfusion pump and catheter system for clinical application.

The currently available retroperfusion system design features a double lumen coronary venous catheter and employs a retroperfusion pump capable of maximum retrograde flows of approximately 120 ml/min. Preliminary studies using such a system in patients with unstable angina and during PTCA have demonstrated that retroperfusion can be performed in 90% of patients and without significant complications. Moreover, therapy can be instituted rapidly, in less than 15 min in the majority of patients by physicians who have had a minimum of experience (approximately five cases) with the technique. Thus, retroperfusion treatment of patients with unstable angina can be provided within a clinically relevant period of time. On the other hand, retroperfusion protection during PTCA does prolong the procedure by approximately 15 min, a fact which must be considered in determining the value of retroperfusion during PTCA in an individual patient. The preliminary data demonstrates some amelioration of ischemia with diastolic retroperfusion in the two patient groups, however the PTCA studies employing 2-D echocardiography demonstrate that although ischemia is reduced, it is not eliminated.

### Future system development

One reason for the limited efficacy may be an inadequate blood flow with the currently available retroperfusion pump-catheter systems. Although antegrade blood flow to the left anterior descending coronary artery at rest is often less than 100 ml/ min, it is recognized that with retrograde treatment probably much less than 50% of the blood flow actually reaches the capillaries and is available for oxygen exchange. The remaining retrograde blood flow is shunted through intramyocardial Thebesian channels and epicardial veno-venous collaterals to the right heart. Thus, greater blood flows must be used with retrograde compared to antegrade delivery to provide similar net capillary blood delivery.

In an attempt to improve retrograde myocardial protection, a new triple lumen catheter is being developed. Using a separate gas inflation balloon system, the effective blood delivery through the catheter is increased by eliminating the need for blood flow into the balloon. In addition, a new pump cassette is being developed to permit retrograde blood flows of up to 250 ml/min. The increase in retrograde flow is deemed feasible because the coronary venous pressures produced with lower retrograde flow rates are well below the safe maximum upper limits. It is anticipated that with these modifications the degree of retrograde myocardial protection will be increased.

Moreover, all of the studies have been performed in patients with left anterior descending coronary artery lesions. This patient subset was chosen primarily because the coronary venous anatomy is best suited for retroperfusion into the great cardiac vein which drains the myocardium supplied by the LAD artery and therefore evaluation of retroperfusion efficacy is most easily performed in these patients. However, if retroperfusion is found to be effective in patients with LAD disease, it should be possible to extend its application to patients with left circumflex and possibly right coronary artery disease – although this will require changes in the coronary sinus catheter design.

### Clinical applications

Further evaluation of the efficacy of diastolic retroperfusion is required before its role in the treatment of patients with ischemic heart disease may be determined.

Preliminary experience in the setting of medically refractory unstable angina suggests that retroperfusion provides a means to stabilize patients prior to cardiac catheterization and eventual definitive therapy. If these preliminary results are reproduced in larger clinical trials, retroperfusion may be of value in the treatment of patients with unstable angina, particularly in hospitals without cardiac catheterization facilities to allow patient stabilization and safe transfer to a referral center for coronary arteriography and definitive therapy. This would be possible since the retroperfusion pump system does have an auxillary battery power supply and is portable. Diastolic retroperfusion may therefore play a role similar to that of the intra-aortic balloon pump. Clinical trials will be necessary to determine the relative efficacy and safety of the two treatments in this setting, however, diastolic retroperfusion does have the theoretic advantage of improving blood delivery to the ischemic myocardium, whereas the intra-aortic balloon pump is effective mainly by decreasing myocardial demand.

The role of diastolic retroperfusion during PTCA is also yet to be determined. Balloon angioplasty is being performed in an ever increasing number of patients with a high initial success rate, however restenosis is common, approximately 30% at 6 months. Diastolic retroperfusion may be beneficial in decreasing the incidence of restenosis by permitting longer balloon inflations. This will require a physiologically significant reduction in angioplasty balloon-induced ischemia with retroperfusion therapy. Retroperfusion support may also permit safe application of newer intracoronary interventions such as laser angioplasty, percutaneous coronary atherectomy and intracoronary stents which necessarily involve prolonged interruption of antegrade blood flow.

Moreover, although balloon angioplasty is associated with a low morbidity and mortality, 2–6% of patients suffer acute coronary occlusion during PTCA requiring emergency coronary bypass graft surgery. Because of the delay between coronary occlusion and surgery there is a significant incidence of myocardial infarction (approximately 40%) and a mortality of approximately 6% [9]. Diastolic retroperfusion has been shown to have been effective in one patient with acute coronary occlusion secondary to PTCA and it may provide the necessary myocardial protection to allow angioplasty operators time to attempt to reestablish antegrade flow with repeated coronary dilations when the procedure is complicated by an acute occlusion. On the other hand, should bypass graft surgery be necessary, retroperfusion treatment may preserve jeopardized myocardium while the patient is prepared for surgery, thereby reducing the incidence of myocardial infarction and death in this setting.

Synchronized retroperfusion may also be of value for the treatment of patients with acute myocardial infarction since the ultimate patho-physiologic consequences of thrombotic coronary occlusions are similar to that of angioplasty-related acute coronary occlusion. However, as most patients who present to hospitals early after acute myocardial infarction are treated with thrombolytic agents, the central venous and arterial catheterizations required by coronary venous retroperfusion presents a problem for combined retroperfusion and thrombolytic therapy. Development of new femoral vein approach catheter systems are contemplated which would significantly reduce the risk of hemorrhage compared to the internal jugular introduction method. Moreover, patients with acute myocar-

206

dial infarction who are not eligible for thrombolytic therapy might be candidates for retroperfusion, providing there are no contraindications to heparinization.

## Conclusion

Clinical trials to determine the safety and efficacy of diastolic retroperfusion are in progress. Preliminary results suggest that the technique is feasible, without significant complications, and effective in stabilizing patients with unstable angina, and ameliorating ischemia in patients undergoing PTCA of the left anterior descending coronary artery. Retroperfusion has also been found to be effective in stabilizing one patient with cardiogenic shock secondary to an abrupt coronary occlusion during PTCA. Further studies will be necessary to determine the ultimate role for diastolic retroperfusion in the treatment of patients with acute myocardial ischemia, however the preliminary results are encouraging.

## References

1. Meerbaum S, Lang T-W, Osher JV, Hashimoto K, Lewis GW, Feldstein C, Corday E (1976) Diastolic retroperfusion of acutely ischemic myocardium. Am J Cardiol 37:588–598
2. Drury JK, Yamazaki S, Fishbein MC, Meerbaum S, Corday E (1985) Synchronized diastolic coronary venous retroperfusion: results of a preclinical safety and efficacy study, J Am Col Cardiol 6:328
3. Gore JM, Weiner BH, Benotti JR, Sloan KM, Okike ON, Cuenoud HF, Gaca JMJ, Alpert JS, Dalen JE (1986) Preliminary experience with synchronized coronary sinus retroperfusion in humans. Circulation 74:381–388
4. Farcot J-C, Berland J, Cribier A, Letac B, Bourdarias J-P (1985) Diastolic synchronized retroperfusion in the coronary sinus during percutaneous transluminal angioplasty: preliminary experience (abstract). Circulation 72 [Suppl III]:III-470
5. Weiner BH, Gore JM, Sloan KM, Benotti JR, Gaca JMJ, Okike ON, Vandersalm TJ, Ball SP, Jeanne Corrao J, Alpert JS, Dalen JE (1986) Synchronized coronary sinus retroperfusion (SCSR) during LAD angioplasty (abstract). JACC 7:64A
6. Berland J, Farcot J-C, Barrier A, Derumeaux G, Cribier A, Bourdarieas J-P, Corday E, Letac B (1988) ECG and 2D echo assessment of myocardial protection achieved by diastolic synchronized coronary sinus retroperfusion (DSR) during LAD angioplasty (abstract). JACC 11:132A
7. Kar S, Drury JK, Eigler N, Buchbinder N, Litvack F, Thessomboon S, Hajduczki I, Corday E (1988) Amelioration of ischemia during PTCA with diastolic coronary venous retroperfusion (abstract). JACC 2:64A
8. Beatt KJ, Patrick WS, de Feyter P, Van Den Brand M, Verdouw PD, Hugenholtz PG (1988) Haemodynamic observations during percutaneous transluminal coronary angioplasty in the presence of synchronized diastolic coronary sinus retroperfusion. Br Heart J 59:159–167
9. Cowley MJ, Dorros G, Kelsey SF, Raden MV, Detre KM (1984) Emergency coronary bypass surgery after coronary angioplasty: the national heart, lung and blood institute's percutaneous transluminal coronary angioplasty registry experience. Am J Cardiol 53:22C–26C

# Coronary venous myocardial contrast echocardiography

S. Meerbaum

This chapter will describe a proposed diagnostic application of the coronary veins, which was first described by Maurer et al. [1], and has been further advanced by a more recent study of Punzengruber et al. (unpublished). The particular technique combines the principles of myocardial contrast two-dimensional echocardiography (MC-2-DE) with retrograde coronary vein echo contrast infusion. The initial studies did not resolve questions about pertinent mechanisms, and the potentials of the method are also not clear. It is thought possible that the very recent development of safe echo contrast agents containing extremely small and adequately persisting microbubbles (about 3–5 microns in diameter), may help overcome some of the methodologic limitations and lead to substantial applications of the proposed technique. We will first briefly describe the experimentally validated and clinically corroborated antegrade coronary artery echo contrast injection technique, and will then proceed to describe and discuss the retrograde MC-2-DE approach. As will be apparent, this retrograde method shares many of the features and issues previously discussed in relation to coronary vein occlusion and retroperfusion.

## Antegrade myocardial contrast 2-DE

In the MC-2-DE method, which is now being successfully applied in clinical studies, special echo contrast agents (e.g., sonicated renografin with less than 10-micron diameter gaseous bubbles) are injected into coronary arteries to evaluate the perfusion in corresponding myocardial regions visualized in 2-DE cross-sectional images of the heart [2–10]. Figure 1 is an early example of experimental delineation of regional myocardium by intracoronary MC-2-DE. Other experimental MC-2-DE modalities applied aortic root, pulmonary artery or intravenous echo contrast injections. It has been demonstrated that intracoronary and aortic root injection of several available echo contrasts is safe and that it effectively outlines the severely underperfused myocardium constituting the ischemic risk area, and may also correlate with the eventual extent of infarction [11]. A left main coronary echo contrast injection normally results in significant echo enhancement of the entire left ventricular wall in a 2-DE short axis cross-section. When the left anterior descending or the left circumflex coronary artery is occluded (e.g., experimentally and also during PTCA balloon inflations), the resulting severely underperfused myocardial zone is clearly outlined (Fig. 1) and its size can be reproducibly measured.

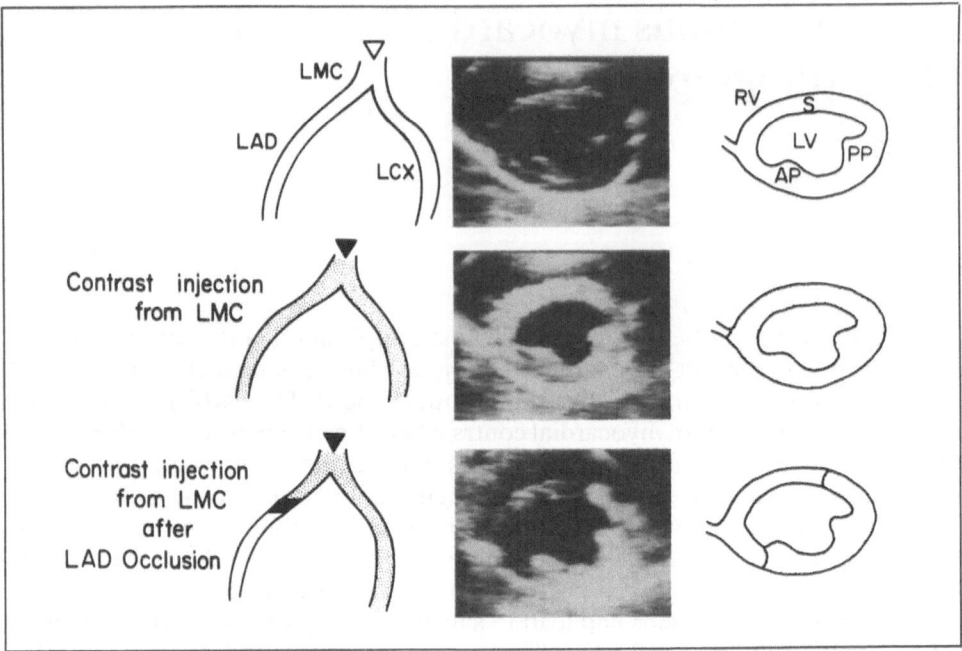

**Fig. 1.** This illustration indicates the quality of myocardial contrast echocardiographic opacification in two-dimensional echocardiographic cross-sections after left main coronary artery injection of saline-Renografin echo contrast agent. The **top panel** shows a midpapillary level short-axis view of the left ventricle before coronary artery occlusion, and before contrast injection. Note the epicardial and endocardial outlines, with relatively echolucent intervening myocardium. The same cross-section appears in the **midpanel** after contrast agent injection from the left main coronary artery, in the preocclusion control state. The entire circumference of the left ventricular myocardium is opacified. The **lower panel** shows the effect of a left main coronary artery contrast agent injection after left anterior descending coronary artery occlusion. Note that a substantial portion of the interventricular septum and a part of the left ventricular anterior wall are devoid of contrast echo. This negative echo contrast area represents the underperfused myocardium during the coronary artery occlusion. AP = anterior papillary muscle; LAD = left anterior descending coronary artery; LCx = left circumflex coronary artery; LMC = left main coronary artery; LV = left ventricle; PP = posterior papillary muscle; RV = right ventricle; S = interventricular septum. (By permission, from: Sakamaki T et al. (1984) Verification of myocardial contrast 2-dimensional echocardiographic assessment of perfusion defects in ischemic myocardium. J Am Col Cardiol 3:34–38.)

Sophisticated computerized MC-2-DE analysis, including derivation of regional myocardial echo intensity-time curves, allow characterization of myocardial perfusion and its distribution [12–13], potentially even selectively in subepicardial and subendocardial layers. Presently, there are limitations and distinct variability of measurement, due to errors related to ultrasound factors, echo contrasts, and injection techniques. Thus, absolute MC-2-DE quantitation of regional myocardial blood flow is yet to be fully demonstrated. Nonetheless, one can now envision an exciting new technique which may allow simultaneous echocardiographic measurement of regional (and global) cardiac function along with assessment of the region's myocardial blood flow. The method is simpler than other techniques, and it can be readily repeated for sequential study. Recent clinical evidence indicates that spe-

210

cial echo contrast agents are both effective and quite safe when applied in very small quantities (e.g., 1–2 cc) into the coronary arteries, causing only minimal transient toxicity.

In contrast with the radionuclide microsphere method (the experimental "gold standard") for quantitation of regional myocardial perfusion, which features microsphere entrapment in the capillary or arteriolar vessels, most MC-2-DE efforts are aimed at using echogenic solutions with tracer microbubbles of a size compatible with unhindered transcapillary passage, allowing application of dilution theory analysis. The intracoronary carrier solutions are also tailored to maintain a low osmolality and viscosity, thus preventing significant alterations of the coronary blood flow [15, 16]. The solutions are prepared in a manner to provide extremely small and persisting gaseous microbubbles, on which the myocardial echo enhancement depends. There are several means for producing uniform and very small microbubbles of sufficient concentration, including the simple procedure of ultrasonic agitation of the solution (using a commercial sonicator). Details of this and other procedures, echo contrast agents, and generally of MC-2-DE development and validations, have been amply reported.

As to the physiologic model of antegrade intracoronary MC-2-DE, small and persisting microbubbles are carried along with the red cells in the blood stream, hopefully distributing in a uniform manner to all myocardial regions. Much of the left ventricular myocardial blood drains via the coronary sinus, the LAD branch of the left coronary artery being largely subserved by the great cardiac vein. With echo contrast injected in a normal state into the main left coronary artery, microbubbles are channeled via both left anterior and left circumflex branches into the intramural vessels, the capillaries, venules, and finally, into the epicardial coronary veins, emptying through the great cardiac vein and coronary sinus, into the right atrium. During a coronary artery obstruction, e.g., the LAD occlusion, the echo contrast is nonuniformly distributed from the left main coronary artery to the nonoccluded vessels. Some extremely small microbubbles might enter the myocardial region beyond the coronary artery occlusion via arterial collaterals. This does not detract much from the MC-2-DE ability to reliably delineate the acutely ischemic risk zones, but it is well to point to the variable nature of coronary anatomy and of collateral blood flow.

**Retrograde coronary venous mode of MC-2-DE**

A retrograde MC-2-DE [1] was initially investigated for its feasibility, and potentially as an alternate for assessing myocardial risk zones, possibly also for study of myocardial perfusion. This method is, however, viewed less as a replacement for the antegrade MC-2-DE technique, and more as an adjunct capable of elucidating mechanisms and effectiveness of retrograde interventions, and possibly as a diagnostic modality for detection and study of underperfused myocardial regions whenever the antegrade approach is more difficult. It was recognized that in all retrograde MC-2-DE applications, the particular coronary venous anatomy would play a crucial role, involving both primary coronary vein variability and profuse coronary veno-venous shunting.

The first effort by investigators was to standardize the retrograde echo contrast agent injections, e.g., by maintaining its pressure and flow. Depending on the position of the intravascular occlusive catheter balloon and the particular site of echo contrast injection, coronary venous shunting is known to variably affect the pressures developed within the coronary sinus and the great cardiac vein. Thus, low resistance shunting from the coronary veins of the LAD territory to coronary veins draining into the ostium of the coronary sinus, occasionally led to rather low developed pressures (e.g., in the presence of a mid coronary sinus site of balloon occlusion and contrast infusion). A substantial mean pressure potential (30−50 mmHg) was needed to cause retrograde echo contrast penetration into the small intramural coronary veins. When "successfully" delivered, the corresponding regional myocardium exhibited enhanced echo intensities. But it is by no means evident from the retrograde MC-2-DE images how deep into the microcirculation the echo contrast microbubbles penetrated, and whether capillaries were involved. Further investigation with the newest microbubble echo contrasts is therefore needed to demonstrate retrograde contrast delivery into the capillary bed, and even beyond into the arterial vessels in an ischemic zone. Nonetheless, given adequate epicardial coronary vein pressures, the retrograde MC-2-DE study documented, in most instances, a distinct regional myocardial opacification, and also demonstrated the Thebesian drainage, primarily into the right chambers of the heart [1].

Before turning to the coronary venous MC-2-DE experiments, let us briefly consider an antegrade intracoronary MC-2-DE during simultaneous, induced coronary vein occlusion (i.e., without the retrograde echo contrast infusion). Analogous to the previously described physiologic effects, coronary sinus occlusion will (in general) cause a significant elevation of the coronary vein blood pressure, and some redistribution of the myocardial blood flow, particularly when a coronary artery is occluded. In the latter case, echo contrast from the nonischemic zone should drain into the corresponding coronary veins, and from these some of the contrast will be shunted to the ischemic region, altering its contrast microbubble concentration. The degree of this redistributive retrograde microbubble delivery into the ischemic zone will depend (among other factors) upon the severity of the coronary artery obstruction. In the absence of a full understanding of the mechanism during antegrade perfusion with coronary vein occlusion, it is not clear whether such a modified MC-2-DE study would be helpful in evaluating the residual circulation in myocardium beyond a coronary artery occlusion.

**Initial retrograde MC-2-DE investigation**

An otherwise routine closed-chest dog preparation featured coronary sinus or great cardiac vein occlusion by means of a catheter balloon [1]. The LAD coronary artery could also be occluded by an appropriately positioned intravascular balloon. A 90% mechanical sector scanner with a 3 MHz transducer provided the cross-sectional images, largely short axis sections of the dog's left ventricle. A 2 cc bolus of sonicated renografin (mean bubble diameter 10 micron) had been found adequate for left main coronary artery antegrade echo contrast injections. Based on prior antegrade vs retrograde rationales (involving the known coronary veno-venous shunting), a larger 5 cc echo contrast bolus was injected retrogradely during 5 s into the balloon-occluded GCV. This latter balloon occlusion was released after 10 s, so as not to significantly interfere with coronary venous drainage. The LAD coronary artery was occluded by a 2-F balloon catheter just below the first septal branch. End-diastolic 2-DE freeze frame images of LV short axis cross-sections were analyzed for the regional extent of myocardial opacification or perfusion deficiency, using antegrade or retrograde MC-2-DE. Reproducibility of the MC-2-DE procedures was assessed. The site and observed amount of coronary venous − Thebesian shunting into the left and right cardiac chambers was estimated in vivo, and validated in postmortem examinations of the dog's heart.

The circumferential extent of the left ventricular myocardial opacification following great cardiac vein echo contrast injection was measured in short axis 2-DE sections. Whereas echo contrast injection into the left main coronary artery in normal states generally opacified the entire cross-sectional myocardium, great cardiac vein injection enhanced only the anatomically corresponding myocardial region (anterior and anteroseptal wall), often in a patchy nontransmural manner (Fig. 2). Thus, of 12 retrograde echo contrast injections, there was no significant myocardial enhancement in two cases, patchy opacification in seven instances, and confluent nontransmural opacification in four cases. Following LAD coronary artery occlusion, retrograde MC-2-DE resulted in confluent and transmural myocardial opacification, encompassing a somewhat larger area than the risk zone perfusion defect previously outlined by means of main coronary artery (antegrade) MC-2-DE (Fig. 3), $42.8 \pm 8.6\%$ ($26-54\%$) vs $30 \pm 6.3\%$ ($21-37\%$) of the short axis section circumference. Injection-to-injection error of retrograde MC-2-DE was found to be $7.2 \pm 8\%$. Coronary venous echo contrast injections were also performed while studying the 2-DE parasternal four-chamber view. The study revealed a sub-

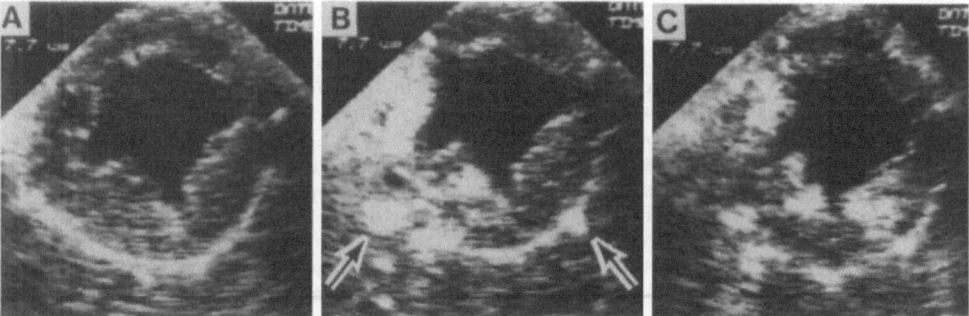

**Fig. 2.** Retrograde coronary venous contrast injections without coronary artery occlusion. **A,** Baseline image. **B,** Great cardiac vein injection results in patchy myocardial opacification. The bright rounded structures (**arrows**) presumably represent epicardial coronary veins containing contrast material. **C,** Coronary sinus injection also results in patchy but more extensive myocardial opacification which now also involves the posterior papillary muscle. (By permission, from: Maurer G et al. (1984) Retrograde coronary venous contrast echocardiography: assessment of shunting and delineation of regional myocardium in the normal and ischemic canine heart. J Am Col Cardiol 4(3):577−586.)

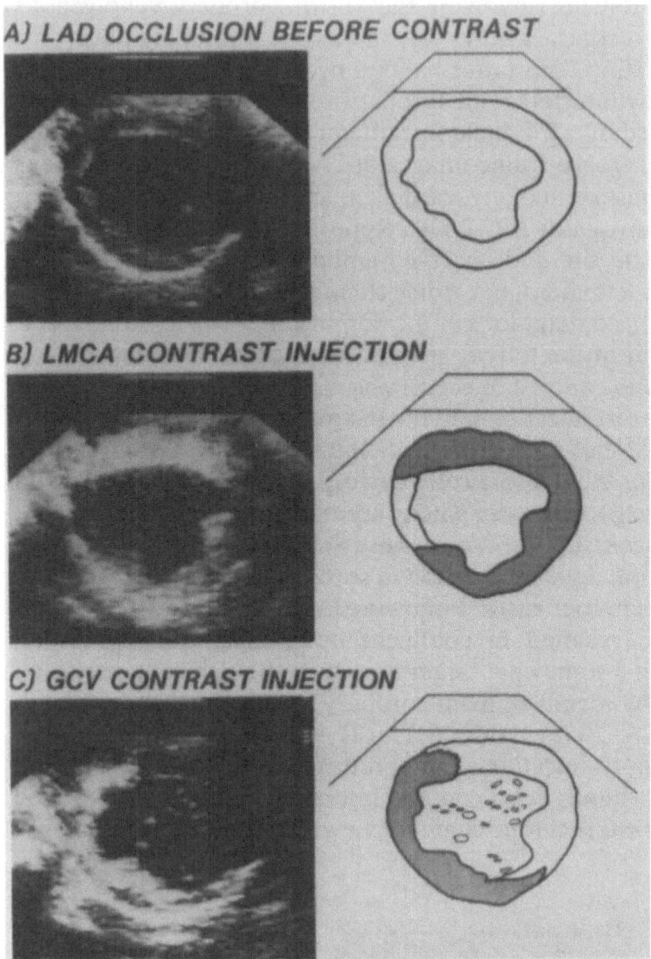

**A) LAD OCCLUSION BEFORE CONTRAST**

**B) LMCA CONTRAST INJECTION**

**C) GCV CONTRAST INJECTION**

**Fig. 3.** Antegrade and retrograde echocardiographic contrast studies after occlusion of the left anterior descending (LAD) coronary artery. **A,** Baseline image before contrast injection. **B,** Injection into the left main coronary artery (LMCA) results in myocardial opacxification with a filling defect in the underperfused region (approximately 8 to 10 o'clock). The diminished contrast intensity adjacent to the posteriomedial papillary muscle (approximately 2 to 4 o'clock) is due to echo dropout, which is most likely caused by lung interference. **C,** Retrograde great cardiac vein (GCV) contrast injection results in transmural myocardial opacification of the underperfused zone as well as of adjacent myocardium (approximately 5 to 11 o'clock). Contrast can also be noted within the left ventricular cavity. (By permission, from: Maurer G ewt al. (1984) Retrograde coronary venous contrast echocardiography: assessment of shunting and delineation of regional myocardium in the normal and ischemic canine heart. J Am Col Cardiol 4(3):577–586.)

stantial Thebesian shunting of the retrogradely injected echo contrast to the right cardiac chambers (grade $3.5 \pm 0.6$ of a maximum of 4), while the left chamber exhibited a lesser degree of Thebesian drainage ($1.5 \pm 0.6$).

Thus, this limited investigation indicated that the echo contrast injected into an occluded cardiac vein could be propelled into the intramural veins, and resulted in

214

regional opacification of myocardium in 2-DE short axis sections. This opacified myocardial region corresponded in general to the anatomic territory subserved by the great cardiac vein, and became confluent and transmural when the proximal LAD coronary artery was occluded. However, the opacified myocardial region included some myocardium adjacent to the perfusion defect. Posterior or inferior walls were never enhanced by GCV MC-2-DE. Failures of retrograde MC-2-DE were in part found to be associated with coronary veno-venous shunting, from the great cardiac and anterior interventricular veins to the posterior left ventricular and middle cardiac veins, causing some drainage into the coronary sinus at its ostium and lowering of the developed coronary venous blood pressure. In most cases, MC-2-DE with adequate pressure great cardiac vein injections delivered the echo contrast into the acutely ischemic myocardial region.

**Corroborative experimental study of retrograde MC-2-DE**

A second (unpublished) study further standardized the retrograde MC-2-DE procedure, extended the investigation by including coronary sinus injections as well as circumflex coronary artery occlusions, provided more phasic analysis of regional myocardial echo intensities, and performed simultaneous measurements of regional function.

This experimental study used two 4-F balloon catheters, inserted under fluoroscopic control. The first catheter had a balloon at its tip which served to occlude the coronary sinus as close as possible to its ostium (but, for practical reasons, still missing the above mentioned substantial venous shunts located at the ostium), as well as a center lumen for the retrograde echo contrast injection into the coronary sinus. The other balloon catheter was used to either occlude the great cardiac vein (near its junction with the anterior interventricular veins) during a regional retrograde MC-2-DE, or else to measure (through its center lumen) the coronary venous blood pressure produced during coronary sinus balloon occlusion. The input pressure to these 4-F balloon catheters was kept constant. Either the proximal LAD or the circumflex coronary artery was occluded with a 2-F balloon catheter.

The 2-DE LV short axis imaging and the echo contrast used during MC-2-DE were similar to those described above. Retrograde renografin contrast injections were also used as part of digital subtraction coronary venography study of the epicardial coronary venous anatomy. The diastolic 2-DE stop frame images of short axis LV cross-sections (at the lower level of the left ventricle) were analyzed to assess the regional extent of myocardial contrast enhancement, and the regional left ventricular function. In addition, a computerized temporal analysis of myocardial contrast enhancement (pixel brightness) was performed, primarily to evaluate the changes from end-diastole to end-systole. These differences were observed to be significant prior to and tended to disappear following coronary artery occlusion. Contractile function was determined in the same myocardial regions.

Consistent regional myocardial opacification could be achieved in normal hearts by retrograde injection of 2.5 cc sonicated renografin (over a period of 2.5−8 s, at a rate of 0.3−1 cc/s). The echo cotrast injection pressure was 2000 mmHg (±10%), this high pressure level being due to the small 4-F catheter lumen and its length.

Measurement of great cardiac vein pressures indicated that coronary sinus balloon occlusion alone caused a significant rise from $7 \pm 3/1 \pm 0.6$ mmHg (systolic/diastolic) to $29 \pm 11/5 \pm 3$ mmHg. The coronary vein pressure then increased significantly more to $55 \pm 2.3/15 \pm 12$ mmHg as a result o the superimposed echo contrast injection. With this coronary venous pressure level, (achieved in spite of the known coronary veno-venous shunting), coronary sinus MC-2-DE produced distinct regional MC-2-DE opacification in eight of 13 dogs. It appeared that in the remaining five dogs, in which retrograde coronary sinus MC-2-DE was ineffective, there may have been incomplete coronary vein occlusion or excessive shunting, resulting in a lower developed mean coronary venous blood pressure of 31/6.3 mmHg as compared to 70/18 mmHg in the above eight dogs. On the other hand, in 15 out of 15 dogs, echo contrast injections into the balloon-occluded great cardiac vein led to myocardial uptake in portions of the anterior left ventricle and the anteroseptal region of the short axis myocardial cross-section. As percentage of the total section circumference, this extent of myocardial opacification in the control state was $28.4 \pm 11.9\%$ (10–40%), compared to $35.9 \pm 17$ (13–58%) with the coronary sinus contrast injections. Because of substantial variability, the differences were not statistically significant, and in spite of coronary sinus retroinjection, no echo contrast opacification could be discerned in the posterior aspects of the LV.

Following LAD coronary artery occlusions, the circumferential extent of opacification with occluded great cardiac vein injections was slightly (but not significantly) greater than in control states, i.e., $36.6 \pm 9.7\%$ (26–60%). Upon examination, not all of the severely asynergic ischemic segments were opacified. During left circumflex occlusion, echo contrast injections into the coronary sinus or the GCV opacified no more of the myocardium than had been opacified in the control state, i.e., enhancement remained restricted to the septal and anterior wall. The posterior septum or posterior wall of the LV were not opacified in spite of asynergy clearly seen in the 2-DE images during left circumflex coronary artery occlusion.

Prior to coronary artery occlusion, retrograde coronary venous MC-2-DE was found to enhance myocardial brightness (in a corresponding region) significantly more in diastole as compared to systole. This may be due to the forceful systolic squeezing out of myocardial blood. After coronary occlusion, this significant phasic difference was abolished, and evidence points to an association between the sharply reduced diastolic-to-systolic ratio in regional contrast enhancement with major dysfunction in the zone distal to the occluded circumflex vessel.

Digital venography performed in control states revealed the epicardial coronary venous pathways. Injection of contrast into the occluded great cardiac vein always led to opacification of the two anterior interventricular veins. Beyond these, in nine of 10 dogs, contrast was rapidly cleared around the left ventricular apex via large coronary veno-venous shunts to an obtuse marginal vein and the posterior interventricular coronary vein, both of which drained into the coronary sinus practically at its ostium. In addition, in seven of the 10 dogs, it was apparent that contrast was also shunted to the anterior right ventricle myocardial veins, separately draining into the right atrium. Occasionally, there was evidence of Thebesian drainage into the right ventricular chamber.

## Significance of coronary venous MC-2-DE during coronary artery occlusion

It would appear that whenever coronary vein occlusion led to an epicardial coronary venous pressure of the order of 25 mmHg in diastole, retrograde MC-2-DE provided adequate regional myocardial opacification. Occasionally low coronary vein blood pressures (e.g., less than 10 mmHg in diastole) failed to produce significant myocardial echo enhancement. The above is in accord with retroperfusion experience, indicating (for example) a satisfactory synchronized retroperfusion effectiveness when the diastolic pressure in the regional coronary vein was augmented to about 25 mmHg.

Any major shunting of retroinfusate back to the coronary sinus ostium and the right atrium would, of course, reduce the desired retrograde penetration into the microcirculation. Also, whenever flow during a great cardiac vein or coronary sinus occlusion is shunted back through the posterior coronary veins, the resulting lowering of developed blood pressure evidently prevents retrograde penetration into the posterior aspects of the LV. Thus, experimental coronary sinus injections of echo contrast generally did not opacify more myocardial territory than when the injection was into the great cardiac vein. It appeared that the small obtuse marginal veins (which drain left circumflex flow into the coronary sinus) offer a much greater flow resistance than the anterior interventricular veins. One would presumably have to fully sequestrate the coronary sinus vessel to achieve with coronary sinus injections a sufficient retroperfusion of the posterior LV.

Retrograde injections performed during experimental obstruction of the left circumflex coronary artery did not result in contrast appearance in posterolateral left ventricular segments, whereas LAD occlusions invariably led to homogeneous opacification and increased uptake of retrograde venous injectates within the LAD subserved myocardium. Thus, using current retrograde MC-2-DE methods, great cardiac venous echo contrast infusion during LAD occlusion propels the retroinfusate deep into small intramural coronary vessels, whereas coronary sinus occlusion and echo contrast injection does not provide effective delivery to myocardium distal to a left circumflex coronary artery occlusion.

The physico-chemical properties of the retroinfusate can significantly affect the extent of retrograde delivery into the corresponding mural microcirculation. Thus, echo contrast agents featuring air bubbles of 10 micron (or larger) diameters may not provide adequate potential retroinfusate penetration into the fine intramyocardial vasculature, even though regional opacification may delineate the general retroinfused myocardial zone. It is possible that the newer and very small microbubbles (of the order of 5 micron diameters) may be significantly more effective, e.g., in passing retrogradely from the coronary sinus via the small obtuse cardiac veins, thus also outlining in echo images the regional myocardium distal to a left circumflex coronary artery.

## Potential application for study of coronary venous interventions

To achieve an effective arterial blood or drug retroinfusion treatment in the highly variable anatomy and clinical setting, it seems desirable to develop for the retro-

217

grade interventions a flexible mode of coronary vein occlusion. Thus, various balloon catheter configurations can be specifically designed for coronary sinus occlusion, either with or without concomittant obstruction of the coronary venous branches in the immediate vicinity of the coronary sinus ostium, and alternately for simultaneous coronary sinus and great cardiac vein occlusion. Functional shunting could thus be minimized and developed coronary vein pressures may be increased, improving the retrograde infusate delivery into the ischemic risk zone. As previously intimated, a double balloon occlusion would enable directionally favorable retrograde delivery into the coronary veins which subserve an occluded circumflex coronary artery.

The previously discussed aspects of retrograde intervention safety set limits on permissible coronary vein pressures. These limits also apply to coronary venous MC-2-DE. Proper positioning of the occlusive balloon within the epicardial coronary veins remains important, taking into account the coronary veno-venous shunting, effects of redistribution or competitive antegrade-retrograde perfusion, and the appreciable coronary vein capacitance due to vessel distensibility. In general, the coronary vein retroinfusion path should be as short as possible, and the retroinfusion site should be positioned as close as practicable to the location of coronary artery obstruction.

In any particular retrograde MC-2-DE study which might be performed in the catheterization laboratory, after a coronary arteriography, limits on retrograde catheter positioning will become apparent to the investigator. Contrast injection into the occluded coronary vein should reveal the retrogradely induced coronary vein pressures, and also define the individual coronary venous configuration available for retroinfusions. These preliminary indications will help decide if supplemental retrograde MC-2-DE diagnosis might be usefully applied.

Retrograde MC-2-DE studies are clearly relevant to assessment of coronary venous interventions, such as the clinically oriented synchronized retroperfusion and pressure controlled intermittent coronary sinus occlusion treatments. It would seem that all data pertaining to the location and manner of coronary vein obstruction and retrograde coronary venous infusion, are very meaningful. Thus, it was encouraging to note in the above MC-2-DE studies that only limited diastolic coronary venous pressure elevation (i.e., to 25 mmHg) was required, although even this moderate blood pressure augmentation could require a substantial retroinfusion flow rate along with a total coronary vein occlusion. In the presence of an elevated (arterial) diastolic pressure distal to the coronary artery obstruction, e.g., due to a significant coronary collateral blood supply to an ischemic zone, a proportionate increase in the augmented diastolic retroperfusion pressure in the corresponding coronary vein is required. It appears from some past experiments that the effectiveness of retroinfusion depends importantly on the pressure differential between the infused coronary vein and the distal coronary artery. A diastolic differential of at least 20–30 mmHg seemed required; a veno-arterial differential of 10 mmHg or less was found to greatly diminish the effectiveness of myocardial retroperfusion. A rational use of coronary venous interventions could well include an initial diagnostic antegrade as well as retrograde MC-2-DE, along with coronary angiography and conventional baseline measurements of hemodynamics, ECG and cardiac function. The first objective may well be to determine (in each particu-

218

lar case) what would be the best possible retrogradely augmented coronary vein pressure and flow. Low resistance coronary veno-venous shunts could make it difficult to achieve the minimally required coronary vein pressure. The converse challenge is not to exceed the safe pressure limits and to maintain adequate systolic coronary venous blood drainage. The degree of this drainage may also be assessed in the catheterization laboratory, by applying contrast echo or digital angiography analysis in a manner previously applied in animal experiments of synchronized retroperfusion [17]. That study yielded a threshold of retrogradely pumped flow, below which there simply was no treatment effectiveness. In a similar application, echo contrast might be injected (through the catheter center lumen) beyond a PTCA balloon occlusion so as to analyze "washout" from the ischemic zone. This may contribute to evaluation of the effectiveness of PTCA support treatments, such as antegrade autoperfusion or retrograde coronary venous interventions.

## References

1. Maurer G, Punzengruber C, Haendchen RV, Torres MAR, Heublein B, Meerbaum S, Corday E (1984) Retrograde coronary venous contrast echocardiography: assessment of shunting and delineation of regional myocardium in the normal and ischemic canine heart. J Am Col Cardiol 4(3):577–586
2. Tei C, Sakamaki T, Shah PM et al. (1983) Myocardial contrast echocardiography: a reproducible technique of myocardial opacification for identifying regional perfusion defects. Circulation 67:585–593
3. Sakamaki T, Tei C, Meerbaum S et al. (1984) Verification of myocardial contrast 2-dimensional echocardiographic assessment of perfusion defects in ischemic myocardium. J Am Coll Cardiol 3:34–38
4. Armstrong W, Mueller T, Kinney E, Tickner G, Dillon J, Feigenbaum H (1982) Assessment of myocardial perfusion abnormalities with contrast enhanced 2-dimensional echocardiography. Circulation 66:166–174
5. Kemper AJ, O'Boile JE, Sharma S, Cohen CA, Kloner RA, Khuri SF, Parisi AF (1983) Hydrogen peroxide contrast-enhanced 2-dimensional echocardiography: real time in vivo delineation of regional myocardial perfusion. Circulation 68:603–611
6. Kaul S, Glasheen W, Ruddy TD, Pandian NG, Weyman AE, Okada RD (1987) The importance of defining left ventricular area at risk in vivo during acute myocardial infarction: an experimental evaluation with myocardial contrast 2-dimensional echocardiography. Circulation 75:1249–1260
7. Feinstein SB, Lang RN, Dick C, Neuman A, Al-Sadir J, Chua KG, Carroll J, Feldman T, Borrow KM (1986) Contrast echocardiographic perfusion studies in humans. Am J Cardiol Imaging 1:29–37
8. Lang RM, Feinstein SB, Feldman T et al. (1986) Contrast echocardiography for evaluation of myocardial perfusion effects of coronary angiography. J Am Col Cardiol 8:232–235
9. Cheirif J, Zoghby WA, Raizner AE et al. (1988) Assessment of myocardial perfusion in humans by contrast echocardiography. I. Evaluation of regional coronary reserve by peak contrast intensity. J Am Col Cardiol 11:735–743
10. Keller MW, Glasheen W, Smucker ML, Burwell LR, Watson DD, Kaul S (1988) Myocardial contrast echocardiography in humans. II. Assessment of coronary blood flow reserve. J Am Col Cardiol 12:925–934
11. Armstrong WF, West SR, Mueller T, Dillon JC, Feigenbaum H (1983) Assessment of location and size of myocardial infarction with contrast enhanced echocardiography. J Am Col Cardiol 2:63–69
12. Kaul S, Chesler DA, Boucher CA, Okada RD (1987) Quantitative aspects of myocardial perfusion imaging. Sem Nucl Med 17:131–144

219

13. Ten Cate FJ, Comel JH, Serruys TW, Vletter WB, Roelandt J, Mittertremer WH (1987) Quantitative assessment of myocardial blood flow by contrast 2-dimensional echocardiography: initial clinical observation. Am J Physiol Imaging 2:56−60
14. Moore CA, Smucker ML, Kaul S (1986) Myocardial contrast echocardiography in humans. I. Safety − a comparison with routine coronary arteriography. J Am Col Cardiol 8:1066−1072
15. Keller MW, Feinstein SB, Watson DD (1987) Successful left ventricular opacification following peripheral venous injection of sonicated contrast agent: an experimental evaluation. Am Heart J 114:570−575
16. Berwing K, Schlepper M (1988) Echocardiographic imaging of the left ventricle by peripheral intravenous injection of echo contrast agent. Am Heart J 115:399−408
17. Chang BL, Drury K, Meerbaum S, Fishbein MC, Whiting JS, Corday E (1987) Enhanced myocardial washout and retrograde blood delivery via synchronized retroperfusion during acute myocardial ischemia. JACC 9(5):1091−1098

# Mathematical modeling of coronary venous interventions

W. Schreiner, F. Neumann

## Introduction

Two basic approaches are open to mathematical modeling of coronary sinus interventions: phenomenological and mechanistic modeling. For each of these approaches we outline the mathematical concepts, computer implementation, results, and final aims.

In phenomenological modeling, parameterized mathematical functions are fitted to systolic and diastolic coronary sinus pressure (CSP) envelopes in order to characterize numerically specific parameters of shape of CSP pressure rise. The mathematical procedures are designed in a way so as to yield numerical results which very closely resemble those features of CSP envelopes (e.g., height of plateau, rise time) which one intuitively looks at when visually inspecting CSP tracings. The modeling results therefore lend themselves as objective and reproducible substitutes for diagnostic information visually obtained from CSP tracings, and in a further step may be used as input of a closed loop regulation of CSI modalities (e.g., occlusion/release time for PICSO[1], SRP[2] flow rate and timing). Mean values, statistical properties and the physiologic interpretation of these derived quantities have been assessed in canine as well as human studies.

Mechanistic modeling is achieved by coupled differential equations describing flows, pressures, and volume changes in a mechanical model designed to resemble the key features of the coronary circulation in the presence of CSIs. The results comprise pulsatile time courses of all calculated quantities (e.g., the forward and backward flow between capillaries and coronary veins, CSP), which can be used to address three main issues. Firstly the modeling allows estimation of intramyocardial hemodynamics which are essentially inaccessible to experimental measurements. Secondly, it provides insights into the performance of different modes of CSI (e.g., PICSO, SRP), in the presence of derangements of perfusion and function. Thus, we have specifically addressed the case of moderate global ischemia with maintained coronary inflow but reduced contractile function (representing the reperfusion phase after surgical cardiac arrest), and also the setting of a single coronary artery obstruction associated with a regional loss of contractile function. Thirdly, the mechanistic model is specifically used to simulate stepwise modifications of CSI patterns (e.g., varying the PICSO occlusion/release timing, or the SRP flow rate), in order to approach an optimal coronary venous intervention.

---

[1] Pressure controlled Intermittent Coronary Sinus Occlusion (no extracorporal pump infusion)
[2] Synchronized Retroperfusion (diastolic retroinfusion of arterial blood – systolic venous drainge)

Results obtained so far demonstrate that phenomenological modeling is a suitable tool to assure precise CSI operation based on predetermined intervention criteria, whereas mechanistic modeling is the method of choice in the search for the optimal intervention mode to be applied, since model computations may yield results useful to minimize the need for costly and time consuming animal experiments.

## Rationales and types of modeling

In addition to experimental and clinical investigations of coronary sinus interventions (CSIs), mathematical modeling may provide an effective tool for elucidation of mechanisms and for the optimization of therapeutically effective techniques. Generally, two types of modeling can be envisaged:
a) phenomenological modeling;
b) mechanistic (or causal) modeling.
The main features, capabilities, and prospects of these two approaches are discussed in the following subsections.

### Phenomenological modeling

Phenomenological modeling emphasizes a mathematical description, e.g., of hemodynamic measurements, and it is to provide quantitative results supplementing the clinicians' intuitive judgement. For example, the rise of the coronary sinus pressure (CSP) after coronary sinus occlusion (CSO) follows a characteristic pattern, from which the clinician may intuitively infer a selection of occlusion and release periods for one of the interventions, viz. pressure controlled intermittent coronary sinus occlusion (PICSO). However, the systolic and diastolic envelopes of CSP may also be subjected to a mathematical fitting procedure, which derives numerical parameters of curve shape (e.g., height of pressure plateau and its rise time). Such results may be used to calculate an apparently optimal occlusion/release timing, which is reproducible and intersubjective. Two basic features apply to phenomenological modeling:
 i) rather than considering mechanisms (e.g., physiologic or hydrodynamic) which are (in a causal sense) responsible for a special form of CSP rise, a quantitative *description* of the experimental measurements is provided;
ii) since no mechanisms are incorporated in the analysis, the functional form of such a phenomenological model remains to some extent arbitrary.
One may argue that little is achieved in this approach as compared to visual inspection since, despite being complex, it still involves the arbitrariness of model selection. However, this argument is really not applicable for the following reason: Even if the initial choice of a model function and the mathematical procedure according to which the intervention pattern is adapted are just as arbitrary as visual inspection, the results become objective and reproducible due to strict algorithms applied to measured data. Thus, while phenomenological modeling does not pretend to extract information principally unavailable to visual inspection, it adds an

222

important quantitative and reproducible feature to various adjustments that must be made during CSI interventions. Moreover, the "calculated diagnostics" approach forms a useful basis for system computerization and automatic closed-loop control desired in routine applications.

## Mechanistic modeling

Whereas phenomenological modeling is mainly applied to measured experimental data (in order to obtain a quantitative assessment or even regulation criteria), mechanistic modeling addresses the important task of achieving an understanding of the influence of principal factors and establishing conditions which will most likely provide optimal therapeutic effects of CSIs.

In vivo studies carried out to evaluate the therapeutic relevance of coronary sinus interventions are based on specific protocols which select the mode of the intervention, and also specify the criteria according to which a possible therapeutic effect is to be evaluated. In this context, the broad term "mode of intervention" describes among others, the occlusion/release phasing of PICSO, the synchronized retroperfusion (SRP) flow rates/pressures and timing within the cardicac cycle, as well as adjustments of these procedures to account for changing physiological needs in the course of the intervention. The characteristic conditions during an experimental investigation are usually kept constant, so that the results (e.g., improvement observed in the treated group, vs control) refers to a particular mode of intervention under the selected circumstances. If the mode of intervention was inappropriately chosen, a positive effect will in all probability not be achieved, and the only information which can be derived from the study is that "this *particular mode* of intervention (say a specific mode of PICSO cycling) *is suboptimal*". In this case the result of a presumably complex and difficult study is quite unsatisfactory. In other words, when an experimental investigation fails to provide evidence of benefits, one might argue that the *particular mode of intervention* was the reason for the negative result, and the intervention could have been beneficial provided "it had been applied appropriately". One can then suggest a modification of the intervention and redo the study. But each study usually provides only a single answer to a particular question, e.g., "does PICSO, applied with a constant occlusion/release timing of 14/4 s, reduce the area at risk or the degree of infarction during an acute LAD coronary artery occlusion in dogs?". If the answer is negative one selects, for example, another phasing ratio, out of an infinite number of possibilities, a rather costly and time-consuming approach. Mathematically speaking, such an in vivo study is only a "point evaluation" and is not by itself capable of indicating the "direction" toward optimization of the mode of intervention.

It therefore seems desirable to develop a tool capable of anticipating trends or even relations between the mode of intervention and at least certain parameters (e.g., washout or redistribution of venous blood) which may greatly influence the effectiveness of CSIs. It is the purpose of mechanistic modeling to provide such insights which could be very useful in setting up experimental protocols or clinical studies. In contrast to the above in vivo studies, a mathematical model which simulates key features of the coronary circulation and myocardial physiology in the presence of CSIs, can be run repeatedly to simulate a large number of intervention modalities, at very low cost and expense of time. Those particular conditions, which have already been studied experimentally are very useful as "reference conditions", according to which the model has to be adjusted so as to reproduce actual study findings. Each set of experimental data constitutes a "point in a grid of boundary conditions". It is the purpose of the model to interpolate between such experimental points and even enable extrapolations to conditions under which the interventions have not yet been experimentally investigated. Moreover, model results are not affected by experimental spread (fully deterministic) and hence very small changes of the intervention's mode can readily be simulated without loosing trace of the changes' effects (due to uncontrolled superimposed physiological changes as seen in vivo). It is thus possible to derive that combination of changes in parameters which is most promising in terms of increasing the beneficial effects (i.e., the gradient in the direction of a therapeutic optimum).

The major drawback of mechanistic modeling is obviously the need to resort to simplifications (of, for example, the real coronary circulation) required if one is to arrive at a manageable quantitative and numerical model. Therefore, questions remain as to how close the applied model resembles reality, and whether the computational results will be confirmed in subsequent experimental studies.

## Model applications

### PICSO coronary sinus pressure rise (phenomenological)

Numerous studies have been carried out to evaluate the value of PICSO as an intervention to salvage myocardium at risk. Specific applications envisaged comprise acute infarction settings, and open heart surgery support during the reperfusion phase after cardiac arrest [1-3, 8-10, 21]. Some studies were carried out with constant occlusion/release timing [16] (intermittent coronary sinus occlusion, ICSO), others used adaptive cycling adjusted according to the *visual* inspection of the coronary sinus pressure rise (*pressure controlled intermittent coronary sinus occlusion*, PICSO) [10, 17]. Some of these studies did find a beneficial effect [10], others did not [21, 23]. The poor outcome of some of the latter investigations was attributed by the Vienna PICSO working group to "improper" PICSO cycling, resulting from inadequate adjustment based on visual inspection of the CSP tracings. To improve the adaption of PICSO cycling, mathematical models have been established, which put the shape of coronary sinus pressure rise on a quantitative basis.

### Models for coronary sinus pressure rise

We used a three-parameter double exponential function (model 1, see Appendix 1, and [14]) to describe coronary sinus pressure rise (Fig. 1). The same form of function was used for systoles and diastoles, the differences between systolic and diastolic envelopes being reflected in different values for the fitted parameters. Other investigators [20] used a monoexponential function (model 2, Appendix 2).

224

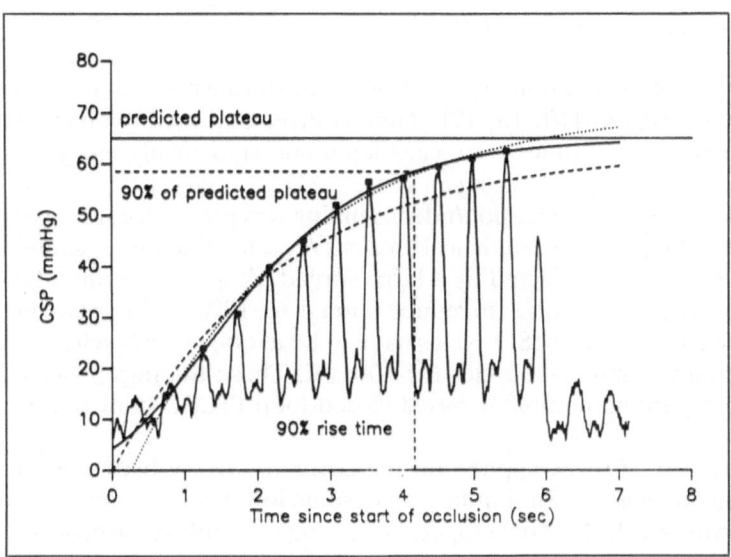

**Fig. 1.** Fitting properties of different models for CSP rise and derived diagnostic quantities. CSP data are taken from a representative canine PICSO cycle. Full squares: peak-detected systolic values. Solid curve: model 1, dotted: model 2 (with non-linear fitting), dashed: model 2 with the plateau being forced through the last peak and fitted upon logarithmic transform. Note the differences in onset (near t = 0) and extrapolation properties between the models. Only the systolic envelope is shown. The predicted plateau and T90_SY (dashed reference line) are calculated from model 1.

*Derived quantities for model 1:* Following the fit, the plateau, P_SY, is calculated as the asymptotic height (mmHg) of the fitted systolic CSP envelope (extrapolated toward $t = \infty$), and may well exceed the last systolic peak within the respective PICSO cycle (Fig. 1). This extrapolation feature of model 1 becomes prominent for short occlusion times, where the envelope still rises near $t = t_o$. The diastolic plateau, P_DI, is obtained along the same lines. The *systolic pressure (ENDP_SY) at the end of CSO,* defined as the height of the fitted envelope at $t = t_o$, is (particularly for short occlusion times) lower than P_SY, and *not necessarily identical* to the last systolic peak detected. The *systolic 90% rise time* of CSP, (T90_SY), is defined as the time period required (counting from the beginning of the respective PICSO cycle) for the fitted envelope to reach 90% of P_SY. T90_DI is calculated in a similar manner. The rise times, in relation to the coronary sinus occlusion time, $t_o$, provide information as to the $P_{CSP}$ developed within the occlusion time. Furthermore, the following other quantities may be constructed from model 1: mean integral under the envelope, (maximum) slope at the inflection point and rise times to 70% or 80% (or any percentage) of the height of the plateau (for details see [14]).

*Derived quantities for model 2:* In the form originally proposed for model 2 [20], the *calculated plateau,* P_SY, is set equal to the height of the last systolic peak. This calculation method is fast but fails to account for a rising tendency of $P_{CSP}$ near $t_o$ (no extrapolation capability). Consequently, ENDP_SY is systematically below the last systolic peak, since ENDP_SY < P_SY = (height of last peak).

225

The above outlined concepts were applied to data obtained during both canine and human investigations of PICSO [10, 16, 17]. Time courses and mean values of plateaus, end-pressures and rise times were calculated and statistically analyzed (Figs. 2a and b). In the dog study (featuring an acute LAD coronary artery occlusion followed by reperfusion), the occlusion/release timing was preset according to the particular experimental protocol. Statistical analysis revealed that plateaus and even more so, the rise times are affected by a large spread, despite experimental conditions remaining constant. Thus, to reduce spread to 10% SEM, plateau values from about five successive PICSO cycles have to be averaged; 10 cycles are required for T90_SY and as much 40 cycles for T90_DI. These findings provide essential information regarding computer based closed-loop PICSO control systems.

In the human study, PICSO was applied during coronary artery bypass graft surgery in the reperfusion phase. $t_o$ was initially set to some low value (5s) and then gradually increased until systolic $P_{CSP}$ developed something resembling the onset of a plateau. Subsequently, $t_o$ was readjusted according to visual inspection according to the following criteria: Should an initial setting of $t_o$ prove insufficient (due to physiological changes) to produce the onset of a plateau, $t_o$ was carefully increased, not exceeding an upper limit of 14s which was deemed potentially hazardous. On the other hand, $t_o$ was *reduced immediately* if a *prolonged systolic CSP plateau* appeared. The release time was kept constant at 4 s. One physician was entirely occupied with monitoring $P_{CSP}$ tracings and adjusting the PICSO timing. The human study data were mainly used to compare the visually adjusted coronary sinus occlusion time with the rise times calculated from the mathematical model. The postoperative computerized data reduction revealed that visual adjustment did not always properly keep pace with the rapid physiologic changes observed during the surgical procedure (Fig. 2b). Opening of the grafts entailed an immediate decrease in rise time (T90_SY), calculated from the model. Interestingly, the physician operating the PICSO controller also reduced $t_o$ (based on his visual inspection), however, with a significant delay in time. These findings suggest that modeling results represent useful aids for CSI control. Considering the spread of numerical diagnostics, as investigated in the dog study, one has to be aware, however, that a reliable adjustment of PICSO phasing requires an averaging of rise times over at least 10 PICSO cycles. The question still remains about how rapidly an automatic regulation should follow the changes in calculated diagnostics, without sacrificing a good deal of regulation security and stability for which time-consuming averaging proved mandatory (tuning PDI-regulation characteristics[3]).

Finally, both phenomenological models (see appendices 1 and 2) were compared using the very same (human or canine) data. The fitting algorithm of model 2 in its original form [20] proved unstable due to the fact that two out of three parameters (A and α) are derived from single $P_{CSP}$ data points. Therefore, we replaced the model 2 fitting method by a true non-linear least squares procedure for all three parameters. Mean derived quantities then turned out to be quite simi-

---

[3] PDI = Proportional Differential Integral regulation algorithm.

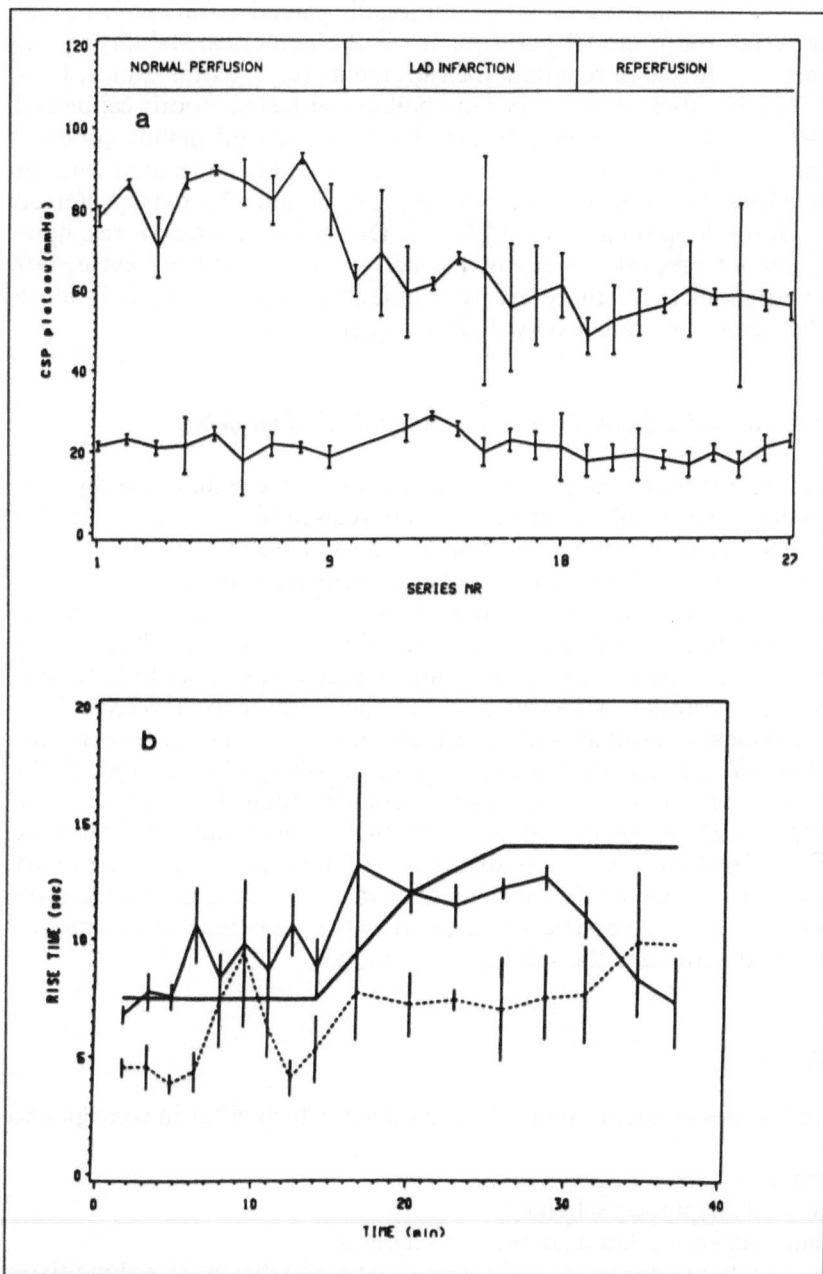

**Fig. 2.** Intraoperative time courses of derived quantities. Panel a: CSP plateau (subaverages and SEM over 20 PICSO cycles) calculated from a canine study featuring LAD ligation [16] (systoles = upper, diastolic = lower graph). Panel b: CSP 90% rise time (systolic = solid, diastolic = dashed) subaverages over eight PICSO cycles with bars denoting SEM, calculated from a human study [17]. The heavy line shows the (manually adjusted) occlusion time actually used. Note the decline in calculated rise time after opening of the bypass grafts, around 30 min post aortic declamp. (Courtesy of Cardiovascular Research.)

lar for both models, and model 2 thereby additionally gained extrapolation capability. Another criterion of model performance is the numerical stability in the presence of certain artifacts in real data measurements (e.g., extrasystoles, temporary loss of data, insufficient coronary sinus balloon occlusion, poorly calibrated signals). Model 2, with the improved fitting algorithm, proved slightly superior when calculating the fraction of PICSO cycles which could be evaluated without numerical difficulties. We concluded that the improved model 2 is to be preferred to model 1 for closed-loop regulation of PICSO. Due to its analytical form, however, model 2 cannot properly represent the early part of a PICSO cycle. For theoretical investigations, we therefore recommend model 1, since it is more realistic and the "drop-out" rate is only slightly higher.

## Interpretation of quantities derived from phenomenological modeling

Derived diagnostic quantities can provide information on the state of the myocardium. For example, the systolic plateau is closely related to contractile function which, during systole, is largely responsible for the squeezing of blood into the venous bed. During the CSO, blood accumulates and distends the coronary veins. Thus, the systolic rise and plateau of venous pressure directly reflect the elastic recoil of the venous bed against the input of additional volume, the latter being directly linked to myocardial contraction. Similar arguments apply to the mean integrals (systolic, diastolic) and even more to their difference (INTM_SY − INTM_DI). Statistically, estimates for P_SY are affected by larger spread than those for INTM_SY. In fact, during experimental LAD ligation in dogs, P_SY declined in one dog while practically no reaction could be found in another animal. INTM_SY − INTM_DI declined in all cases [16]. In the human study [17] it was the 90% rise time which showed a remarkable decline following enhanced coronary artery perfusion after opening of the bypass grafts. The detailed mechanisms governing these processes have to be investigated by the use of mechanistic modeling techniques, as described in the subsequent section.

## Modes of adaption

Three principal modes of adaption may be envisaged, which differ in concept and complexity:
  i) fixed adaptation schemes for $t_o$ and $t_r$;
 ii) disease-related adaptation schemes;
iii) disease- and recovery-related adaptation schemes.
A fixed adaption scheme derives $t_o$ and $t_r$ from calculated diagnostics along fixed lines (i.e., inserts T90_SY, P_SY, etc. into fixed mathematical formulas to get $t_o$ and $t_r$). As P_SY, T90_SY, etc. change with changing physiological states, $t_o$ and $t_r$ will track these changes and may thus be called adaptive (following a *fixed* adaptation mechanism).

We may speculate that an *acute regional infarction* actually requires an algorithm (for deriving optimal $t_o$ and $t_r$) which is different from the algorithm for *moderate global ischemia* (i.e., disease-related adaptation). As yet it is not clear whether, or which mode of adaptation is most effective for the above mentioned two main types of ischemia.

Finally, with either type of ischemia, even a disease-related adaptation does not account for the fact that the adaptation most favorable during the initial phase of treatment (e.g., set $t_o$ = T90_SY) could be suboptimal (and should therefore be changed, e.g., to set $t_o$ = T70_SY) in an advanced state of recovery (similar to respirator modes and settings in to CCU). Such a "disease-and-recovery-related" adaption scheme additionally has to incorporate criteria for switching algorithms during recovery.

It is important to point out that *experimental studies* are only useful to *validate* (evaluate) the efficacy of a given adaptation scheme, they are an inappropriate tool however, *in the search* for an adaptation scheme.

*Relationship between coronary sinus pressure and coronary artery flow*
*(phenomenological modeling with limited experimental data)*

In modeling the release time regulation of PICSO, we assume that peak flow in the coronary sinus during its release occurs well before a maximum of antegrade flow (reactive hyperemia) can be observed in the coronary arteries [6, 7]. Therefore, the occurrence of peak arterial flow is believed to be a suitably conservative indicator of the duration of the coronary sinus release phase in order to achieve sufficient drainage of venous blood into the right atrium. It is thus possible to both avoid a reduction of overall coronary perfusion, and to allow for toxic metabolites to be washed out from jeopardized regions of the myocardium [4]. In humans arterial flow cannot as yet be measured directly and CSP is the only quantity generally available during PICSO application, from which the arterial flow rate might be inferred. We investigated in an experimental study [11] the relationship between CSP and arterial flow and tried to predict the occurrence of the reactive hyperemic response from CSP measurements.

**Methods**

In three dogs, left ventricular pressure (LVP), CSP and arterial flow in the left anterior descending (LAD) and in the circumflex artery (CX) were monitored continuously. PICSO was applied during normal perfusion, LAD-infarction, and reperfusion (1 h each), in 3 × 9 different series, each consisting of 20−25 uniform occlusion and release cycles, characterized by systematically varied values of cycle length and occlusion/release ratio. Fourier analysis was applied to averages of 10 superimposed PICSO cycles of each series. CX and LAD flows were added to yield the total left coronary blood flow.

Fourier coefficients for 40 frequencies of total arterial flow, $\Phi(t)$, and of the LVP-CSP pressure difference (pressure gradient), $\Delta p(t)$, were calculated for all such averages of PICSO cycles (cf. Eqs. A3.1a and A3.1b in Appendix 3).

It was evident from the spectra that 4−5 frequencies are sufficient to study the pressure-flow relationship in the low-frequency range. Assuming, as a first approximation, a generalized linear relationship between arterial flow and pressure gradient, i.e., assuming that for each frequency the respective Fourier coefficients of pressure and flow are proportional to each other, $\hat{\Phi}_n$ is given by Eq. A3.2 (see Appendix 3).

Since the average flow was found to be almost constant (probably due to a non-coronary sinus flow $\hat{\Phi}_0^{res}$, such as Thebesian flow, lymphatic flow, or collateral venous flow), and independent of the average pressure gradient, proportionality holds for non-zero frequencies only. Experimental proportionality coefficients $\hat{y}_n'$, were calculated by dividing the five lowest flow Fourier coefficients obtained from each series of 10 superimposed PICSO cycles by the corresponding pressure Fourier coefficient, i.e.,

$$\hat{y}_n' = \hat{\Phi}_n/\Delta p_n; n = 0, 1, \ldots, 4.$$

Inspection of the experimental $\hat{y}_n'$ shows that for $n \neq 0$ the transfer function [12] can be roughly approximated by Eq. A3.3, i.e., arterial flow is essentially the convolution in the time domain of the pressure gradient with an exponential $A.\exp(-t/\tau)$. The parameters can be determined by minimizing the mean square difference between predicted and observed arterial flow (cf. Appendix 3, Eq. A3.4).

For reasons to be discussed later, the average (= net) flow $\hat{\Phi}_0 = T^{-1} \int dt\, \Phi(t)$ is subtracted from the observed flow values, i.e., it is only the deviation of flow from its average value, or the shape of the flow curve and not the net amount of flow, which is optimized.

Multiplying the experimental CSP Fourier coefficients of any PICSO cycle by $\hat{y}(\omega)$, and inserting the resulting coefficients into the (truncated) Fourier transform, yields the predicted estimate of the flow curve for that PICSO cycle. The time lag between balloon deflation (at the end of the occlusion period) and the occurrence of the predicted flow maximum can then be used to regulate the length of the release period of the following PICSO cycle.

In our study the transfer function was only fitted to the experimental data of two dogs, and applied to the pressure data of the third dog, thus trying a first validation of our method by predicting the occurrence of reactive hyperemia for a third dog from its CSP data, and comparing the predictions with the experimental data.

## Results

Figures 3a and b depict the real and imaginary parts of the proportionality coefficients $\hat{y}_n'$ at the five lowest frequencies $\omega_0, \ldots, \omega_4$ for all types of PICSO cycles applied in the two dogs during normal perfusion, LAD-infarction, and reperfusion. While the $\hat{y}_n'$-coefficients at $\omega \neq 0$ show a common trend toward zero, a tremendous spread of the data was found at $\omega = 0$. The reason for this is that for each set of experimental conditions (normal perfusion, LAD-infarction, reperfusion) the average total flow was found to remain relatively stable, irrespective of the type of PICSO cycle applied. Obviously, slow processes, corresponding to $\omega \rightarrow 0$, allow for compensatory reactions of the heart to the intermittent flow depression, so that

230

a certain level of flow can be sustained by non-coronary sinus pathways. Therefore, it is not possible to maintain that the average value of flow $\hat{\Phi}_0$ is proportional to the average LVP-CSP pressure difference $\Delta \hat{p}_0$, when PICSO cycles with different occlusion periods are used. Formally, this is indicated in Eq.(A3.2) by adding to $\hat{y}_0$. $\Delta p_0$ a quantity $\hat{\Phi}_0^{res}$, which is supposed to account for the cycle type, the experimental condition, and the dog under consideration.

Within the experimental error the $\hat{y}_n'$-coefficients for non-zero frequencies of both dogs appear to be lying on a single curve, regardless of cycle type and experimental condition. Therefore, when fitting the function $\hat{y}(\omega)$ to $\hat{y}_n'$, $n = 1, \ldots, 4$, (thus excluding $\omega = 0$) we summed Eq.(A3.4) over all PICSO cycles of both dogs to obtain a single transfer function, representative of the data of both dogs. The parameter estimates for the transfer function were $A = 0.50$ ml/(min $\times$ mmHg) and $\tau = 0.86$ s. The RMS-deviation of the fit was 0.985 ml/min. The results of the fit are shown as continuous curves in Figs. 3a and b. By means of this function it was possible to predict flow curves for the third dog using only its measured CSP and to determine the points of maximum flow on these curves. Figure 4 shows to what extent this prediction is in agreement with the actual experimental data of one PICSO cycle.

## Conclusion

For several reasons, it might be questionable to assume a simple linear relationship between coronary venous pressure and arterial flow: Firstly, non-coronary sinus flow sustaining a certain level of total perfusion is thought to become increasingly important as a compensatory mechanism when coronary sinus flow is heavily obstructed; secondly, the distensibility of the vessels and the pressure in the surrounding tissues will change when PICSO is applied, so that the distribution of impedances in the micro-circulation cannot be assumed to remain constant [5, 18]. We believe these mechanisms to be responsible for the long-time stability of the flow level observed at a given experimental condition, irrespective of the PICSO cycle length and occlusion time applied. Consequently, we dropped the assumption of linearity for zero frequency. For non-zero frequencies, however, we could show that a suitable transfer function can be found, by means of which it is possible to predict at least the shape of the flow curve from CSP measurements.

Flow prediction was possible, even though in this first approach we simply chose the Fourier transform on an exponential for the transfer function. With more experimental data it should be possible to fit more than two parameters and to use a more sophisticated function which could represent the data more accurately. The question remains, however, whether a single transfer function, i.e., a single set of parameters can be applied to different individuals at all. In order to adjust parameters individually, one could try to relate them to other variables (such as heart rate, arterial and venous pressure, etc.) and to use these quantities as estimators for the parameters of the transfer function.

Thus, further experiments are needed to validate and improve the flow prediction, to fit the parameters of the transfer function to a larger sample, and to re-validate the new parameters.

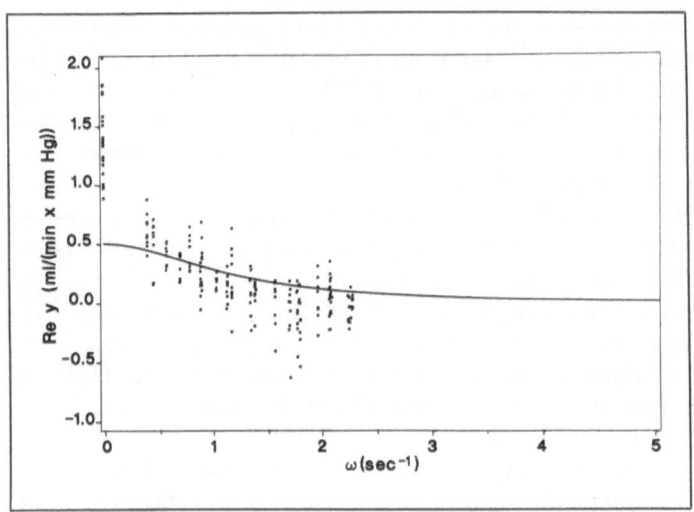

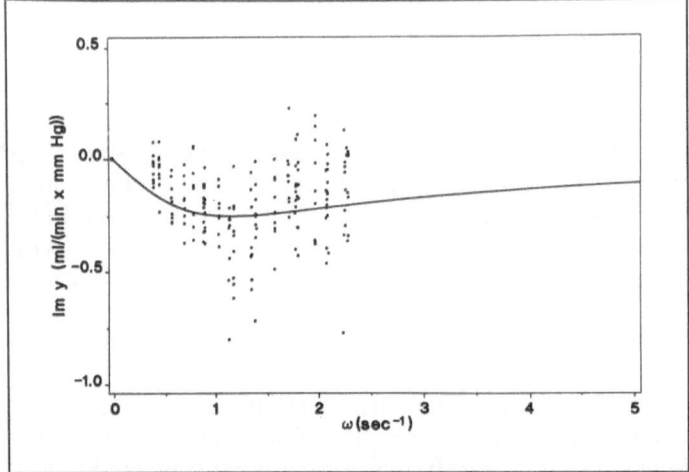

**Fig. 3.** Real (panel A) and imaginary part (panel B) of the transfer function $\hat{y}(\omega)$. Squares: experimental proportionality coefficients $\hat{y}_n'$, with $n < 5$ for all PICSO cycles applied in two dogs. Solid line: fitted analytical transfer function, Eq. (A3.3). At $\omega = 0$, 28 of the $\hat{y}_0'$-coefficients are in the range of $2 < \hat{y}_n' < 5$ ml/(min $\times$ mmHg), and thus are not shown in the figure. Note that at $\omega = 0$ Im $\hat{y}_0' =$ Im $\hat{y}(\omega) = 0$ by definition. (By permission, from: Neumann F et al. (1989) Coronary sinus pressure and arterial flow during intermittent coronary sinus occlusion. Am J Physiol.)

*Coronary circulation during coronary sinus interventions (CSIs) (mechanistic)*

It is evident that the full complexity of coronary circulation is beyond the scope of mathematical modeling. Therefore, simplifications have to be introduced *in any model* designed to give a mechanistic description of cardiovascular physiology. It is mainly up to the skill of the investigator to introduce these simplifications and at the same time retain those important features of reality which have direct impact on the

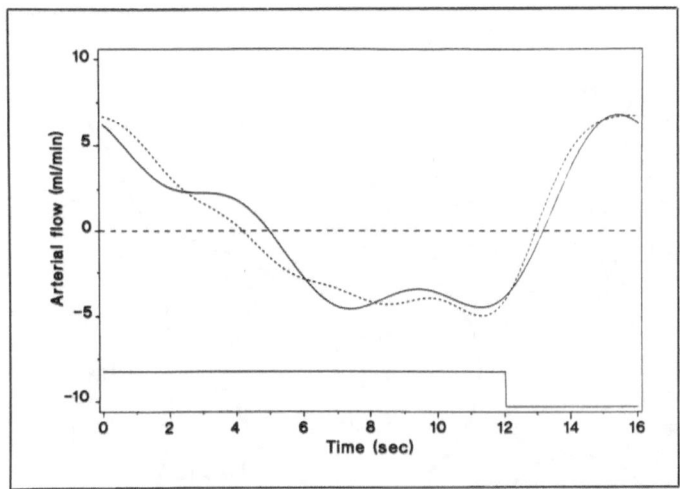

**Fig. 4.** Filtered experimental flow curve (solid line) of one dog during LAD-infarction, when PICSO with an occlusion/release ratio of 12:4 s was applied, and predicted flow (dashed line) using the transfer function fitted to the data of two other dogs. The average value of flow has been subtracted from the experimental flow curve, i.e., only the deviation from the average is shown. (By permission, from: Neumann F et al. (1989) Coronary sinus pressure and arterial flow during intermittent coronary sinus occlusion. Am J Physiol.)

quantities evaluated from the model. As part of simplification in the model, parts of detailed structures (say sections of the vascular tree) are lumped together into single volumes, resistors and compliances. Searching for "the best possible model", one is always tempted to retain as much complexity as possible in order to make the model "realistic". However, this approach is prone to involve more and more parameters for which experimental data are simply not available, thus introducing many unknowns to be adjusted "realistically". Similar to simple systems of equations, the solution may become arbitrary (or manifold) if there are more unknowns (model parameters) than equations (experimental tests). It was therefore a major aim to restrict a model for the hemodynamics of coronary sinus interventions to a form of minimal complexity, which is nevertheless still sufficiently realistic to reproduce key features observed in measurements.

*Lumped parameter model of myocardial circulation*

The model proposed [15] comprises several compartments as shown in Fig. 5.

Starting from the aortic root and a main left coronary artery perfusion pressure $p_{LCA}$ (Fourier expansion), the flow $Q_{LCA}$ traverses a small resistance $R_{LCA}$ before dividing at a bifurcation pressure $p_{bif}$ into the LCX- and the LAD coronary artery branches. Crossing the corresponding two major regional coronary resistances (from epicardial vessels up to the capillaries, $R_{art \to cap}^{LCX}$ and $R_{art \to cap}^{LAD}$, the model shows two distensible capillary compartments (volumes $V_{cap}^{LCX}$ and $V_{cap}^{LAD}$, compliances $C_{cap}^{LCX} = C_{cap}^{LAD}$) which are filled, and are subjected to extramural myocardial squeezing pressures ($p_{sqz}^{LCX}$, $p_{sqz}^{LAD}$) the latter being set proportional to left ventricular pressure, $p_{LV}(t)$, (Fourier expansion). The outflows from both LCX and LAD

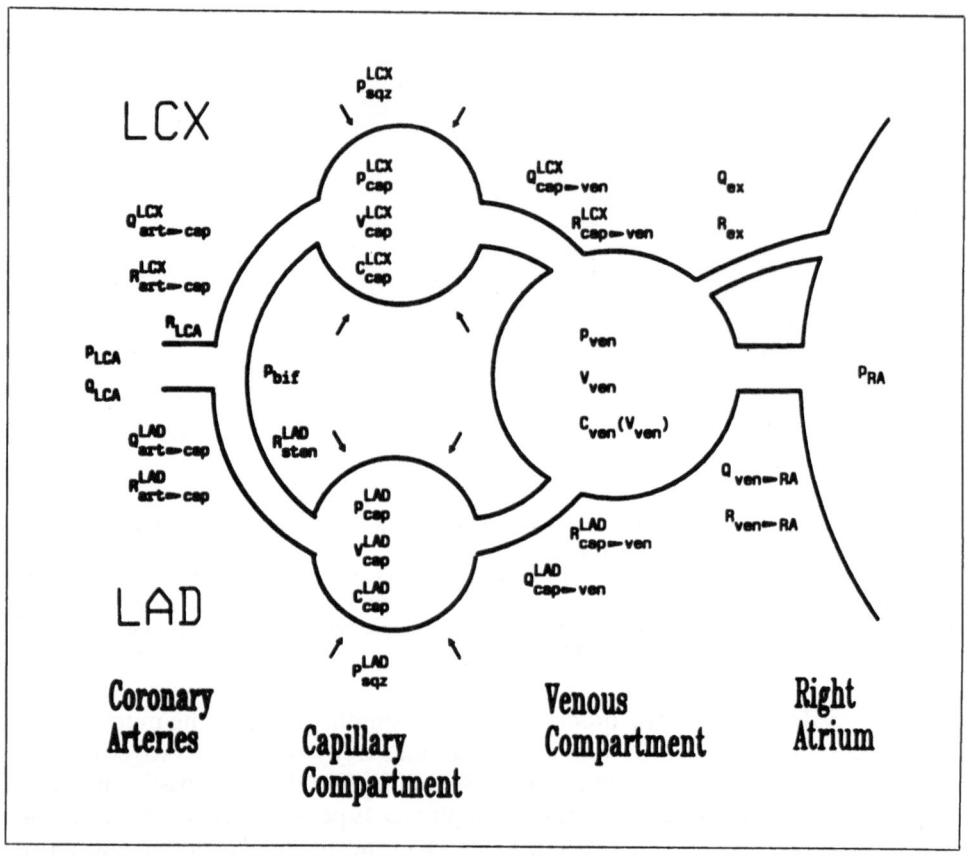

**Fig. 5.** Schematic diagram of the model. The following components (from left to right) constitute the vascular bed of the left coronary artey in the model: From the left coronary artery ostium (driving pressure $p_{LCA}(t)$), the left common artery ($R_{LCA}$, $Q_{LCA}$), leads to the bifurcation ($p_{bif}$). LCX- and LAD branches are modeled similarly: The LCX coronary artery ($R_{art \to cap}^{LCX}$, $Q_{art \to cap}^{LCX}$) leads to its capillary compartment ($C_{cap}^{LCX}$, $V_{cap}^{LCX}$, $p_{cap}^{LCX}$) which is subjected to extravascular squeezing pressure ($p_{sqz}^{LCX}$). The venules draining the LCX bed ($R_{cap \to ven}^{LCX}$, $Q_{cap \to ven}^{LCX}$) and the LAD bed ($R_{cap \to ven}^{LAD}$, $Q_{cap \to ven}^{LAD}$) lead to one single coronary venous compartment (volume-dependent compliance $C_{ven}(V_{ven})$, volume $V_{ven}$, $p_{ven}$). The venous compartment drains to the right atrium ($p_{RA}$ = const.) via the coronary sinus ($R_{ven \to RA}$, $Q_{ven \to RA}$) and the extra outlet ($R_{ex}$, $Q_{ex}$), shown as narrow outlet above the coronary sinus.

capillary compartments cross resistances $R_{cap \to ven}^{LCX}$ and $R_{cap \to ven}^{LAD}$, respectively (capillaries and venules), and both circulations drain into a single venous compartment (volume $V_{ven}$, compliance $C_{ven}$, pressure $p_{ven}$). Venous egress occurs via the coronary sinus (resistance $R_{ven \to RA}$, flow $Q_{ven \to RA}$) into the right atrium (pressure $p_{RA}$). Simultaneously, provision is made in the model to account for coronary veno-venous shunting and for Thebesian channels by an extra outlet ($R_{ex}$, $Q_{ex}$) directly from the venous microcirculation into the cavities of the heart.

A rather important part of the model is a volume-dependent coronary venous compliance, which decreases as the venous bed is distended. A pressure-volume relation of this kind proved to be mandatory in order to reproduce the CSP rise, with its characteristic shapes of systolic and diastolic envelopes. Further (non-standard) elements of the model are the volume-dependent resistances governing (antegrade or retrogradely induced) egress from the capillary compartment. In particular, the total resistance to flow leaving the capillaries was assumed to increase with the inverse square of capillary volume, i.e., effectively becomes infinite as the capillary bed is totally emptied. This assumption mimics the increased resistance during vessel "collapse" due to systolic squeezing. The rest of the model employs ordinary linear relations between pressure gradients, resistances and flows.

Pulsatile time courses of all coronary hemodynamic quantities are obtained by numerical integration (time step 0.001 s) of the differential equations used to describe the balance between change in volume, inflow and egress of blood for each compartment. Each model simulation of the PICSO intervention consisted of an equilibration (control) phase ($0 < t < 20$ s), a 10-s-phase of coronary sinus occlusion (CSO) ($20 < t < 30$ s) and another 10 s of coronary sinus release (CSR) ($30 < t < 40$ s). Mean values of all quantities were calculated for I) a single heartbeat near the end of the CSO period, and II) a single beat during CSR. Except for the simulation of heart rate variation (see below), the heart rate was generally set at 60 bpm, and the single beat periods for averaging were $28 < t < 29$ s for CSO, and $33 < t < 34$ s for CSR conditions. Forward and retrograde phasic components of flows were averaged separately, and the total mean flow was also calculated. Note that phasic flows are shown in the same time intervals for which the averaging was performed.

Systolic and diastolic envelopes were fitted to the rise of coronary venous pressure (after the onset of CSO), using the very same mathematical function described in the phenomenological modeling section. This offered the possibility to derive analogous diagnostic quantities from the model as those studied in vivo.

*Simulation of PICSO under normal conditions, global loss of contractility, and reduced perfusion pressure*

PICSO is simulated by forcing coronary sinus flow to zero during the balloon inflation periods. Mathematically, this is a boundary condition to the differential equations.

Of all the quantities calculated from the model (e.g., pressures and volumes in each compartment, flows between compartments), the quantity of most interest with regard to CSI effectiveness is the retrograde flow delivered from the coronary venous bed to the capillaries. This computed delivery directly characterizes the retrograde access of the myocardium. For the normal state Fig. 6 displays the time course of the LCX flow from arteries to capillaries (dashed) and from capillaries to veins (solid curve) over a whole PICSO cycle (panel a), and in single beat resolution for CSO (panel b), and CSR conditions (panel c). The area between the baseline and the negative portion of the solid curve reflects the volume of blood redistributed from the coronary venous system toward the capillaries within each heartbeat, and may be considered an important quantity characterizing the degree of PICSO "washout" (of toxic metabolites).

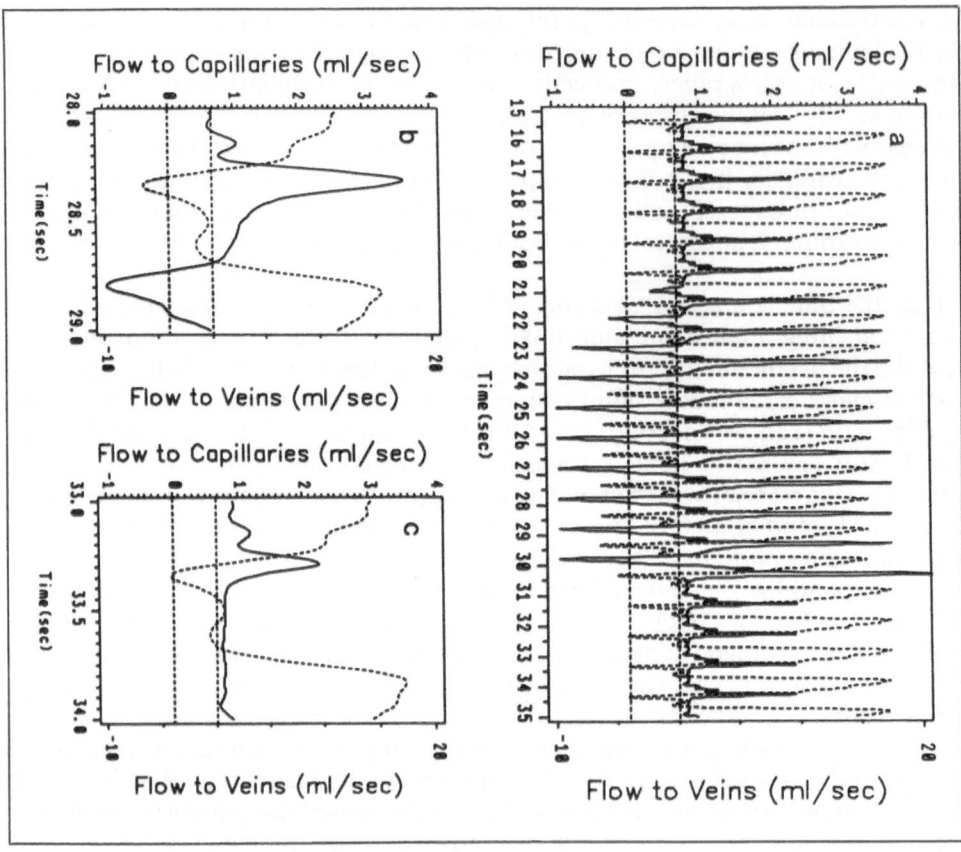

**Fig. 6.** Flow dynamics: a) Full time course of flow from the coronary arteries to the capillaries ($Q_{art \rightarrow cap}^{LCX}$, dashed curve) and of flow from the capillaries to the veinous compartment ($Q_{cap \rightarrow ven}^{LCX}$, solid curve). The occlusion/release pattern is the same as in Fig. 2; b) Single beat resolution during coronary sinus occlusion; c) Single beat resolution during coronary sinus release.

Just like any other quantity obtained from the mathematical model, the redistributed blood volume may be investigated as a function of any other model parameter(s) to mimic the effects observed under pathological condition. For the sake of briefness only the effects of globally reduced myocardial contractility and changes in arterial perfusion pressure will be illustrated here.

The joint variation of myocardial squeezing (parameter $\gamma_{norm}$) and of arterial perfusion pressure ($p_{LCA}$) was simulated on a grid defined by $0 \leq \gamma_{norm} \leq 1$ and $10 \leq p_{LCA} \leq 140$ mmHg. The mean retrograde flow per heartbeat from veins to capillaries was calculated in each of these simulations during a reference beat under CSO and CSR conditions, respectively. The results are displayed as surface charts in Fig. 7. During CSO (left panel) maximum redistribution is found around a rather linear relation between $\gamma_{norm}$ and $p_{LCA}$, cf. the valley extending from ($\gamma_{norm}$ = 0.25, $p_{LCA}$ = 40 mmHg) to ($\gamma_{norm}$ = 0.95, $p_{LCA}$ = 120 mmHg). Moreover, it becomes evident that no redistribution is observed in the absence of squeezing ($\gamma_{norm}$ = 0). This seems to be an important result since one may conclude that

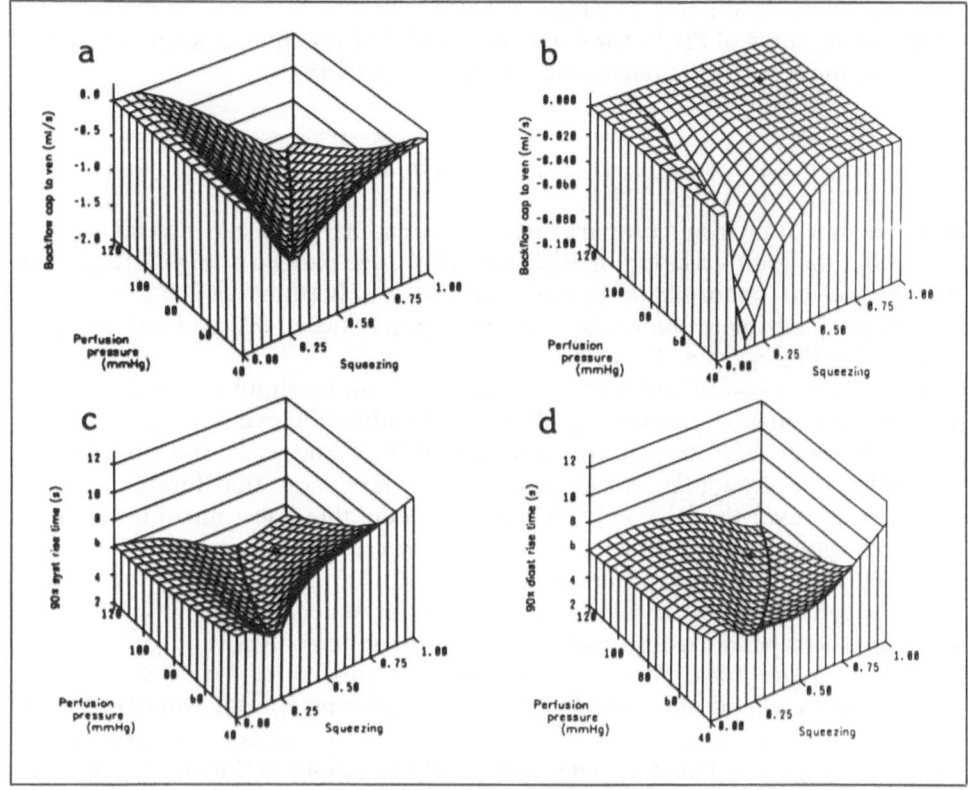

**Fig. 7.** Joint variation of squeezing and perfusion pressure. Each intersection of the grid lines corresponds to a separate simulation with the respective values of $\gamma_{norm}$ and $p_{LCA}$, the other parameters have their default values. Note that results along lines for $p_{LCA}$ = const. differ from pure variations of squeezing since a constant perfusion pressure is used here. Nevertheless, the approximate location of the default parameter set ($p_{LCA}$ = 100 mmHg, $\gamma_{norm}$ = 0.75) is indicated by a rhomboid symbol. The heavy curves running across the surface charts indicate the relations between $\gamma_{norm}$ and $p_{LCA}$ for which local extrema of the calculated quantities are observed.
a) Retrograde component of flow from capillaries to veins under CSO[4];
b) Retrograde component of flow from capillaries to veins under CSR;
c) 90% systolic rise time calculated from a 10 s period of CSO;
d) 90% diastolic rise time calculated from a 10 s period of CSR.

severely ischemic areas (or arrested myocardium), in which myocardial contraction has ceased, are inaccessible to assistance by PICSO.

During CSR, no redistribution is found in the vicinity of ordinary parameter conditions (the rhomboid symbol indicates the location of $\gamma_{norm}$ and $p_{LCA}$ within the default parameter set). Only for very faint squeezing does redistribution occur (see the deeply cut valley in the right panel of Fig. 7). In a similar way, the derived diagnostic quantities, systolic and diastolic 90% rise times can be obtained for every combination of squeezing and perfusion pressure (see Fig. 7, lower panels).

---

[4] All flows count positive in the direction from arteries to veins. "Backflow" in the axis label stands short for the retrograde (i.e., negative) component of the flow from capillaries to veins.

Physiologically, the line along $p_{LCA}$ = const = 100 mmHg may be interpreted as showing the effect of PICSO under moderate global ischemia, as seen, for example, during the reperfusion phase after open heart surgery.

### Simulation of PICSO during LAD stenosis and infarction

The retrograde flow from coronary veins to the capillary bed can also be investigated under the joint variation of $R_{sten}^{LAD}$ and $p_{sqz}^{LAD}$, see panel c of Fig. 8. $R_{sten}^{LAD}$ was varied between 0 (= normal perfusion) and 5000 mmHg (total LAD obstruction), extramural squeezing in the LAD area was varied between 0% and 100% of the normal value ($0 \leq \gamma_{sten}^{LAD} \leq 1$).

Whereas every combination of $R_{sten}^{LAD}$ and $\gamma_{sten}^{LAD}$ can be simulated in the model (yielding a full surface graph in Fig. 8), an LAD infarction event in a patient will run only through characteristic combinations of $R_{sten}^{LAD}$ and $\gamma_{sten}^{LAD}$, since the loss of contractility is a physiologic consequence of flow reduction, thus forming an "infarction trace" over the $R_{sten}^{LAD}$-$\gamma_{sten}^{LAD}$-plane. To model this trace, mean LAD flow, $Q_{art \rightarrow cap}^{LAD}$, was calculated for each step of $R_{sten}^{LAD}$ increase and expressed as fraction of normal LAD flow. Results reported in literature [13, 19, 22] were used to estimate the corresponding reduction in myocardial squeezing ($\gamma_{sten}^{LAD}$). Then the simulation was repeated using this particular combination of $R_{sten}^{LAD}$ and $\gamma_{sten}^{LAD}$ to yield one point on the trace shown. Repeating this procedure for $0 \leq R_{sten}^{LAD} \leq$ 5000 mmHg · s/ml adds up to the full trace. An infarction episode would run from location 1 (healthy) via locations 2 to 5 toward the 68% end (zero flow) of the full heavy trace. Figure 8, panel a, gives single beat resolutions of the intramyocardial flow between capillaries and veins, as calculated in the points along the trace.

The relation between $R_{sten}^{LAD}$ and $\gamma_{sten}^{LAD}$ inferred from [19, 22] terminates at $\gamma_{sten}^{LAD}$ = 0.68, since experimental measurements [13] indicate that even in the case of total loss of *regional contractility*, intramyocardial pressure is maintained around 68% of normal values, due to the wall tension within the bulging segment. However, since measurements of intramyocardial pressure are very controversial [13] two other assumptions for the reduction in squeezing (to 30% and to 0% of normal values) have additionally been investigated in the model (dashed traces in Fig. 8). Obviously, these lead to combinations of $R_{sten}^{LAD}$ and $\gamma_{sten}^{LAD}$ much less favorable for retrograde redistribution during PICSO. See panels a and b of Fig. 8 for calculated pulsatile flows between capillaries and veins for six points along each trace.

Since $R_{sten}^{LAD}$ and $\gamma_{sten}^{LAD}$ refer to the LAD bed, the model results are non-symmetric for the LAD- and LCX beds. Besides drastic changes in the LAD region, LCX-hemodynamics are also (slightly) affected, due to redistribution phenomena. For reasons of briefness, only the (more important) LAD results are shown here.

### Further concepts

The variations of model parameters may be applied in support of the search for a direction toward an optimal intervention mode, as outlined in section 1. In fact, the "valleys" observed in Fig. 7 indicate such a direction. However, the parameters

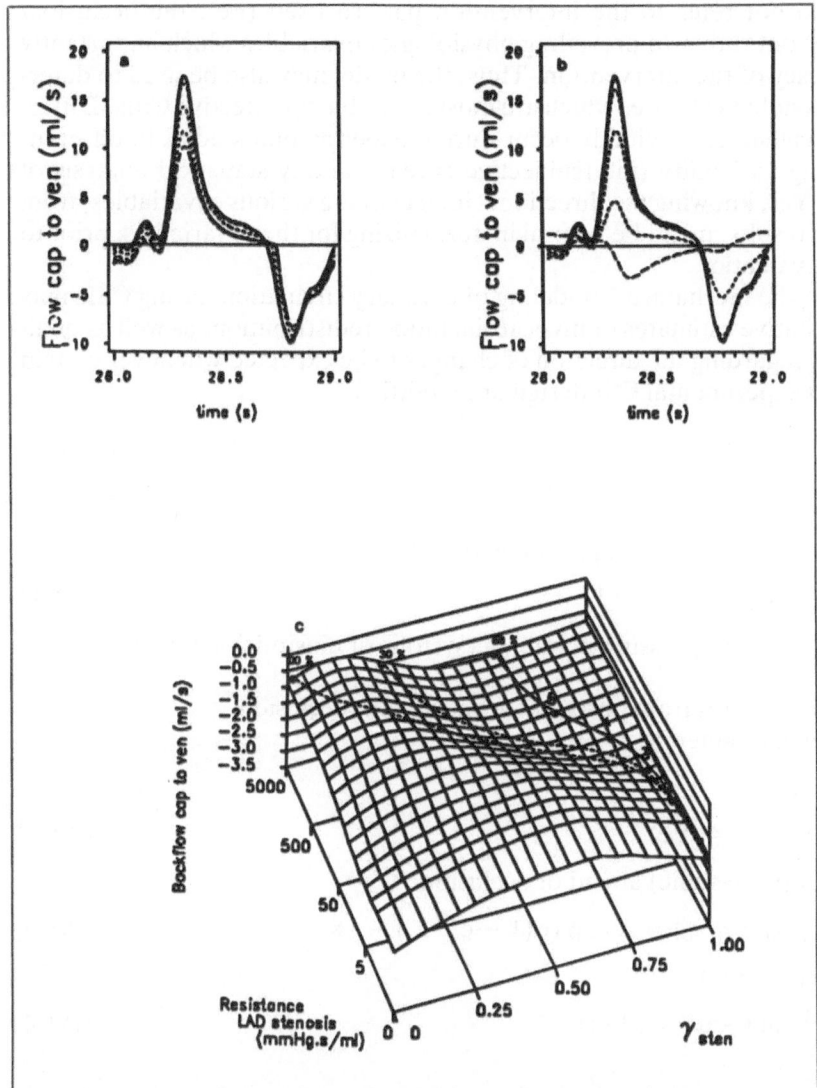

**Fig. 8.** Flow between capillaries and coronary veins for different degrees of LAD stenosis. Panel a: Pulsatile single beat resolution of $Q^{LAD}_{cap \rightarrow ven}$ during coronary sinus occlusion. The solid curve refers to location labelled [1] in panel c (healthy state, $R^{LAD}_{sten} = 0$, $\gamma^{LAD}_{sten} = 1$). Dashed curves display changes along the infarction trace ending at $\gamma^{LAD}_{sten} = 0.68$. Panel b: Like panel a but for the infarction trace ending at $\gamma^{LAD}_{sten} = 0$. Panel c: Retrograde flow (averaged over one beat during CSO) from coronary veins to capillaries (negative in sign) shown for a full grid of combinations of LAD stenosis ($R^{LAD}_{sten}$) and loss of contractile function ($\gamma^{LAD}_{sten}$) in the LAD bed. Lower parts of the surface correspond to pronounced backflow. Infarction traces are modeled to display the physiologic link between stenosis and the impairment of contractile function.

investigated do not refer to the intervention pattern itself (i.e., the occlusion/release timing) but rather to prevailing physiologic covariables which importantly affect the efficacy of the intervention. Thus, the model may also be used to determine those physiological states which are most favorable for effective CSIs. Different physiological states inevitably occur during experimental studies in different animals, adding variability (intersubjective spread) in any statistical analysis of CSIs efficacy. Yet, knowing the directional impact of the various covariables, from the simulation results, might be helpful in normalizing for the covariables, prior to the statistical evaluation.

In summary, the mechanistic modeling of coronary circulation during CSIs may provide quantitative estimates of myocardial blood redistribution, as well as qualitative insights regarding the direction of changes to be expected whenever certain elements of an experimental CSI design are modified.

## Appendix 1

Mathematical representation of $p_{CSP}$ envelope in model 1:

$$p_{CSP}(t) = A \exp\left(\alpha(1 - e^{-\beta t}) - 1\right) \tag{A1.1}$$

where    coronary sinus pressure in mmHg (systolic or diastolic) at time t
$p_{CSP}(t)$
t        time in seconds from onset of coronary sinus occlusion
$A, \alpha, \beta$  model parameters

Systolic plateau:

$$P\_SY = p_{CSP}(t = \infty) = A\, e^{\alpha - 1} \tag{A1.2}$$

Pressure envelope (systolic) at end of occlusion:

$$ENDP\_SY = p_{CSP}(t = t_0) = A \exp\left(\alpha(1 - e \quad) - 1\right) \tag{A1.3}$$

90% rise time (systolic):

$$T90\_SY = \beta^{-1} \cdot \ln\left(-\alpha / \ln(0.9)\right) \tag{A1.4}$$

## Appendix 2

Mathematical representation of $p_{CSP}$ envelope in model 2:

$$p_{CSP}(t) = A + \alpha\,(1 - e^{-\beta t}) \tag{A.21}$$

Single point assignments in original version of model 2:

Parameter A: $A = p_{CSP}(t = 0)$ (A2.2)

Parameter α: $\alpha = p_{CSP}(\text{last peak}) - A$ (A2.3)

In the improved version, with A, α and β fitted by least squares, the plateau becomes a calculated rather than assigned quantity:

$$P\_SY = p_{CSP}(t = \infty) = A + \alpha \tag{A2.4}$$

Pressure at end of occlusion:

$$ENDP\_SY = p_{CSP}(t = t_0) = A + \alpha\,(1 - e^{-\beta t_0}) \tag{A2.5}$$

90% rise time (systolic):

$$T90\_SY = -\beta^{-1} \ln\left((A + \alpha) \cdot (1-0.9)/\alpha\right) \tag{A2.6}$$

## Appendix 3

Complex Fourier coefficients

$$\hat{\Phi}_n = \frac{1}{T} \int_0^T dt\, e^{-i\omega_n t}\, \Phi(t) \tag{A3.1a}$$

$$\Delta p_n = \frac{1}{T} \int_0^T dt\, e^{-i\omega_n t}\, \Delta p(t) \tag{A3.1b}$$

T is the total length of a PICSO cycle and

$$\omega_n = 2\,\pi n/T; \quad n = 0, \pm 1, \pm 2, \ldots$$

is the discrete set of frequencies compatible with the cycle length T.

$$\hat{\Phi}_n = \begin{cases} \hat{y}_0\,\Delta p_0 + \hat{\Phi}_0^{res} & n = 0 \\ \hat{y}_n\,\Delta p_n & n \neq 0 \end{cases} \tag{A3.2}$$

Transfer function:

$$\hat{y}(\omega) = A/(1 + i\omega\tau) \tag{A3.3}$$

Minimizing criterium:

$$\frac{1}{T} \int_0^T dt \left\{ \sum_{n \neq 0} \hat{y}(\omega_n)\,\Delta p_n\, e^{i\omega_n t} - [\bar{\Phi}(t) - \hat{\Phi}_0] \right\}^2 \tag{A3.4}$$

$$= 2 \sum_{n=1}^{n_{max}} |\Delta p_n|^2\, |\hat{y}(\omega_n) - \hat{y}_n'|^2 \longrightarrow \min.$$

## References

1. Chang BL, Drury JK, Meerbaum S, Fishbein MC, Whiting JS, Corday E (1987) Enhanced myocardial washout and retrograde blood delivery with synchronized retroperfusion during acute myocardial ischemia. J Am Coll Cardiol 9:1091
2. Guerci AD, Ciuffo AA, Dipaula AF (1987) Intermittent coronary sinus occlusion in dogs: reduction of infarct size 10 days after reperfusion. J Am Coll Cardiol 9:1075
3. Jacobs AK, Faxon DP, Mohl W, Coats WD, Gottsman SB, Ryan TJ (1985) Pressure control-led intermittent coronary sinus occlusion (PICSO) during reperfusion markedly reduces infarct size. Clin Res 33:197A

4. Kenner T, Moser M, Mohl W (1984) Wave reflections and pressure flow relations in the coronary circulation. In: Mohl W, Wolner E, Glogar D (eds) The coronary sinus. Steinkopff, Darmstadt, pp 60−72
5. Kenner T, Moser M, Mohl W, Tiedt N (1986) Inflow, outflow and pressures in the coronary circulation. In: Mohl W, Faxon D, Wolner E (eds) CSI − A new approach to interventional cardiology. Steinkopff, Darmstadt/Springer, New York, pp 15−26
6. Mohl W (1987) Coronary sinus interventions: From concepts to clinics. J Cardiac Surg 2:467−493
7. Mohl W (1988) The momentum of coronary sinus interventions clinically. Circulation 77:6−12
8. Mohl W, Glogar D, Mayr H, Losert U, Sochor H, Pachinger O, Kaindl F, Wolner E (1984) Reduction of infarct size induced by pressure controlled intermittent coronary sinus occlusion. Am J Cardiol 53:923−928
9. Mohl W, Punzengruber C, Moser M, Kenner T, Heimisch W, Haendchen R, Meerbaum S, Maurer G, Corday E (1985) Effects of pressure controlled intermittent coronary sinus occlusion on regional ischemic myocardial function. J Am Coll Cardiol 5:939−947
10. Mohl W, Simon P, Neumann F, Schreiner W, Punzengruber C (1988) Clinical evaluation of pressure-controlled intermittent coronary sinus occlusion: Randomized trial during coronary artery surgery. Ann Thor Surg 46:192−201
11. Neumann F, Mohl W, Schreiner W (1989) Coronary sinus pressure and arterial flow during intermittent coronary sinus occlusion. Am J Physiol 256:H906−H915
12. Priestley MB (1981) Spectral analysis and time series. Academic Press, London
13. Sabbah HN, Stein PD (1982) Effect of acute regional ischemia on pressure in the subepicardium and subendocardium. Am J Physiol 242:H240−H244
14. Schreiner W, Mohl W, Neumann F, Schuster J (1987) Model of the hemodynamic reactions to intermittent coronary sinus occlusion. J Biomed Eng 9:141−147
15. Schreiner W, Neumann F, Mohl W (1989) Modeling of the left coronary circulation in normal states and during coronary sinus occlusion. J. Biomed. Eng. (submitted)
16. Schreiner W, Neumann F, Schuster J, Froehlich KC, Mohl W (1988) Computation of derived diagnostic quantities during intermittent coronary sinus occlusion in dogs. Cardiovasc Res 22:265−276
17. Schreiner W, Neumann F, Schuster J, Simon P, Froehlich KC, Mohl W (1988) Intermittent coronary sinus occlusion in humans: pressure dynamics and calculation of diagnostic quantities. Cardiovasc Res 22:277−286
18. Sethna DH, Moffit EA (1986) An appreciation of the coronary circulation. Anesth Analg 65:294−305
19. Stein PD, Sabbah HN, Marzilli HN, Boick EF (1980) Comparison of the distribution of intramyocardial pressure across the canine left ventricular wall in the beating heart during diastole and in the arrested heart. Circ Res 47:258−267
20. Sun Y, Mohl W (1986) Characterization of the reactive hyperemic response time during intermittent coronary sinus occlusion. In: Mohl W, Faxon D, Wolner E (eds) Clinics of CSI. Steinkopff, Darmstadt, pp 355−362
21. Toggart EJ, Nellis SH, Liedke AJ (1987) The efficacy of intermittent coronary sinus occlusion in the absence of coronary artery collaterals. Circulation 76:667−677
22. Vatner SF (1980) Correlation between acute reductions in myocardial blood flow and function in concious dogs. Circ Res 47:201−207
23. Zalewski A, Goldberg S, Slysh S, Maroko PR (1985) Myocardial protection via coronary sinus interventions: superior effects of arterialization compared with intermittent occlusion. Circulation 71:1215

# Subject Index